Number Theory and Symmetry

Number Theory and Symmetry

Editor

Michel Planat

MDPI • Basel • Beijing • Wuhan • Barcelona • Belgrade • Manchester • Tokyo • Cluj • Tianjin

Editor
Michel Planat
Institut FEMTO-ST
France

Editorial Office
MDPI
St. Alban-Anlage 66
4052 Basel, Switzerland

This is a reprint of articles from the Special Issue published online in the open access journal *Symmetry* (ISSN 2073-8994) (available at: https://www.mdpi.com/journal/symmetry/special_issues/Number_Theory_Symmetry).

For citation purposes, cite each article independently as indicated on the article page online and as indicated below:

LastName, A.A.; LastName, B.B.; LastName, C.C. Article Title. *Journal Name* **Year**, *Article Number*, Page Range.

ISBN 978-3-03936-686-6 (Hbk)
ISBN 978-3-03936-687-3 (PDF)

© 2020 by the authors. Articles in this book are Open Access and distributed under the Creative Commons Attribution (CC BY) license, which allows users to download, copy and build upon published articles, as long as the author and publisher are properly credited, which ensures maximum dissemination and a wider impact of our publications.

The book as a whole is distributed by MDPI under the terms and conditions of the Creative Commons license CC BY-NC-ND.

Contents

About the Editor vii

Preface to "Number Theory and Symmetry" ix

Germán Sierra
The Riemann Zeros as Spectrum and the Riemann Hypothesis
Reprinted from: *Symmetry* 2019, *11*, 494, doi:10.3390/sym11040494 1

Michel Planat, Raymond Aschheim, Marcelo M. Amaral and Klee Irwin
Universal Quantum Computing and Three-Manifolds
Reprinted from: *Symmetry* 2018, *10*, 773, doi:10.3390/sym10120773 39

Torsten Asselmeyer-Maluga
Braids, 3-Manifolds, Elementary Particles: Number Theory and Symmetry in Particle Physics
Reprinted from: *Symmetry* 2019, *11*, 1298, doi:10.3390/sym11101298 55

Bruno Aiazzi, Stefano Baronti, Leonardo Santurri and Massimo Selva
An Investigation on the Prime and Twin Prime Number Functions by Periodical Binary Sequences and Symmetrical Runs in a Modified Sieve Procedure
Reprinted from: *Symmetry* 2019, *11*, 775, doi:10.3390/sym11060775 81

S.T. Ishmukhametov, B.G. Mubarakov and R.G. Rubtsova
On the Number of Witnesses in the Miller–Rabin PrimalityTest
Reprinted from: *Symmetry* 2020, *12*, 890, doi:10.3390/sym12060890 103

Atsushi Yamagami and Kazuki Taniguchi
On a Generalization of a Lucas' Resultand an Application to the 4-Pascal's Triangle
Reprinted from: *Symmetry* 2020, *12*, 288, doi:10.3390/sym12020288 115

Pavel Trojovský
Algebraic Numbers as Product of Powers of Transcendental Numbers
Reprinted from: *Symmetry* 2019, *11*, 887, doi:10.3390/sym11070887 123

Atsushi Yamagami and Yūki Matsui
On Some Formulas for Kaprekar Constants
Reprinted from: *Symmetry* 2019, *11*, 885, doi:10.3390/sym11070885 129

Ilwoo Cho
Asymptotic Semicircular Laws Induced by p-Adic Number Fields Q_p and C^*-Algebras over Primes p
Reprinted from: *Symmetry* 2019, *11*, 819, doi:10.3390/sym11060819 161

About the Editor

Michel Planat was a researcher at the National Center of Scientific Research in France from 1982 to 2018. From 1980 to 2001, he researched nonlinear waves in piezoelectric crystals and 1/f noise in quartz resonators. He established links between 1/f noise and number theory. He also researched the Riemann hypothesis. He discovered Ramanujan sums signal processing. From 2002 to 2018, he was interested in quantum information theory with work concerning mutually unbiased bases, quantum entanglement and contextuality, and quantum computing, using mathematical tools such as finite geometries, number theory, dessin d'enfants, and free group theory. Since 2019, he is a visiting scientist at FEMTO-ST Institute in Besançon, France. He is also an associate research scientist at Quantum Gravity Research in Los Angeles. His current research is about topological quantum computing from three- and four-manifolds to establish bridges between quantum computing and quantum gravity.

Preface to "Number Theory and Symmetry"

"Number Theory and Symmetry" deals with topics connecting numbers (integers, algebraic integers, transcendental numbers, p-adic numbers) and symmetries. First of all, symmetry became part of number theory when Riemann investigated the distribution of prime numbers and for that purpose introduced the complex functional equation and the related Riemann hypothesis (RH) that non-trivial zeros of the Riemann zeta function lie on the symmetry axis $s = 1/2$. Then, in a quest to justify RH on physical grounds, the Hilbert–Polya conjecture claimed that the imaginary part of the Riemann zeros on the symmetry axis should correspond to the eigenvalues of a Hermitian operator. This topic is covered by German Sierra.

Besides these classical areas, number fields offer clues to the connection between numbers and symmetries through arithmetic groups, geometry, and topology. I have in mind the Poincaré conjecture and the whole work of Thurston about 3-manifolds. This topic is the kernel of the two papers by Michel Planat and co-authors and Torsten Asselmeyer Maluga. The aforementioned three papers highlight a strong connection between number theory and quantum physics.

The range of the three subsequent papers in this series is about more standard topics of number theory. A modified Sieve procedure by Bruno Aiazzi and coauthors, the Miller–Rabin primality test by Shamil Ishmukhametov and co-authors, and the 4-Pascal's triangle by Atsushi Yamagami and Kazuki Taniguchi are investigated.

The paper by Pavel Trojovsky offers clues to the relation between algebraic and transcendental numbers through polynomials. Atsushi Yamagami and Yuki Matsui's paper is in the field of b-adic numbers. The last paper by Ilwoo Cho covers the topic of p-adic numbers thanks to C*-algebras and Banach*-probability spaces.

The rich panel of mathematical concepts involved in this Special Issue illustrates the continuous interest of scholars in the relationship between numbers, their symmetries, and physics.

Michel Planat
Editor

Article

The Riemann Zeros as Spectrum and the Riemann Hypothesis

Germán Sierra

Instituto de Física Teórica UAM/CSIC, Universidad Autónoma de Madrid, Cantoblanco, 28049 Madrid, Spain; german.sierra@uam.es

Received: 31 December 2018; Accepted: 26 March 2019; Published: 4 April 2019

Abstract: We present a spectral realization of the Riemann zeros based on the propagation of a massless Dirac fermion in a region of Rindler spacetime and under the action of delta function potentials localized on the square free integers. The corresponding Hamiltonian admits a self-adjoint extension that is tuned to the phase of the zeta function, on the critical line, in order to obtain the Riemann zeros as bound states. The model suggests a proof of the Riemann hypothesis in the limit where the potentials vanish. Finally, we propose an interferometer that may yield an experimental observation of the Riemann zeros.

Keywords: zeta function; Pólya-Hilbert conjecture; Riemann interferometer

1. Introduction

One of the most promising approaches to prove the Riemann Hypothesis [1–7] is based on the conjecture, due to Pólya and Hilbert, that the Riemann zeros are the eigenvalues of a quantum mechanical Hamiltonian [8]. This bold idea is supported by several results and analogies involving Number Theory, Random Matrix Theory and Quantum Chaos [9–17]. However, the construction of a Hamiltonian whose spectrum contains the Riemann zeros, has eluded researchers for several decades. In this paper we shall review the progress made along this direction starting from the famous xp model proposed in 1999 by Berry, Keating and Connes [18–20] that inspired many works [21–45], some of them will be discuss below. See [46] for a general review on physical approaches to the RH. Other approaches to the RH and related material can be found in [47–63].

To relate xp with the Riemann zeros, Berry, Keating and Connes used two different regularizations. The Berry and Keating regularization led to a discrete spectrum related to the smooth Riemann zeros [18,19], while Connes's regularization led to an absorption spectrum where the *zeros* are missing spectral lines [20]. A physical realization of the Connes model was obtained in 2008 in terms of the dynamics of an electron moving in two dimensions under the action of a uniform perpendicular magnetic field and an electrostatic potential [29]. However this model has not been able to reproduce the exact location of the Riemann zeros. On the other hand, the Berry–Keating xp model was revisited in 2011 in terms of the classical Hamiltonians $H = x(p + 1/p)$, and $H = (x + 1/x)(p + 1/p)$ whose quantizations contain the smooth approximation of the Riemann zeros [32,36]. Later on, these models were generalized in terms of the family of Hamiltonians $H = U(x)p + V(x)/p$ that were shown to describe the dynamics of a massive particle in a relativistic spacetime whose metric can be constructed using the functions U and V [35]. This result suggested a reformulation of $H = U(x)p + V(x)/p$ in terms of the massive Dirac equation in the aforementioned spacetimes [38]. Using this reformulation, the Hamiltonian $H = x(p + 1/p)$ was shown to be equivalent to the massive Dirac equation in Rindler spacetime that is the natural arena to study accelerated observers and the Unruh effect [42]. This result provides an appealing spacetime interpretation of the xp model and in particular of the smooth Riemann zeros.

To obtain the exact *zeros*, one must make further modifications of the Dirac model. First, the fermion must become massless. This change is suggested by a field theory interpretation of the Pólya's ζ function and its comparison with the Riemann's ζ function. On the other hand, inspired by the Berry's conjecture on the relation between prime numbers and periodic orbits [12,14] we incorporated the prime numbers into the Dirac action by means of Dirac delta functions [42]. These delta functions represent moving mirrors that reflect or transmit massless fermions. The spectrum of the complete model can be analyzed using transfer matrix techniques that can be solved exactly in the limit where the reflection amplitudes of the mirrors go to zero that is when the mirrors become transparent. In this limit we find that the *zeros* on the critical line are eigenvalues of the Hamiltonian by choosing appropriately the parameter that characterizes the self-adjoint extension of the Hamiltonian. One obtains in this manner a spectral realization of the Riemann zeros that differs from the Pólya and Hilbert conjecture in the sense that one needs to fine tune a parameter to *see* each individual *zero*. In our approach we are not able to find a single Hamiltonian encompassing all the *zeros* at once. Finally, we propose an experimental realization of the Riemann zeros using an interferometer consisting of an array of semitransparent mirrors, or beam splitters, placed at positions related to the logarithms of the square free integers.

The paper is organized in a historical and pedagogical way presenting at the end of each section a summary of achievements (✓), shortcomings/obstacles (✗) and questions/suggestions (?).

2. The Semiclassical XP Berry, Keating and Connes Model

In this section, we review the main results concerning the classical and semiclassical xp model [18–20]. A classical trajectory of the Hamiltonian $H = xp$, with energy E, is given by

$$x(t) = x_0 e^t, \qquad p(t) = p_0 e^{-t}, \qquad E = x_0 p_0, \tag{1}$$

that traces the parabola $E = xp$ in phase space plotted in Figure 1. E has the dimension of an action, so one should multiply xp by a frequency to get an energy, but for the time being we keep the notation $H = xp$. Under a time reversal transformation, $x \to x, p \to -p$ one finds $xp \to -xp$, so that this symmetry is broken. This is why reversing the time variable t in (1) does not yield a trajectory generated by xp. As $t \to \infty$, the trajectory becomes unbounded that is $|x| \to \infty$, so one expects the semiclassical and quantum spectrum of the xp model to form a continuum. To get a discrete spectrum Berry and Keating introduced the constraints $|x| \geq \ell_x$ and $|p| \geq \ell_p$, so that the particle starts at $t = 0$ at $(x,p) = (\ell_x, E/\ell_x)$ and ends at $(x,p) = (E/\ell_p, \ell_p)$ after a time lapse $T = \log(E/\ell_x \ell_p)$ (we assume for simplicity that $x, p > 0$). The trajectories are now bounded, but not periodic. A semiclassical estimate of the number of energy levels, $n_{\rm BK}(E)$, between 0 and $E > 0$ is given by the formula

$$n_{\rm BK}(E) = \frac{A_{\rm BK}}{2\pi \hbar} = \frac{E}{2\pi \hbar} \left(\log \frac{E}{\ell_x \ell_p} - 1 \right) + \frac{7}{8}, \tag{2}$$

where $A_{\rm BK}$ is the phase space area below the parabola $E = xp$ and the lines $x = \ell_x$ and $p = \ell_p$, measured in units of the Planck's constant $2\pi\hbar$ (see Figure 1). The term $7/8$ arises from the Maslow phase [18]. In the course of the paper, we shall encounter this equation several times with the constant term depending on the particular model.

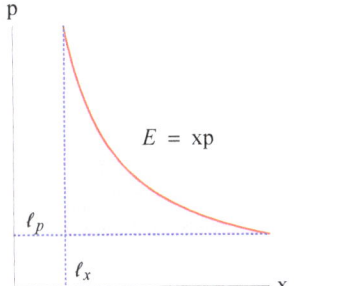
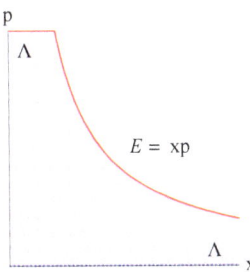

Figure 1. (**Left**): The region in shadow describes the allowed phase space with area A_{BK} bounded by the classical trajectory (1) with $E > 0$ and the constraints $x \geq \ell_x, p \geq \ell_p$. (**Right**): Same as before with the constraints $0 < x, p < \Lambda$.

Berry and Keating compared this result with the average number of Riemann zeros, whose imaginary part is less than t with $t \gg 1$,

$$\langle n(t) \rangle \simeq \frac{t}{2\pi}\left(\log\frac{t}{2\pi} - 1\right) + \frac{7}{8} + O(1/t), \qquad (3)$$

finding an agreement with the identifications

$$t = \frac{E}{\hbar}, \qquad \ell_x \ell_p = 2\pi\hbar. \qquad (4)$$

Thus, the semiclassical energies E, expressed in units of $\hbar$, are identified with the Riemann zeros, while $\ell_x \ell_p$ is identified with the Planck's constant. This result is remarkable given the simplicity of the assumptions. However, one must observe that the derivation of Equation (2) is heuristic, so one goal is to find a consistent quantum version of it.

Connes proposed another regularization of the xp model based on the restrictions $|x| \leq \Lambda$ and $|p| \leq \Lambda$, where Λ is a common cutoff, which is taken to infinity at the end of the calculation [20]. The semiclassical number of states is computed as before yielding (see Figure 1, we set $\hbar = 1$)

$$n_C(E) = \frac{A_C}{2\pi} = \frac{E}{2\pi}\log\frac{\Lambda^2}{2\pi} - \frac{E}{2\pi}\left(\log\frac{E}{2\pi} - 1\right). \qquad (5)$$

The first term on the RHS of this formula diverges in the limit $\Lambda \to \infty$, which corresponds to a continuum of states. The second term is minus the average number of Riemann zeros, which according to Connes, become missing spectral lines in the continuum [17,20]. This is called the *absorption* spectral interpretation of the Riemann zeros, as opposed to the standard *emission* spectral interpretation where the *zeros* form a discrete spectrum. Connes, relates the minus sign in Equation (5) to a minus sign discrepancy between the fluctuation term of the number of zeros and the associated formula in the theory of Quantum Chaos. We shall show below that the negative term in Equation (5) must be seen as a finite size correction of discrete energy levels and not as an indication of missing spectral lines.

Let us give for completeness the formula for the exact number of *zeros* up to t [2,3]

$$n_R(t) = \langle n(t) \rangle + n_{fl}(t), \qquad (6)$$
$$\langle n(t) \rangle = \frac{\theta(t)}{\pi} + 1, \qquad n_{fl}(t) = \frac{1}{\pi}\text{Im}\log\zeta\left(\frac{1}{2} + it\right),$$

where $\langle n(t) \rangle$ is the Riemann–von Mangoldt formula that gives the average behavior in terms of the function $\theta(t)$

$$\theta(t) = \text{Im} \log \Gamma\left(\frac{1}{4} + \frac{it}{2}\right) - \frac{t}{2} \log \pi \stackrel{t \to \infty}{\longrightarrow} \frac{t}{2} \log \frac{t}{2\pi} - \frac{t}{2} - \frac{\pi}{8} + O(1/t), \quad (7)$$

that can also be written as

$$e^{2i\theta(t)} = \pi^{-it} \frac{\Gamma\left(\frac{1}{4} + \frac{it}{2}\right)}{\Gamma\left(\frac{1}{4} - \frac{it}{2}\right)}. \quad (8)$$

$\theta(t)$ is the phase of the Riemann zeta function on the critical line, that can be expressed as

$$\zeta\left(\frac{1}{2} + it\right) = e^{-i\theta(t)} Z(t), \quad (9)$$

where $Z(t)$ is the Riemann-Siegel zeta function, or Hardy function, that on the critical line satisfies

$$Z(t) = Z(-t) = Z^*(t), \quad t \in \mathbb{R}. \quad (10)$$

Summary:

> ✓ The semiclassical spectrum of the xp Hamiltonian reproduces the average Riemann zeros.
> ✗ There are two schemes leading to opposite physical realizations: emission vs absorption.
> ? Quantum version of the semiclassical xp models.

3. The Quantum XP Model

To quantize the xp Hamiltonian, Berry and Keating used the normal ordered operator [18]

$$\hat{H} = \frac{1}{2}(x\hat{p} + \hat{p}x) = -i\hbar\left(x\frac{d}{dx} + \frac{1}{2}\right), \quad x \in \mathbb{R}, \quad (11)$$

where x belongs to the real line and $\hat{p} = -i\hbar d/dx$ is the momentum operator. We shall show below that despite of being a natural quantization of the classical xp Hamiltonian, it does not reproduce the semiclassical spectrum obtained in the previous section. It is, however, of great interest to study it in detail since it is the basis of the rest of the work.

It is convenient to restrict x to the positive half-line, then (11) is equivalent to the expression

$$\hat{H} = \sqrt{x}\,\hat{p}\,\sqrt{x}, \quad x \geq 0. \quad (12)$$

$\hat{H}$ is an essentially self-adjoint operator acting on the Hilbert space $L^2(0, \infty)$ of square integrable functions in the half-line $\mathbb{R}_+ = (0, \infty)$ [23,24,30]. The eigenfunctions, with eigenvalue E, are given by

$$\psi_E(x) = \frac{1}{\sqrt{2\pi\hbar}} x^{-\frac{1}{2} + \frac{iE}{\hbar}}, \quad x > 0, \quad E \in \mathbb{R}, \quad (13)$$

and the spectrum is the real line $\mathbb{R}$. The normalization of (13) is given by the Dirac's delta function

$$\langle \psi_E | \psi_{E'} \rangle = \int_0^\infty dx\, \psi_E^*(x)\, \psi_{E'}(x) = \delta(E - E'). \quad (14)$$

The eigenfunctions (13) form an orthonormal basis of $L^2(0, \infty)$, that is related to the Mellin transform in the same manner that the eigenfunctions of the momentum operator $\hat{p}$, on the real line, are related to the Fourier transform [24]. If one takes x in the whole real line, then the spectrum of the Hamiltonian (11) is doubly degenerate. This degeneracy can be understood from the invariance of xp

under the parity transformation $x \to -x, p \to -p$, which allows one to split the eigenfunctions with energy E into even and odd sectors

$$\psi_E^{(e)}(x) = \frac{1}{\sqrt{2\pi\hbar}} |x|^{-\frac{1}{2}+\frac{iE}{\hbar}}, \quad \psi_E^{(o)}(x) = \frac{\text{sign } x}{\sqrt{2\pi\hbar}} |x|^{-\frac{1}{2}+\frac{iE}{\hbar}}, \quad x \in \mathbb{R}, \quad E \in \mathbb{R}. \quad (15)$$

Berry and Keating computed the Fourier transform of the even wave function $\psi_E^{(e)}(x)$ [18]

$$\begin{aligned}
\hat{\psi}_E^{(e)}(p) &= \frac{1}{\sqrt{2\pi\hbar}} \int_{-\infty}^{\infty} dx \, \psi_E^{(e)}(x) \, e^{-ipx/\hbar} \quad (16) \\
&= \frac{1}{\sqrt{2\pi\hbar}} |p|^{-\frac{1}{2}-\frac{iE}{\hbar}} (2\hbar)^{iE/\hbar} \frac{\Gamma\left(\frac{1}{4}+\frac{iE}{2\hbar}\right)}{\Gamma\left(\frac{1}{4}-\frac{iE}{2\hbar}\right)},
\end{aligned}$$

which means that the position and momentum eigenfunctions are each other's time reversed, giving a physical interpretation of the phase $\theta(t)$, see Equation (8). Choosing odd eigenfunctions leads to an equation similar to Equation (16) in terms of the gamma functions $\Gamma(\frac{3}{4} \pm \frac{iE}{2})$ that appear in the functional relation of the odd Dirichlet L-functions. Equation (16) is a consequence of the exchange $x \leftrightarrow p$ symmetry of the xp Hamiltonian, which is an important ingredient of the xp model.

Comments:

- Removing Connes's cutoff, i.e., $\Lambda \to \infty$, gives the quantum Hamiltonians (11) or (12), whose spectrum is a continuum. This shows that the negative term in Equation (5) does not correspond to missing spectral lines. In the next section we give a physical interpretation of this term in another context.
- xp is invariant under the scale transformation (dilations) $x \to Kx, p \to K^{-1}p$, with $K > 0$. An example of this transformation is the classical trajectory (1), whose infinitesimal generator is xp. Under dilations, $\ell_x \to K\ell_x$, $\ell_p \to K^{-1}\ell_p$, so, the condition $\ell_x \ell_p = 2\pi\hbar$ is preserved. Berry and Keating suggested to use integer dilations $K = n$, corresponding to evolution times $\log n$, to write [18]

$$\psi_E(x) \to \sum_{n=1}^{\infty} \psi_E(nx) = \frac{1}{\sqrt{2\pi\hbar}} x^{-\frac{1}{2}+\frac{iE}{\hbar}} \sum_{n=1}^{\infty} \frac{1}{n^{\frac{1}{2}-\frac{iE}{\hbar}}} = \frac{1}{\sqrt{2\pi\hbar}} x^{-\frac{1}{2}+\frac{iE}{\hbar}} \zeta(1/2 - iE/\hbar). \quad (17)$$

If there exists a physical reason for this quantity to vanish one would obtain the Riemann zeros E_n. Equation (17) could be interpreted as the breaking of the continuous scale invariance to discrete scale invariance.

Summary:

> ✗ The normal order quantization of xp does not exhibit any trace of the Riemann zeros.
> ✓ The phase of the zeta function appears in the Fourier transform of the xp eigenfunctions.

4. The Landau Model and XP

Let us consider a charged particle moving in a plane under the action of a perpendicular magnetic field and an electrostatic potential $V(x,y) \propto xy$ [29]. The Langrangian describing the dynamics is given, in the Landau gauge, by

$$\mathcal{L} = \frac{\mu}{2}(\dot{x}^2 + \dot{y}^2) - \frac{eB}{c}\dot{y}x - e\lambda xy, \quad (18)$$

where μ is the mass, e the electric charge, B the magnetic field, c the speed of light and λ a coupling constant that parameterizes the electrostatic potential. There are two normal modes with real, ω_c, and imaginary, ω_h, angular frequencies, describing a cyclotronic and a hyperbolic motion respectively. In the limit where $\omega_c \gg |\omega_h|$, only the Lowest Landau Level (LLL) is relevant and the effective Lagrangian becomes

$$\mathcal{L}_{\text{eff}} = p\dot{x} - |\omega_h|xp, \quad p = \frac{\hbar y}{\ell^2}, \quad \ell = \left(\frac{\hbar c}{eB}\right)^{1/2}, \tag{19}$$

where ℓ is the magnetic length, which is proportional to the radius of the cyclotronic orbits in the LLL. The coordinates x and y, which commute in the 2D model, after the projection to the LLL, become canonical conjugate variables, and the effective Hamiltonian is proportional to the xp Hamiltonian with the proportionality constant given by the angular frequency $|\omega_h|$ (this is the missing frequency factor mentioned in Section 2). The quantum Hamiltonian associated with the Lagrangian (18) is

$$\hat{H} = \frac{1}{2\mu}\left[\hat{p}_x^2 + \left(\hat{p}_y + \frac{\hbar}{\ell^2}x\right)^2\right] + e\lambda xy, \tag{20}$$

where $\hat{p}_x = -i\hbar\partial_x$ and $\hat{p}_y = -i\hbar\partial_y$. After a unitary transformation (20) becomes the sum of two commuting Hamiltonians corresponding to the cyclotronic and hyperbolic motions alluded to above

$$H = H_c + H_h, \tag{21}$$
$$H_c = \frac{\omega_c}{2}(\hat{p}^2 + \hat{q}^2), \quad H_h = \frac{|\omega_h|}{2}(\hat{P}Q + Q\hat{P}).$$

In the limit $\omega_c \gg |\omega_h|$ one has

$$\omega_c \simeq \frac{eB}{\mu c}, \quad |\omega_h| \sim \frac{\lambda c}{B}. \tag{22}$$

The unitary transformation that brings Equation (20) into Equation (21) corresponds to the classical canonical transformation

$$q = x + p_y, \quad p = p_x, \quad Q = -p_y, \quad P = y + p_x. \tag{23}$$

When $\omega_c \gg |\omega_h|$, the low energy states of H are the product of the lowest eigenstate of H_c, namely $\psi = e^{-q^2/2\ell^2}$, times the eigenstates of H_h that can be chosen as even or odd under the parity transformation $Q \to -Q$

$$\Phi_E^+(Q) = \frac{1}{|Q|^{\frac{1}{2}-iE}}, \quad \Phi_E^-(Q) = \frac{\text{sign}(Q)}{|Q|^{\frac{1}{2}-iE}}. \tag{24}$$

The corresponding wave functions are given by (we choose $|\omega_h| = 1$)

$$\psi_E^\pm(x,y) = C\int dQ\, e^{-iQy/\ell^2} e^{-(x-Q)^2/2\ell^2} \Phi_E^\pm(Q), \tag{25}$$

where C is a normalization constant, which yields

$$\psi_E^+(x,y) = C_E^+ e^{-\frac{x^2}{2\ell^2}} M\left(\frac{1}{4} + \frac{iE}{2}, \frac{1}{2}, \frac{(x-iy)^2}{2\ell^2}\right), \tag{26}$$
$$\psi_E^-(x,y) = C_E^-(x-iy)e^{-\frac{x^2}{2\ell^2}} M\left(\frac{3}{4} + \frac{iE}{2}, \frac{3}{2}, \frac{(x-iy)^2}{2\ell^2}\right),$$

where $M(a,b,z)$ is a confluent hypergeometric function [64]. Figure 2 shows that the maximum of the absolute value of ψ_E^+ is attained on the classical trajectory $E = xy$ (in units of $\hbar = \ell = 1$). This 2D representation of the classical trajectories is possible because in the LLL x and y become canonical conjugate variables and consequently the 2D plane coincides with the phase space (x,p).

To count the number of states with an energy below E one places the particle into a box: $|x| < L, |y| < L$ and impose the boundary conditions

$$\psi_E^+(x,L) = e^{ixL/\ell^2} \psi_E^+(L,x), \qquad (27)$$

which identifies the outgoing particle at $x = L$ with the incoming particle at $y = L$ up to a phase. The asymptotic behavior $L \gg \ell$ of (26) is

$$\psi_E^+(L,x) \simeq e^{-ixL/\ell^2 - x^2/2\ell^2} \frac{\Gamma\left(\frac{1}{2}\right)}{\Gamma\left(\frac{1}{4} + \frac{iE}{2}\right)} \left(\frac{L^2}{2\ell^2}\right)^{-\frac{1}{4} + \frac{iE}{2}}, \qquad (28)$$

$$\psi_E^+(x,L) \simeq e^{-x^2/2\ell^2} \frac{\Gamma\left(\frac{1}{2}\right)}{\Gamma\left(\frac{1}{4} - \frac{iE}{2}\right)} \left(\frac{L^2}{2\ell^2}\right)^{-\frac{1}{4} - \frac{iE}{2}},$$

that plugged into the BC (27) yields

$$\frac{\Gamma\left(\frac{1}{4} + \frac{iE}{2}\right)}{\Gamma\left(\frac{1}{4} - \frac{iE}{2}\right)} \left(\frac{L^2}{2\ell^2}\right)^{-iE} = 1, \qquad (29)$$

or using Equation (8)

$$e^{2i\theta(E)} \left(\frac{L^2}{2\pi\ell^2}\right)^{-iE} = 1. \qquad (30)$$

Hence the number of states $n(E)$ with energy less that E is given by

$$n(E) \simeq \frac{E}{2\pi} \log\left(\frac{L^2}{2\pi\ell^2}\right) + 1 - \langle n(E) \rangle, \qquad (31)$$

whose asymptotic behavior coincides with Connes's Formula (5) for a cutoff $\Lambda = L/\ell$. In fact, the term $\langle n(E) \rangle$ is the exact Riemann–von Mangoldt Formula (6).

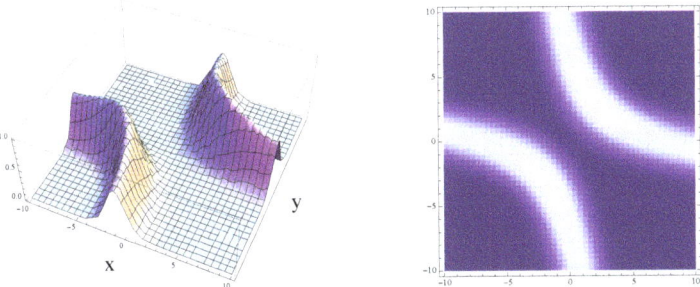

Figure 2. Plot of $|\psi_E^+(x,y)|$ for $E = 10$ in the region $-10 < x, y < 10$. **Left**: 3D representation, **Right**: density plot.

Summary:

> ✓ The Landau model with a xy potential provides a physical realization of Connes's xp model.
> ✓ The finite size effects in the spectrum are given by the Riemann–von Mangoldt formula.
> ✗ There are no missing spectral lines in the physical realizations of xp à la Connes.

5. The XP Model Revisited

An intuitive argument of why the quantum Hamiltonian $(x\hat{p} + \hat{p}x)/2$ has a continuum spectrum is that the classical trajectories of xp are unbounded. Therefore, to have a discrete spectrum one should modify xp to bound the trajectories. This is achieved by the classical Hamiltonian [32]

$$H_I = x\left(p + \frac{\ell_p^2}{p}\right), \quad x \geq \ell_x. \tag{32}$$

For $|p| \gg \ell_p$, a classical trajectory with energy E satisfies $E \simeq xp$, but for $|p| \sim \ell_p$, the coordinate $|x|$ slows down, reaches a maximum and goes back to the value ℓ_x, where it bounces off starting again at high momentum. In this manner one gets a periodic orbit (see Figure 3)

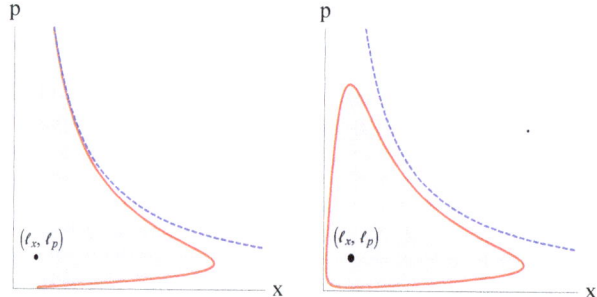

Figure 3. Classical trajectories of the Hamiltonians (32) (**left**) and (35) (**right**) in phase space with $E > 0$. The dashed lines denote the hyperbola $E = xp$. (ℓ_x, ℓ_p) is a fixed-point solution of the classical equations generated by (32) and (35).

$$\begin{aligned} x(t) &= \frac{\ell_x}{|p_0|} e^{2t} \sqrt{(p_0^2 + \ell_p^2) e^{-2t} - \ell_p^2}, \quad 0 \leq t \leq T_E, \\ p(t) &= \pm\sqrt{(p_0^2 + \ell_p^2) e^{-2t} - \ell_p^2}, \end{aligned} \tag{33}$$

where T_E is the period given by (we take $E > 0$)

$$T_E = \cosh^{-1}\frac{E}{2\ell_x \ell_p} \to \log\frac{E}{\ell_x \ell_p} \quad (E \gg \ell_x \ell_p). \tag{34}$$

The asymptotic value of T_E is the time lapse it takes a particle to go from $x = \ell_x$ to $x = E/\ell_p$ in the xp model.

The exchange symmetry $x \leftrightarrow p$ of xp is broken by the Hamiltonian (32). To restore it, Berry and Keating proposed the $x - p$ symmetric Hamiltonian [36]

$$H_{II} = \left(x + \frac{\ell_x^2}{x}\right)\left(p + \frac{\ell_p^2}{p}\right), \quad x \geq 0. \tag{35}$$

Here the classical trajectories turn clockwise around the point (ℓ_x, ℓ_p), and for $x \gg \ell_x$ and $p \gg \ell_p$, approach the parabola $E = xp$ (see Figure 3). The semiclassical analysis of (32) and (35) reproduce the asymptotic behavior of Equation (2) to leading orders $E \log E$ and E, but differ in the remaining terms.

The two models discussed above have the general form

$$H = U(x)p + \ell_p^2 \frac{V(x)}{p}, \quad x \in D, \tag{36}$$

where $U(x)$ and $V(x)$ are positive functions defined in an interval D of the real line. H_I corresponds to $U(x) = V(x) = x, D = (\ell_x, \infty)$, and H_{II} corresponds to $U(x) = V(x) = x + \ell_x^2/x, D = (0, \infty)$. The classical Hamiltonian (36) can be quantized in terms of the operator

$$\hat{H} = \sqrt{U}\, \hat{p}\, \sqrt{U} + \ell_p^2 \sqrt{V}\, \hat{p}^{-1}\, \sqrt{V}, \tag{37}$$

where $\hat{p}^{-1}$ is pseudo-differential operator

$$\left(\hat{p}^{-1}\psi\right)(x) = -\frac{i}{\hbar}\int_x^\infty dy\, \psi(y), \tag{38}$$

which satisfies that $\hat{p}\, \hat{p}^{-1} = \hat{p}^{-1}\hat{p} = \mathbf{1}$ acting on functions which vanish sufficiently fast in the limit $x \to \infty$. The action of $\hat{H}$ is

$$(\hat{H}\psi)(x) = -i\hbar\sqrt{U(x)}\frac{d}{dx}\left\{\sqrt{U(x)}\psi(x)\right\} - \frac{i\ell_p^2}{\hbar}\int_x^\infty dy\, \sqrt{V(x)V(y)}\,\psi(y). \tag{39}$$

The normal order prescription that leads from (36) to (39) will be derived in Section 7 in the case where $U(x) = V(x) = x$, but holds in general [38]. We want the Hamiltonian (37) to be self-adjoint, that is [65,66]

$$\langle \psi_1 | \hat{H} | \psi_2 \rangle = \langle \hat{H}\psi_1 | \psi_2 \rangle. \tag{40}$$

When the interval is $D = (\ell_x, \infty)$, Equation (40) holds for wave functions that vanishes sufficiently fast at infinity and satisfy the non-local boundary condition

$$\hbar\, e^{i\vartheta} \sqrt{U(\ell_x)}\, \psi(\ell_x) = \ell_p \int_{\ell_x}^\infty dx\, \sqrt{V(x)}\, \psi(x), \tag{41}$$

where $\vartheta \in [0, 2\pi)$ parameterizes the self-adjoint extensions of $\hat{H}$. The quantum Hamiltonian associated with (32) is

$$\hat{H}_I = \sqrt{x}\, \hat{p}\, \sqrt{x} + \ell_p^2 \sqrt{x}\, \hat{p}^{-1}\, \sqrt{x}, \quad x \geq \ell_x, \tag{42}$$

and its eigenfunctions are proportional to (see Figure 4)

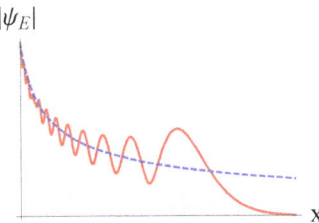

Figure 4. Absolute values of the wave function $\psi_E(x)$, given in Equation (43) (continuous line), and $x^{-\frac{1}{2}}$ (dashed line).

$$\psi_E(x) = x^{\frac{iE}{2\hbar}} K_{\frac{1}{2} - \frac{iE}{2\hbar}}\left(\frac{\ell_p x}{\hbar}\right) \propto \begin{cases} x^{-\frac{1}{2} + \frac{iE}{\hbar}} & x \ll \frac{E}{2\ell_p}, \\ x^{-\frac{1}{2} + \frac{iE}{2\hbar}} e^{-\ell_p x/\hbar} & x \gg \frac{E}{2\ell_p}, \end{cases} \quad (43)$$

where $K_\nu(z)$ is the modified K-Bessel function [64]. For small values of x, the wave functions (43) behave as those of the xp Hamiltonian, given in Equation (13), while for large values of x they decay exponentially giving a normalizable state. The boundary condition (41) reads in this case

$$\hbar\, e^{i\vartheta}\,\sqrt{\ell_x}\,\psi(\ell_x) = \ell_p \int_{\ell_x}^\infty dx\,\sqrt{x}\,\psi(x), \quad (44)$$

and substituting (43) yields the equation for the eigenenergies E_n,

$$e^{i\vartheta} K_{\frac{1}{2} - \frac{iE}{2\hbar}}\left(\frac{\ell_x \ell_p}{\hbar}\right) - K_{\frac{1}{2} + \frac{iE}{2\hbar}}\left(\frac{\ell_x \ell_p}{\hbar}\right) = 0. \quad (45)$$

For $\vartheta = 0$ or π, the eigenenergies form time reversed pairs $\{E_n, -E_n\}$, and for $\vartheta = 0$, there is a zero-energy state $E = 0$. Considering that the Riemann zeros form pairs $s_n = 1/2 \pm i t_n$, with t_n real under the RH, and that $s = 1/2$ is not a zero of $\zeta(s)$, we are led to the choice $\vartheta = \pi$. On the other hand, using the asymptotic behavior

$$K_{a + \frac{it}{2}}(z) \longrightarrow \sqrt{\frac{\pi}{t}}\left(\frac{t}{z}\right)^a e^{-\pi t/4}\, e^{\frac{i\pi}{2}(a - \frac{1}{2})}\left(\frac{t}{ze}\right)^{it/2}, \quad a > 0, t \gg 1, \quad (46)$$

one derives in the limit $|E| \gg \hbar$,

$$K_{\frac{1}{2} + \frac{iE}{2\hbar}}\left(\frac{\ell_x \ell_p}{\hbar}\right) + K_{\frac{1}{2} - \frac{iE}{2\hbar}}\left(\frac{\ell_x \ell_p}{\hbar}\right) = 0 \longrightarrow \cos\left(\frac{E}{2\hbar}\log\frac{E}{\ell_x \ell_p e}\right) = 0, \quad (47)$$

hence the number of eigenenergies in the interval $(0, E)$ is given asymptotically by

$$n(E) \simeq \frac{E}{2\pi\hbar}\left(\log\frac{E}{\ell_x \ell_p} - 1\right) - \frac{1}{2} + O(E^{-1}). \quad (48)$$

This equation agrees with the leading terms of the semiclassical spectrum (2) and the average Riemann zeros (3) under the identifications (4). Concerning the classical Hamiltonian (35), Berry and Keating obtained, by a semiclassical analysis, the asymptotic behavior of the counting function $n(E)$

$$n(t) \simeq \frac{t}{2\pi}\left(\log\frac{t}{2\pi} - 1\right) - \frac{8\pi}{t}\log\frac{t}{2\pi} + \ldots, \quad t \gg 1, \quad (49)$$

where $t = E/\hbar$ and $\ell_x \ell_p = 2\pi\hbar$. Again, the first two leading terms agree with Riemann's Formula (3), while the next leading corrections are different from (48). In both cases, the constant $7/8$ in Riemann's Formula (3) is missing.

Summary:

✓ The Berry–Keating xp model can be implemented quantum mechanically.
✗ The classical xp Hamiltonian must be modified with ad-hoc terms to have bounded trajectories.
✗ In the quantum theory the latter terms become non-local operators.
✗ The modified xp quantum Hamiltonian related to the average Riemann zeros is not unique.
✗ There is no trace of the exact Riemann zeros in the spectrum of the modified xp models.

6. The Spacetime Geometry of the Modified XP Models

In this section, we show that the modified xp Hamiltonian (36) is a disguised general theory of relativity [35]. Let us first consider the Langrangian of the xp model,

$$L = p\dot{x} - H = p\dot{x} - xp. \tag{50}$$

In classical mechanics, where $H = p^2/2m + V(x)$, the Lagrangian can be expressed solely in terms of the position x and velocity $\dot{x} = dx/dt$. This is achieved by writing the momentum in terms of the velocity by means of the Hamilton equation $\dot{x} = \partial H/\partial p = p/m$. However, in the xp model the momentum p is not a function of the velocity because $\dot{x} = \partial H/\partial p = x$. Hence the Lagrangian (50) cannot be expressed uniquely in terms of x and $\dot{x}$. The situation changes radically for the Hamiltonian (36) whose Lagrangian is given by

$$L = p\dot{x} - H = p\dot{x} - U(x)p - \ell_p^2 \frac{V(x)}{p}. \tag{51}$$

Here the equation of motion

$$\dot{x} = \frac{\partial H}{\partial p} = U(x) - \ell_p^2 \frac{V(x)}{p^2}, \tag{52}$$

allows one to write p in terms of x and $\dot{x}$,

$$p = \eta \ell_p \sqrt{\frac{V(x)}{U(x) - \dot{x}}}, \qquad \eta = \text{sign}\, p, \tag{53}$$

where $\eta = \pm 1$ is the sign of the momentum that is a conserved quantity. The positivity of $U(x)$ and $V(x)$, imply that the velocity $\dot{x}$ must never exceed the value of $U(x)$. Substituting (53) back into (51), yields the action

$$S_\eta = -\ell_p \eta \int \sqrt{-ds^2}, \tag{54}$$

which, for either sign of η, is the action of a relativistic particle moving in a 1+1 dimensional spacetime metric

$$ds^2 = 4V(x)(-U(x)dt^2 + dtdx). \tag{55}$$

The parameter ℓ_p plays the role of mc where m is the mass of the particle and c is the speed of light. This result implies that the classical trajectories of the Hamiltonian (36) are the geodesics of the metric (55). The unfamiliar form of (36) is due to a special choice of spacetime coordinates where the component g_{xx} of the metric vanishes. A diffeomorphism of x permits to set $V(x) = U(x)$. The scalar curvature of the metric (55), in this *gauge*, is

$$R(x) = -2\frac{\partial_x^2 V(x)}{V(x)}, \tag{56}$$

and vanishes for the models $V(x) = x$ and $V(x) = $ constant. For the Hamiltonian (35) one obtains $R(x) = -4\ell_x^2/(x(x^2 + \ell_x^2))$ which vanishes asymptotically.

The flatness of the metric associated with the Hamiltonian (32) implies the existence of coordinates x^0, x^1 where (55) takes the Minkowski form

$$ds^2 = \eta_{\mu\nu} dx^\mu dx^\nu, \qquad \text{diag}\, \eta_{\mu\nu} = (-1, 1). \tag{57}$$

The change of variables is given by

$$t = \frac{1}{2}\log(x^0 + x^1), \qquad x = \sqrt{-(x^0)^2 + (x^1)^2}. \tag{58}$$

Let $\mathcal{U}$ denote the spacetime domain of the model. In both coordinates it reads

$$\mathcal{U} = \{(t,x) \mid t \in (-\infty, \infty), \, x \geq \ell_x\} = \left\{(x^0, x^1) \mid x^0 \in (-\infty, \infty), \, x^1 \geq \sqrt{(x^0)^2 + \ell_x^2}\right\}. \tag{59}$$

The boundary of $\mathcal{U}$, denoted by $\partial \mathcal{U}$, is the hyperbola $x^1 = \sqrt{(x^0)^2 + \ell_x^2}$, that passes through the point $(x^0, x^1) = (0, \ell_x)$, (see Figure 5).

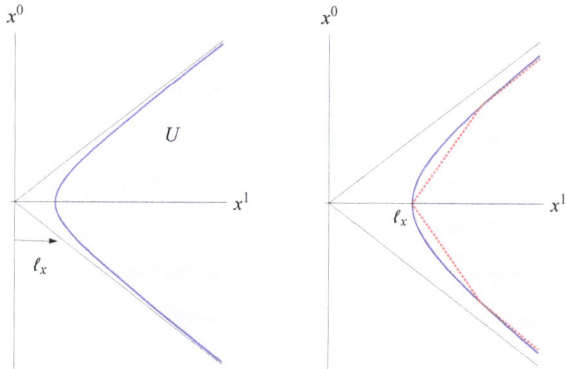

Figure 5. (**Left**): Domain $\mathcal{U}$ of Minkowski spacetime given in Equation (59). (**Right**): The classical trajectory given in Equation (33), and plotted in Figure 3-left, becomes a straight line that bounces off regularly at the boundary (dotted line).

A convenient parametrization of the coordinates x^μ is given by the Rindler variables ρ and ϕ [67]

$$x^0 = \rho \sinh \phi, \qquad x^1 = \rho \cosh \phi, \tag{60}$$

or in light-cone coordinates

$$x^\pm = x^0 \pm x^1 = \pm \rho e^{\pm \phi}, \tag{61}$$

where the Minkowski metric becomes

$$ds^2 = -dx^+ dx^- = d\rho^2 - \rho^2 d\phi^2. \tag{62}$$

These coordinates describe the right wedge of Rindler spacetime in 1+1 dimensions

$$\mathcal{R}_+ = \left\{(x^0, x^1) \mid x^0 \in (-\infty, \infty), \, x^1 \geq |x^0|\right\} = \{(\rho, \phi) \mid \phi \in (-\infty, \infty), \, \rho > 0\}. \tag{63}$$

Notice that $\mathcal{U} \subset \mathcal{R}_+$. The boundary $\partial \mathcal{U}$ corresponds to the hyperbola $\rho = \ell_x$ that is the worldline of a particle moving with uniform acceleration equal to $1/\ell_x$ (in units $c = 1$). The Rindler variables are the ones used to study the Unruh effect [68].

Let us now consider the classical Hamiltonian (35). The underlying metric is given by Equation (55) with $U(x) = V(x) = x + \ell_x^2/x$. The change of variables

$$t = \frac{1}{2}\log(x^0 + x^1), \qquad x = \sqrt{-(x^0)^2 + (x^1)^2 - \ell_x^2}, \tag{64}$$

brings the metric to the form

$$ds^2 = \frac{-(x^0)^2 + (x^1)^2}{-(x^0)^2 + (x^1)^2 - \ell_x^2} \eta_{\mu\nu} dx^\mu dx^\nu = \frac{\rho^2}{\rho^2 - \ell_x^2}(d\rho^2 - \rho^2 d\phi^2), \qquad \rho \geq \ell_x, \qquad (65)$$

which in the limit $\rho \to \infty$ converges to the flat metric (62).

Summary:

> ✓ The classical modified xp models are general relativistic theories in 1+1 dimensions.
> ✓ $H = x(p + \ell_p^2/p)$ is related to a domain $\mathcal{U}$ of Rindler spacetime.
> ✓ l_p is the mass of the particle.
> ✓ $1/\ell_x$ is the acceleration of a particle whose worldline is the boundary of $\mathcal{U}$.
> ? Relativistic quantum field theory of the modified xp models.

7. Diracization of $H = X(P + \ell_p^2/P)$

In this section, we show that the Dirac theory provides the relativistic quantum version of the modified xp models [42]. We shall focus on the classical Hamiltonian $H = x(p + \ell_p^2/p)$ because the flatness of the associated spacetime makes the computations easier, but the result is general: the quantum Hamiltonian (39) can be derived from the Dirac equation in a curved spacetime with metric (55) [38].

The Dirac action of a fermion with mass m in the spacetime domain (59) is given by (in units $\hbar = c = 1$)

$$S = \frac{i}{2} \int_\mathcal{U} dx^0 dx^1 \, \bar{\psi}(\slashed{\partial} + im)\psi, \qquad (66)$$

where ψ is a two-component spinor, $\bar{\psi} = \psi^\dagger \gamma^0$, $\slashed{\partial} = \gamma^\mu \partial_\mu$ ($\partial_\mu = \partial/\partial x^\mu$), and γ^μ are the 2d Dirac matrices written in terms of the Pauli matrices $\sigma^{x,y}$ as

$$\gamma^0 = \sigma^x, \qquad \gamma^1 = -i\sigma^y, \qquad \psi = \begin{pmatrix} \psi_- \\ \psi_+ \end{pmatrix}. \qquad (67)$$

The variational principle applied to (66) provides the Dirac equation

$$(\slashed{\partial} + im)\psi = 0, \qquad (68)$$

and the boundary condition

$$\dot{x}^- \psi_-^\dagger \delta\psi_- - \dot{x}^+ \psi_+^\dagger \delta\psi_+ = 0, \qquad (69)$$

where $\dot{x}^\pm = dx^\pm/d\phi = \ell_x e^{\pm\phi}$ is the vector tangent to the boundary $\partial\mathcal{U}$ in the light-cone coordinates $x^\pm = x^0 \pm x^1$. The Dirac equation reads in components

$$(\partial_0 - \partial_1)\psi_+ + im\psi_- = 0, \qquad (\partial_0 + \partial_1)\psi_- + im\psi_+ = 0. \qquad (70)$$

If $m = 0$ then $\psi_\pm$ depends only $x^\pm$, and so the fields propagate to the left, $\psi_+(x^+)$, or to the right, $\psi_-(x^-)$, at the speed of light. The derivatives in Equation (70) can be written in terms the variables t and x using Equation (58),

$$\partial_0 - \partial_1 = -\frac{2e^{2t}}{x}\partial_x, \qquad \partial_0 + \partial_1 = e^{-2t}(\partial_t + x\partial_x). \qquad (71)$$

Let us denote by $\tilde{\psi}_{\mp}(t,x)$ the fermion fields in the coordinates t,x and by $\psi_{\mp}(x^0,x^1)$ the fields in the coordinates x^0, x^1. The relation between these fields is given by the transformation law

$$\psi_- = \left(\frac{\partial x}{\partial x^-}\right)^{\frac{1}{2}} \tilde{\psi}_- = (2x)^{-\frac{1}{2}} e^t \tilde{\psi}_-, \quad \psi_+ = \left(\frac{\partial x}{\partial x^+}\right)^{\frac{1}{2}} \tilde{\psi}_- = (x/2)^{\frac{1}{2}} e^{-t} \tilde{\psi}_+. \tag{72}$$

Plugging Equations (71) and (72) into (70) gives

$$i\partial_t \tilde{\psi}_- = -i\sqrt{x}\partial_x \left(\sqrt{x}\tilde{\psi}_-\right) + mx\tilde{\psi}_+, \quad \partial_x(\sqrt{x}\tilde{\psi}_+) = im\sqrt{x}\tilde{\psi}_-. \tag{73}$$

The second equation is readily integrated

$$\tilde{\psi}_+(x,t) = -\frac{im}{\sqrt{x}} \int_x^\infty dy \sqrt{y} \tilde{\psi}_-(y,t), \tag{74}$$

and replacing it into the first equation in (73) gives

$$i\partial_t \tilde{\psi}_-(x,t) = -i\sqrt{x}\partial_x \left(\sqrt{x}\tilde{\psi}_-\right) - im^2 \sqrt{x} \int_x^\infty dy \sqrt{y} \tilde{\psi}_-(y,t). \tag{75}$$

This is the Schrödinger equation with Hamiltonian (42) and the relation $m = \ell_p$ found in the previous section. The non-locality of the Hamiltonian (42) is a consequence of the special coordinates t, x where the component $\tilde{\psi}_+$ becomes non-dynamical and depends non-locally on the component $\tilde{\psi}_-$ that is identified with the wave function of the modified xp model. Similarly, the boundary condition (44) can be derived from Equation (69) as follows. In Rindler coordinates the latter equation reads

$$e^{-\phi}\psi_-^\dagger(\ell_x,\phi)\,\delta\psi_-(\ell_x,\phi) = e^\phi \psi_+^\dagger(\ell_x,\phi)\,\delta\psi_+(\ell_x,\phi), \quad \forall \phi, \tag{76}$$

that is solved by

$$-ie^{i\vartheta}e^{-\phi/2}\,\psi_-(\ell_x,\phi) = e^{\phi/2}\,\psi_+(\ell_x,\phi), \quad \forall \phi, \tag{77}$$

where $\vartheta \in [0,2\pi)$. Using Equation (72) this equation becomes

$$-ie^{i\vartheta}\,\tilde{\psi}_-(\ell_x,t) = \tilde{\psi}_+(\ell_x,t), \quad \forall t, \tag{78}$$

that together with Equation (74) yields Equation (44). This completes the derivation of the quantum Hamiltonian and boundary condition associated to $H = x(p + \ell_p^2/p)$. The eigenfunctions and eigenvalue equation of this model were found in Section 5. However, we shall rederive them in alternative way that will provide new insights in the next section.

Let us start by constructing the plane wave solutions of the Dirac Equation (70),

$$\begin{pmatrix} \psi_- \\ \psi_+ \end{pmatrix} \propto \begin{pmatrix} e^{i\pi/4}e^{\beta/2} \\ e^{-i\pi/4}e^{-\beta/2} \end{pmatrix} e^{i(-p^0 x^0 + p^1 x^1)}, \tag{79}$$

where (p^0, p^1) is the energy-momentum vector parameterized in terms of the rapidity variable β

$$(p^0)^2 - (p^1)^2 = m^2, \tag{80}$$
$$p^0 = im\sinh\beta, \quad p^1 = im\cosh\beta, \quad \beta \in (-\infty,\infty).$$

In Rindler coordinates these plane wave solutions decay exponentially with the distance as corresponds to a localized wave function

$$e^{i(-p^0 x^0 + p^1 x^1)} = e^{-m\rho \cosh(\beta - \phi)} \to 0, \quad \text{as} \quad \rho \to \infty. \tag{81}$$

The general solution of the Dirac equation is given by the linear superposition of plane waves (79). The superposition that reproduces the eigenfunctions of the modified xp model is

$$\psi_\mp(\rho, \phi) = e^{\pm i\pi/4} \int_{-\infty}^{\infty} d\beta \, e^{-iE\beta/2} e^{\pm\beta/2} e^{-m\rho \cosh(\beta - \phi)} \tag{82}$$
$$= 2 e^{\pm i\pi/4} e^{(\pm\frac{1}{2} - \frac{iE}{2})\phi} K_{\frac{1}{2} \mp \frac{iE}{2}}(m\rho),$$

that replaced in Equation (77) gives

$$e^{i\vartheta} K_{\frac{1}{2} - \frac{iE}{2}}(m\ell_x) - K_{\frac{1}{2} + \frac{iE}{2}}(m\ell_x) = 0, \tag{83}$$

which coincides with the eigenvalue Equation (45) with $m = \ell_p$. Setting $m\ell_x = 2\pi$ and $\vartheta = \pi$, brings Equation (83) to the form

$$\zeta_H(t) \equiv K_{\frac{1}{2} + \frac{it}{2}}(2\pi) + K_{\frac{1}{2} - \frac{it}{2}}(2\pi) = 0. \tag{84}$$

Summary:

✓ The spectrum of a relativistic massive fermion in the domain $\mathcal{U}$ agrees with the average Riemann zeros.
? Does this result provide a hint on a physical realization of the Riemann zeros.

8. ξ-Functions: Pólya's Is Massive and Riemann's Is Massless

The function $\zeta_H(t)$ appearing in Equation (84) reminds the *fake* ξ function defined by Pólya in 1926 [69,70]

$$\xi^*(t) = 4\pi^2 \left(K_{\frac{9}{4} + \frac{it}{2}}(2\pi) + K_{\frac{9}{4} - \frac{it}{2}}(2\pi) \right). \tag{85}$$

This function shares several properties with the Riemann ξ function

$$\xi(t) = \frac{1}{4} s(s-1) \Gamma\left(\frac{s}{2}\right) \pi^{-s/2} \zeta(s), \quad s = \frac{1}{2} + it, \tag{86}$$

namely, $\xi^*(t)$ is an entire and even function of t, its zeros lie on the real axis and behave asymptotically like the average Riemann zeros, as shown by the expansion obtained using Equation (46)

$$\xi^*(t) \xrightarrow{t \to \infty} 2^{3/4} \pi^{-7/4} t^{7/4} e^{-\pi t/4} \cos\left(\frac{t}{2} \log\left(\frac{t}{2\pi e}\right) + \frac{7\pi}{8}\right). \tag{87}$$

The zeros of $\xi(t), \zeta_H(t)$ and $\xi^*(t)$ are plotted in Figure 6. The slight displacement between the two top curves is due to the constant $7\pi/8$ appearing in the argument of the cosine function in Equation (87) as compared to that in Equation (47).

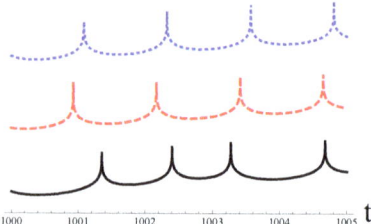

Figure 6. From bottom to top: plot of $-\log|\xi(t)|$ (Riemann zeros), $-\log|\xi_H(t)|$ (eigenvalues of the Hamiltonian (42) with $\ell_x\ell_p = 2\pi$) and $-\log|\xi^*(t)|$ (Pólya zeros). The cusp represents the zeros of the corresponding functions.

The similarity between $\xi_H(t)$ and $\xi^*(t)$, and the relation between $\xi^*(t)$ and $\xi(t)$ provides a hint on the field theory underlying the Riemann zeros. To show this, we shall review how Pólya arrived at $\xi^*(t)$. The starting point is the expression of $\xi(t)$ as a Fourier transform [3]

$$\xi(t) = 4 \int_1^\infty dx \frac{d[x^{\frac{3}{2}}\psi'(x)]}{dx} x^{-\frac{1}{4}} \cos\left(\frac{t \log x}{2}\right), \quad (88)$$

$$\psi(x) = \sum_{n=1}^\infty e^{-n^2\pi x}, \quad \psi'(x) = \frac{d\psi(x)}{dx}.$$

In the variable $x = e^\beta$ these equations become,

$$\xi(t) = \int_0^\infty d\beta \, \Phi(\beta) \cos\frac{t\beta}{2}, \quad (89)$$

$$\Phi(\beta) = 2\pi e^{5\beta/4} \sum_{n=1}^\infty \left(2\pi e^\beta n^2 - 3\right) n^2 e^{-\pi n^2 e^\beta}.$$

The function $\Phi(\beta)$ behaves asymptotically as

$$\Phi(\beta) \to 4\pi^2 e^{9\beta/4} e^{-\pi e^\beta}, \quad \beta \to \infty, \quad (90)$$

which Pólya replaced by the following expression (see Figure 7).

$$\Phi^*(\beta) = 4\pi^2 \left(e^{9\beta/4} + e^{-9\beta/4}\right) e^{-\pi(e^\beta + e^{-\beta})}. \quad (91)$$

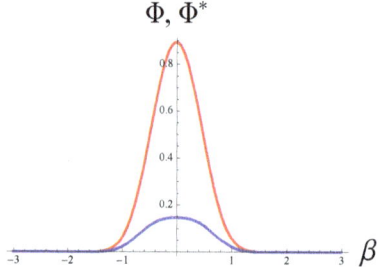

Figure 7. Plot of $\Phi(\beta)$ (red on line), and $\Phi^*(\beta)$ (blue on line). Outside the region $|\beta| < 1$ the difference is very small.

The function $\zeta^*(t)$ is defined as the Fourier transform of $\Phi^*(\beta)$,

$$\zeta^*(t) = \int_0^\infty d\beta \, \Phi^*(\beta) \cos \frac{t\beta}{2}, \qquad (92)$$

which finally gives Equation (85). The function (84) can also be written as the Fourier transform

$$\zeta_H(t) = \int_0^\infty d\beta \, \Phi_H(\beta) \cos \frac{t\beta}{2}, \qquad (93)$$

with

$$\Phi_H(\beta) = (e^{\beta/2} + e^{-\beta/2}) e^{-2\pi \cosh \beta} \qquad (94)$$

Observe that the term $e^{-2\pi \cosh \beta}$ appears in $\Phi_H(\beta)$ and $\Phi^*(\beta)$. The origin of this term in the Dirac theory is the plane wave factor (81) of a fermion with mass m located at the boundary $\rho = \ell_x$ with $m\ell_x = 2\pi$. This observation suggests that the Pólya ζ^* function arises in the relativistic theory of a massive particle with scaling dimension $9/4$, rather than $1/2$, that corresponds to a fermion (this would explain the different order of the corresponding Bessel functions). The approximation $\Phi(\beta) \simeq \Phi^*(\beta)$, that is $e^{-\pi e^\beta} \simeq e^{-2\pi \cosh \beta}$, can then be understood as the replacement of a massless particle by a massive one. Indeed, the energy-momentum of a massless right moving particle is given by $p^0 = p^1 = \Lambda e^\beta$, where Λ is an energy scale. The corresponding plane wave factor is $e^{-\pi e^\beta}$, with $\Lambda = \pi$. For large rapidities, $\beta \gg 1$, a massive particle behaves as a massless one, i.e., $e^{-2\pi \cosh \beta} \simeq e^{-\pi e^\beta}$. However, for small rapidities this is not the case. These arguments suggest that the field theory underlying the Riemann ζ function, if it exists, must associated with a massless particle.

Summary:

> ✓ The zeros of the Polya ζ^* function behave as the spectrum of a relativistic massive particle in the domain $\mathcal{U}$.
> ? Polya's construction of ζ^* suggests that the Riemann's ζ function is related to a massless particle.

9. The Massive Dirac Model in Rindler Coordinates

Let us formulate the Dirac theory in Rindler coordinates. Under a Lorentz transformation with boost parameter λ, the light-cone coordinates $x^\pm$ and the Dirac spinors $\psi_\pm$ transform as

$$x^\pm \to e^{\mp \lambda} x^\pm, \qquad \psi_\pm \to e^{\pm \lambda/2} \psi_\pm, \qquad (95)$$

and the Rindler coordinates (60) as

$$\phi \to \phi - \lambda, \qquad \rho \to \rho. \qquad (96)$$

Hence the new spinor fields $\chi_\pm$ defined as

$$\chi_\pm = e^{\pm \phi/2} \psi_\pm, \qquad (97)$$

remain invariant under (96). The Rindler wedge $\mathcal{R}_+$, and its domain $\mathcal{U}$, are also invariant under Lorentz transformations. The Dirac action (66) written in terms of the spinors $\chi_\pm$ reads

$$S = \frac{i}{2} \int_{-\infty}^\infty d\phi \int_{\ell_x}^\infty d\rho \left[\chi_-^\dagger (\partial_\phi + \rho \partial_\rho + \tfrac{1}{2}) \chi_- + \chi_+^\dagger (\partial_\phi - \rho \partial_\rho - \tfrac{1}{2}) \chi_+ + im\rho \, (\chi_-^\dagger \chi_+ + \chi_+^\dagger \chi_-) \right], \qquad (98)$$

while the Dirac Equation (68) and the boundary condition (77) become

$$(\partial_\phi \pm \rho \partial_\rho \pm \tfrac{1}{2}) \chi_\mp + im\rho \chi_\pm = 0, \qquad (99)$$

and

$$-ie^{i\vartheta}\chi_- = \chi_+ \quad \text{at } \rho = \ell_x. \tag{100}$$

The infinitesimal generator of translations of the Rindler time ϕ, acting on the spinor wave functions, is the Rindler Hamiltonian H_R, which can be read off from (99)

$$i\partial_\phi \chi = H_R \chi, \quad \chi = \begin{pmatrix} \chi_- \\ \chi_+ \end{pmatrix}, \tag{101}$$

$$H_R = \begin{pmatrix} -i(\rho\,\partial_\rho + \tfrac{1}{2}) & m\rho \\ m\rho & i(\rho\,\partial_\rho + \tfrac{1}{2}) \end{pmatrix} = \sqrt{\rho}\,\hat{p}_\rho\sqrt{\rho}\,\sigma^z + m\rho\,\sigma^x, \tag{102}$$

where $\hat{p}_\rho = -i\partial/\partial\rho$, is the momentum operator conjugate to the radial coordinate ρ. Notice that the operator

$$H_{\rho p_\rho} = -i(\rho\,\partial_\rho + \tfrac{1}{2}) = \tfrac{1}{2}(\rho\,\hat{p}_\rho + \hat{p}_\rho \rho) = \sqrt{\rho}\,\hat{p}_\rho\sqrt{\rho}, \tag{103}$$

coincides with Equation (11) with the identification $x = \rho$ (in units $\hbar = 1$). The eigenfunctions of (103) are

$$H_{\rho p_\rho}\psi_E = E\psi_E, \quad \psi_E = \frac{1}{\sqrt{2\pi}}\rho^{-1/2+iE}, \tag{104}$$

with real eigenvalue E for $\rho > 0$ (recall Equation (13)). Thus, H_R consists of two copies of xp, with different signs corresponding to opposite fermion chiralities that are coupled by the mass term $m\rho\sigma^x$.

The scalar product of two wave functions, in the domain $\mathcal{U}$, can be defined as

$$\langle \chi_1 | \chi_2 \rangle = \int_{\ell_x}^\infty d\rho\,(\chi_{1,-}^*\chi_{2,-} + \chi_{1,+}^*\chi_{2,+}). \tag{105}$$

The Hamiltonian H_R is Hermitian with this scalar product acting on wave functions that satisfy Equation (100) and vanish sufficiently fast at infinity, i.e., $\lim_{\rho\to\infty}\rho^{1/2}\chi_\pm(\rho,\phi) = 0$. The eigenvalues and eigenvectors of the Hamiltonian (102), are given by the solutions of the Schrödinger equation

$$H_R\chi = E_R\chi, \quad \chi_\pm(\rho,\phi) = e^{-iE_R\phi \mp i\pi/4}K_{\tfrac{1}{2}\pm iE_R}(m\rho), \quad \rho \geq \ell_x, \tag{106}$$

which coincide with Equation (82) with the identification

$$E_R = \frac{E}{2}. \tag{107}$$

The factor of $1/2$ comes from the relation $e^{2t} = x^0 + x^1 = \rho e^\phi$ (see Equation (58)), that implies $e^{-iE_R\phi} \propto e^{-iEt}$. The Rindler eigenenergies are obtained replacing E by $2E_R$ in Equation (83).

Comments:

- The Dirac Hamiltonian associated with the metric (65) is

$$H = \begin{pmatrix} h & m\rho\Lambda \\ m\rho\Lambda & -h \end{pmatrix}, \quad h = -i\left(\rho\partial_\rho + \tfrac{1}{2} + \tfrac{1}{2}\rho\partial_\rho(\log\Lambda)\right), \quad \Lambda = \frac{\rho}{\sqrt{\rho^2 - \ell_x^2}}. \tag{108}$$

In the limit $\rho \gg \ell_x$ this Hamiltonian converges towards (102).

- Gupta, Harikumar and de Queiroz proposed the Hamiltonian $(x\not{p} + \not{p}x)/2$ as a Dirac variant of the xp Hamiltonian [37]. The Hamiltonian is defined on a semi-infinite cylinder and effectively becomes one dimensional by considering the winding modes on the compact dimension. The eigenfunctions are given by Whittaker functions and the spectrum satisfies an equation similar to Equation (29) in the Landau theory. In the limit where a regularization parameter goes to zero one obtains a continuum spectrum with a correction term related to the Riemann–von Mangoldt formula.

- Bender, Brody and Müller proposed recently a generalization of the xp operator [43]

$$H = \frac{1}{1 - e^{-i\hat{p}}} (x\hat{p} + \hat{p}x)(1 - e^{-i\hat{p}}), \qquad (109)$$

with the property that its eigenvalues E_n give the Riemann zeros as $z_n = \frac{1}{2}(1 - iE_n)$. This interesting result follows from the fact the eigenfunctions of (109) are given in terms of the Hurwitz zeta function as $\psi_z(x) = \zeta(z, x+1)$ and imposing the boundary condition

$$\psi_{z_n}(0) = 0 \rightarrow \zeta(z_n, 1) = \zeta(z_n) = 0. \qquad (110)$$

Unfortunately, the operator (109) is not self-adjoint, so that the reality of its eigenvalues is not guaranteed. However, the authors of [43] found that iH has a PT symmetry which, if it is maximally broken, would imply the reality of the eigenvalues. This property though remains to be proved. Further details can be found in references [44,45].

Summary:

✓ The massless Dirac Hamiltonian in Rindler spacetime is the direct sum of xp and $-xp$.
✓ The mass term couples the left and right modes of the fermions.

10. The Massless Dirac Equation with Delta Function Potentials

From analogies between the Polya ζ^* function, the Riemann ζ function and the ζ_H function of the massive Dirac model, we conjectured in Section 8 the existence of a massless field theory underlying ζ. At first look this idea does not look correct because the Hamiltonian obtained by setting $m = 0$ in Equation (102), is equivalent to two copies of the quantum xp model which has a continuum spectrum. In fact, the mass term in that Hamiltonian is the mechanism responsible for obtaining a discrete spectrum.

To resolve this puzzle, we shall replace the *bulk* mass term in the Dirac action (98) by a sum of ultra-local interactions placed at fixed values ℓ_n of the radial coordinate ρ [42]. These interactions can arise from moving mirrors, or beam splitters, that move with a uniform acceleration $1/\ell_n$ (see Figure 8). The fermion moves freely, until it hits one of the mirrors and it is reflected or transmitted. The moving mirrors are realized mathematically by delta functions added to the massless Dirac action that couple the left and right components of the fermion on both sides of the mirror. These delta functions provide the matching conditions for the wave functions and can be parameterized by a complex number ϱ_n with $n = 2, \ldots, \infty$. The scattering of the fermion at each mirror preserves unitarity that is equivalent to the self-adjointness of the Hamiltonian.

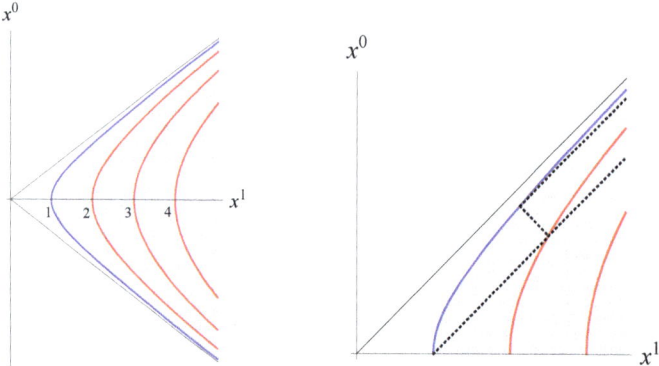

Figure 8. (**Left**): worldlines of the mirrors with accelerations $a_n = 1/\ell_n = 1/n$ ($n = 1, 2, \ldots$). (**Right**): A massless fermion (dotted line) at the point $(x^0, x^1) = (0, 1)$ moves to the right until it hits a moving mirror where it can be reflected or transmitted.

The model is formulated in the spacetime $\mathcal{U}$ defined in Equation (59). We divide $\mathcal{U}$ into an infinite number of domains separated by hyperbolas with constant values of $\rho = \ell_n$, as follows. First we define the intervals (see Figure 9)

$$I_n = \{\rho \mid \ell_n < \rho < \ell_{n+1}\}, \quad n = 1, 2, \ldots, \infty, \tag{111}$$

where using the scale invariance of the model we set $\ell_1 = 1$ (ℓ_1 plays the role of ℓ_x in previous sections).

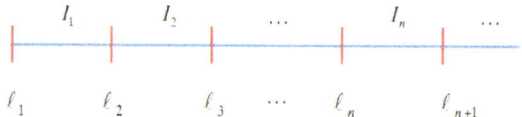

Figure 9. Intervals I_n defined in Equation (111).

The partition of $\mathcal{U}$ is given by

$$\mathcal{U} \to \tilde{\mathcal{U}} = \cup_{n=1}^{\infty} \mathcal{U}_n, \quad \mathcal{U}_n = \mathcal{I}_n \times \mathbb{R}, \tag{112}$$

where the factor $\mathbb{R}$ denotes the range of the Rindler time ϕ. See Figure 8 for an example with $\ell_n = n$. The wave function of the model is the two component Dirac spinor (see Equation (101))

$$\chi(\rho) = \begin{pmatrix} \chi_-(\rho) \\ \chi_+(\rho) \end{pmatrix}, \quad \rho \in \mathcal{I} = \cup_{n=1}^{\infty} I_n, \tag{113}$$

and the scalar product is given by (recall Equation (105))

$$\langle \chi | \chi \rangle = \sum_{n=1}^{\infty} \int_{\ell_n}^{\ell_{n+1}} d\rho \, \chi^\dagger(\rho) \cdot \chi(\rho). \tag{114}$$

The complex Hilbert space is $\mathcal{H} = L^2(\mathcal{I}, \mathbb{C}) \oplus L^2(\mathcal{I}, \mathbb{C})$ and the Hamiltonian is obtained setting $m = 0$ in Equation (102)

$$H = \begin{pmatrix} -i(\rho \, \partial_\rho + \frac{1}{2}) & 0 \\ 0 & i(\rho \, \partial_\rho + \frac{1}{2}) \end{pmatrix}, \quad \rho \notin \mathcal{I}. \tag{115}$$

H is a self-adjoint operator acting on the subspace $\mathcal{H}_\vartheta \subset \mathcal{H}$ of wave functions that satisfy the boundary conditions [42] (see [71] for the relation between self-adjointness of operators and boundary conditions)

$$\chi \in \mathcal{H}_\vartheta: \quad \chi(\ell_n^-) = L(\varrho_n)\chi(\ell_n^+), \quad (n \geq 2), \quad -ie^{i\vartheta}\chi_-(\ell_1^+) = \chi_+(\ell_1^+), \tag{116}$$

where

$$\chi(\ell_n^\pm) = \lim_{\varepsilon \to 0^+} \chi(\ell_n \pm \varepsilon), \tag{117}$$

and

$$\vartheta \in [0, 2\pi), \quad L(\varrho) = \frac{1}{1-|\varrho|^2}\begin{pmatrix} 1+|\varrho|^2 & 2i\varrho \\ -2i\varrho^* & 1+|\varrho|^2 \end{pmatrix}, \quad \varrho \in \mathbb{C}, \quad |\varrho| \neq 1. \tag{118}$$

This means that H satisfies

$$\langle \chi_1 | H\chi_2 \rangle = \langle H\chi_1 | \chi_2 \rangle, \quad \chi_{1,2} \in \mathcal{H}_\vartheta. \tag{119}$$

This condition guarantees that the norm (114) of the state is conserved by the time evolution generated by the Hamiltonian. The subspace $\mathcal{H}_\vartheta$ also depends on ℓ_n and ϱ_n but we shall not write this dependence explicitly. Similarly, we shall also denote the Hamiltonian as H_ϑ. The matching conditions (116) describe a scattering process where two incoming waves χ_n^{in} collide at the nth-mirror and become two outgoing waves χ_n^{out} given by (see Figure 10)

$$\chi_n^{\text{in}} = \begin{pmatrix} \chi_-(\ell_n^-) \\ \chi_+(\ell_n^+) \end{pmatrix}, \quad \chi_n^{\text{out}} = \begin{pmatrix} \chi_-(\ell_n^+) \\ \chi_+(\ell_n^-) \end{pmatrix}, \quad n > 1. \tag{120}$$

At the mirror $n = 1$, the components $\chi_\pm(\ell_1^-)$ of these vectors are null, i.e., there is no propagation at the left of the boundary. The scattering process is described by the matrix S_n

$$\chi_n^{\text{out}} = S_n \chi_n^{\text{in}}, \quad S_n = \frac{1}{1+|\varrho_n|^2}\begin{pmatrix} 1-|\varrho_n|^2 & -2i\varrho_n \\ -2i\varrho_n^* & 1-|\varrho_n|^2 \end{pmatrix}, \quad n > 1, \tag{121}$$

that is unitary,

$$S_n S_n^\dagger = \mathbf{1}. \tag{122}$$

Notice that the boundary condition at $\rho = \ell_1$, is also described by Equation (121) with a parameter ϱ_1

$$\varrho_1 = -e^{-i\vartheta}, \tag{123}$$

that is a pure phase for the Hamiltonian H_ϑ to be self-adjoint. The matrix $L(\varrho)$ satisfies

$$L(1/\varrho^*) = -L(\varrho). \tag{124}$$

Hence, replacing ϱ_n by $1/\varrho_n^*$ gives a unitary equivalent model because the sign changes at $\rho = \ell_n$, given in Equation (124), can be compensated by changing the sign of the wave function in the remaining intervals. Hence, without losing generality, we shall impose the condition $|\varrho_n| < 1$, $\forall n > 1$.

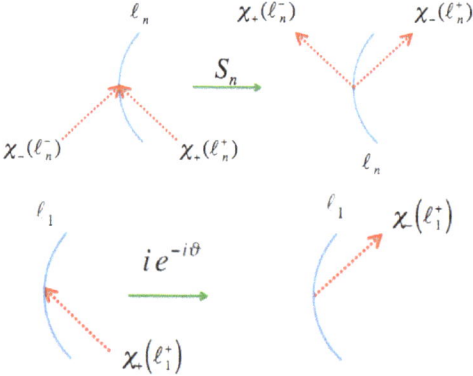

Figure 10. (**Top**): scattering process taking place at the mirror located at $\rho = \ell_n$ for $n > 1$ (Equation (121)). (**Bottom**): reflexion at the perfect mirror located at $\rho = \ell_1$ (Equation (116)).

The eigenfunctions of the Hamiltonian (115) are the customary functions (see Equation (13))

$$H\chi = E\chi \longrightarrow \chi_{\mp} \propto \rho^{-1/2 \pm iE}. \tag{125}$$

From now one, we shall assume that E is a real number which is guaranteed by the self-adjointness of the Hamiltonian H. In the nth interval we take

$$\chi_{\mp,n}(\rho) = e^{\pm i\pi/4} \frac{A_{\mp,n}}{\rho^{1/2 \mp iE}}, \qquad \ell_n < \rho < \ell_{n+1}, \tag{126}$$

where $A_{\mp,n}$ are constants that in general will depend on E. The phases $e^{\pm i\pi/4}$ have been introduced by analogy with those appearing in Equation (106). The boundary values of χ at $\rho = \ell_n^{\pm}$ ($n \geq 1$) are (see Equation (117))

$$\chi_{\mp}(\ell_n^+) = \chi_{\mp,n}(\ell_n) = e^{\pm \frac{i\pi}{4}} \frac{A_{\mp,n}}{\ell_n^{1/2 \mp iE}}, \qquad \chi_{\mp}(\ell_n^-) = \chi_{\mp,n-1}(\ell_n) = e^{\pm \frac{i\pi}{4}} \frac{A_{\mp,n-1}}{\ell_n^{1/2 \mp iE}}. \tag{127}$$

Let us define the vectors

$$|\mathbf{A}_n\rangle = \begin{pmatrix} A_{-,n} \\ A_{+,n} \end{pmatrix}, \qquad n \geq 1. \tag{128}$$

The boundary conditions (116) together with Equation (127) imply

$$|\mathbf{A}_{n-1}\rangle = T_n |\mathbf{A}_n\rangle \quad (n \geq 2), \qquad |\mathbf{A}_1\rangle = |\mathbf{A}_1(\vartheta)\rangle = \begin{pmatrix} 1 \\ e^{i\vartheta} \end{pmatrix}, \tag{129}$$

where the transfer matrix T_n is given by

$$T_n = \frac{1}{1 - |\varrho_n|^2} \begin{pmatrix} 1 + |\varrho_n|^2 & 2\varrho_n \ell_n^{-2iE} \\ 2\varrho_n^* \ell_n^{2iE} & 1 + |\varrho_n|^2 \end{pmatrix} \quad (n \geq 2). \tag{130}$$

The norm of the eigenstate can be computed using Equations (114) and (126)

$$||\chi||^2 = \sum_{n=1}^{\infty} \log \frac{\ell_{n+1}}{\ell_n} \langle \mathbf{A}_n | \mathbf{A}_n \rangle, \qquad \langle \mathbf{A}_n | \mathbf{A}_n \rangle = |A_{-,n}|^2 + |A_{+,n}|^2. \tag{131}$$

The log term comes from the integral of the norm of the wave function in the nth interval, $\int_{\ell_n}^{\ell_{n+1}} d\rho/\rho$ (we used that E is real). If $\varrho_n = 0$ then $T_n = 1$ which implies that $|\mathbf{A}_{n-1}\rangle = |\mathbf{A}_n\rangle$. If this happens for all n, then $|\mathbf{A}_n\rangle = |\mathbf{A}_1\rangle$, in which case the norm of these states diverges, but they can be normalized using Dirac delta functions, so they correspond to scattering states. In the general case, iterating Equation (129) yields $|\mathbf{A}_n\rangle$ in terms of $|\mathbf{A}_1(\vartheta)\rangle$

$$|\mathbf{A}_n\rangle = T_n^{-1}T_{n-1}^{-1}\cdots T_2^{-1}|\mathbf{A}_1(\vartheta)\rangle, \quad n \geq 2. \tag{132}$$

For special values of ℓ_n and ϱ_n one can find the exact expression of these amplitudes. An example is $\ell_n = e^{n/2}$, $\varrho_n =$ cte [42]. To make contact with the Riemann zeros, we shall consider a limit where the reflection coefficients vanish asymptotically.

Summary:

- ✓ The massless Dirac Hamiltonian with delta function potential is solvable by transfer matrix methods.
- ✓ The model is completely characterized by the set of parameters $\{\ell_n, \varrho_n\}_{n=2}^{\infty}$ and ϑ.

11. Heuristic Approach to the Spectrum

Let us replace ϱ_n by $\varepsilon\varrho_n$, and consider the limit $\varepsilon \to 0$ of the transfer matrix (130)

$$T_n \simeq 1 + \varepsilon\,\tau_n + O(\varepsilon^2), \quad \tau_n = \begin{pmatrix} 0 & 2\varrho_n\,\ell_n^{-2iE} \\ 2\varrho_n^*\,\ell_n^{2iE} & 0 \end{pmatrix} \quad (n \geq 2). \tag{133}$$

Plugging this equation into Equation (132) yields

$$|\mathbf{A}_n\rangle \simeq \left(1 - \varepsilon \sum_{m=2}^{n} \tau_m\right)|\mathbf{A}_1(\vartheta)\rangle + O(\varepsilon^2), \quad n \geq 2, \tag{134}$$

and in components

$$A_{-,n} \simeq 1 - 2\varepsilon\,e^{i\vartheta}\sum_{m=2}^{n}\varrho_m\,\ell_m^{-2iE} + O(\varepsilon^2), \quad A_{+,n} \simeq e^{i\vartheta} - 2\varepsilon\sum_{m=2}^{n}\varrho_m^*\,\ell_m^{2iE} + O(\varepsilon^2). \tag{135}$$

For a normalizable state, the amplitudes $A_{\pm,n}$ must vanish as $n \to \infty$. In the next section we shall study in detail the normalizability of the state. We shall make the following choice of lengths and reflection coefficients [42]

$$\ell_n = n^{1/2}, \quad \varrho_n = \frac{\mu(n)}{n^{1/2}}, \quad n > 1, \tag{136}$$

where $\mu(n)$ is the Moëbius function that is equal to $(-1)^r$, with r the number of distinct primes factors of a square free integer n, and $\mu(n) = 0$, if n is divisible by the square of a prime number [4]. See Figures 11 and 12 for a graphical representation of Equations (136) and (135). The Moebius function has been used in the past to provide physical models of prime numbers, most notably in the ideal gas of primons with fermionic statistics [72,73] and a potential whose semiclassical spectrum are the primes [46,74].

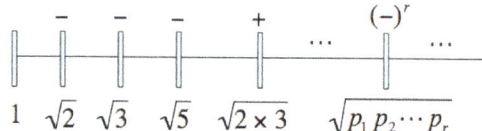

Figure 11. Localization of the mirrors corresponding to the choice (136), together with the values of $\mu(n)$.

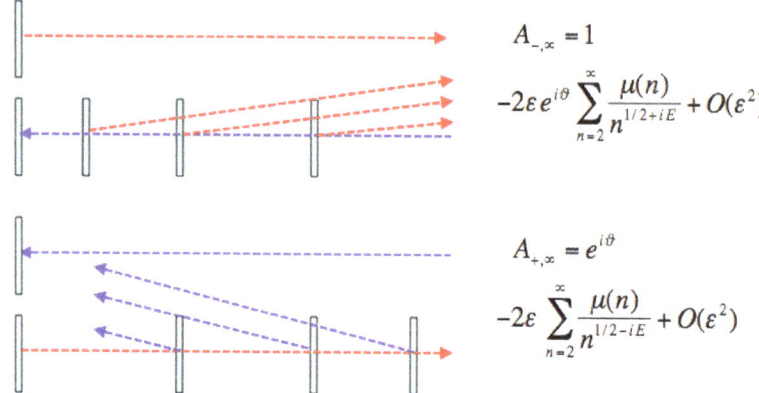

Figure 12. Depiction of the amplitudes $A_{\pm,\infty}$ as the superposition of a principal wave with the waves resulting from the scattering with all the mirrors along its trajectory (see Equation (135)). The terms of higher order in ε correspond to more than one scattering.

Another motivation of the choice (136) is the following [42]. Consider a fermion that leaves the boundary at $\rho = \ell_1$, moves rightwards until it hits the mirror at $\rho = \ell_n$ where it gets reflected and returns to the boundary. The time lapse for the entire trajectory is given by

$$\tau_n = 2\log(\ell_n/\ell_1) \qquad (137)$$

where we used the Rindler metric Equation (62). If the mirror is associated with the prime p, that is $\ell_p = \sqrt{p}$, the time will be given by $\tau_p = \log p$. This result reminds the Berry conjecture that postulates the existence of a classical chaotic Hamiltonian whose primitive periodic orbits are labelled by the primes p, with periods $\log p$, and whose quantization will give the Riemann zeros as energy levels [12]. A classical Hamiltonian with this property has not been found, but the array of mirrors presented above, displays some of its properties. In particular, the trajectory between the boundary and the mirror at ℓ_p, with p a prime number, behaves as a primitive orbit with a period $\log p$. Moreover, the trajectories and periods of these orbits are independent of the energy of the fermion because it moves at the speed of light.

Let us work out the consequences (136). The condition for a normalizable eigenstate, that is $\lim_{n\to\infty} A_{\pm,n} = 0$, is

$$1 \simeq 2\varepsilon\, e^{i\vartheta} \sum_{n=1}^{\infty} \frac{\mu(n)}{n^{\frac{1}{2}+iE}} = \frac{2\varepsilon\, e^{i\vartheta}}{\zeta(\frac{1}{2}+iE(\varepsilon))}, \qquad (138)$$

where we have included the term $n=1$ in the series because it does not modify its value when $\varepsilon \to 0$. We have employed the formula $\sum_{n=1}^{\infty} \mu(n)/n^s = 1/\zeta(s)$ for a value of s where the series may not converge. In the next section we shall compute the value of the finite sum that determines the norm of the state. $E_n(\varepsilon)$ denotes a solution such that $\lim_{\varepsilon \to 0} E_n(\varepsilon) = E_n$, where $\frac{1}{2}+iE_n$ is a zero of the zeta function. All known zeros of $\zeta(s)$ on the critical line are simple, but we shall also consider the case where $\frac{1}{2}+iE_n$ might be a zero of order $r \geq 1$, that is $\zeta^{(r)}(s) \neq 0$. The Taylor expansion of $\zeta(\frac{1}{2}+iE(\varepsilon))$ around $\frac{1}{2}+iE_n$, in Equation (138) yields

$$1 \simeq \frac{2\varepsilon\, r!\, e^{i\vartheta}}{i^r (E_n(\varepsilon)-E_n)^r \zeta^{(r)}(\frac{1}{2}+iE_n)}. \tag{139}$$

Hence $E_n(\varepsilon)-E_n$ is of order $\varepsilon^{1/r}$, as $\varepsilon \to 0$ and

$$\frac{\zeta^{(r)}(\frac{1}{2}+iE_n)}{\zeta^{(r)}(\frac{1}{2}-iE_n)} = (-1)^r e^{2i\vartheta}. \tag{140}$$

On the other hand, from Equation (9) one finds

$$i^r\, \zeta^{(r)}(\frac{1}{2}+iE_n) = e^{-i\theta(E_n)} Z^{(r)}(E_n), \tag{141}$$

that plugged into (140) yields

$$e^{2i(\vartheta+\theta(E_n))} = 1, \qquad \forall r. \tag{142}$$

We can collect these results in the equation

$$\text{If } \zeta(\frac{1}{2} \pm iE_n) = 0 \text{ and } e^{2i(\vartheta+\theta(E_n))} = 1 \iff H_\vartheta\, \chi_{E_n} = E_n\, \chi_{E_n}. \tag{143}$$

Observe that ϑ is fixed mod π. In the next section we shall fix this ambiguity. This equation is heuristic. It has been derived by (i) solving the eigenvalue equation in the limit $\varepsilon \to 0$, (ii) imposing the vanishing of the eigenfunction at infinity and (iii) using the Dirichlet series of $1/\zeta(s)$ in a region where it may not converge. In the next section we shall derive Equation (143) without making the previous assumptions (see Equation (176)). Let us notice that this spectral realization of the *zeros* requires the fine tuning of the parameter ϑ in terms of the phase of the zeta function, $\theta(E_n)$ (see Figure 13). This realization is different from the Pólya-Hilbert conjecture of a single Hamiltonian encompassing all the Riemann zeros at once. This Hamiltonian would exist if $\theta(E_n) = \theta_0, \forall n$, but this is certainly not the case.

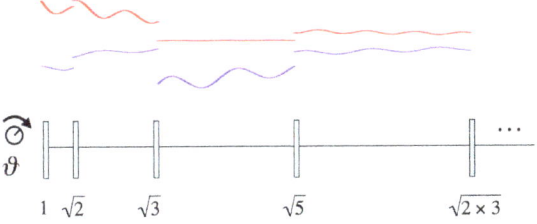

Figure 13. Schematic representation of the array of mirrors that give rise to a spectral realization of the Riemann zeros. The red and blue lines represent the left and right wave functions $\chi_{\pm,n}(\rho)$. The wave functions are discontinuous at the moving mirrors located at the positions $\ell_n = \sqrt{n}$ with n a square free integer. The knob on the left represents the scattering phase at the perfect mirror that is set to minus the phase of the zeta function at the *zero* E_n, namely $\vartheta = -\theta(E_n)$ mod π.

Summary:

✓ A Riemann zero, on the critical line, becomes an eigenvalue of the Hamiltonian H_ϑ by tuning the phase ϑ according to the phase of the zeta function.
✗ The previous result is obtained in the limit $\varepsilon \to 0$ and is heuristic.

12. The Riemann Zeros as Spectrum and the Riemann Hypothesis

In this section, we provide more rigorous arguments that support the heuristic results obtained previously. Let us first review the main properties of the model discussed so far. The Hamiltonian, Equation (115), describes the dynamics of a massless Dirac fermion in the region of Rindler spacetime bounded by the hyperbola $\rho = \ell_1$. The reflection of the wave function at this boundary is characterized by a parameter ϑ, which is real for a self-adjoint Hamiltonian. At the positions $\ell_{n>1}$ the wave function is discontinuous due to the presence of delta function potentials characterized by the reflection amplitudes ϱ_n that provide the matching conditions of the wave function at those sites. An eigenfunction χ, with eigenvalue E, has a simple expression, Equation (126), in terms of the amplitudes $A_{n,\pm}$, which are related by the transfer matrix T_n (130). The norm of χ is given by the sum of the squared length of the vectors $\mathbf{A}_n$, weighted with a factor that depends on the positions ℓ_n, Equation (131). We introduce a scale factor ε in the parameters ϱ_n, which allows us to study the limit $\varepsilon \to 0$, where the mirrors become semitransparent. In this way we found an ansatz for the parameters ℓ_n and ϱ_n that heuristically led to an individual spectral realization of the *zeros* by fine tuning the parameter ϑ.

12.1. Normalizable Eigenstates

Under the choice $\ell_n = n^{1/2}$, Equation (131) becomes

$$||\chi||^2 = \frac{1}{2} \sum_{n=1}^{\infty} \log\left(1 + \frac{1}{n}\right) \langle \mathbf{A}_n | \mathbf{A}_n \rangle. \tag{144}$$

This series can be replaced by

$$||\chi||_c^2 \equiv \sum_{n=1}^{\infty} \frac{1}{n} \langle \mathbf{A}_n | \mathbf{A}_n \rangle, \tag{145}$$

which is convergent if and only if (144) is convergent. The vectors $\mathbf{A}_n$ are obtained by acting on $\mathbf{A}_1(\vartheta)$ with the transfer matrices T_n (see Equation (132)). These matrices have unit determinant and can be written as the exponential of traceless Hermitian matrices, that is,

$$T_n = e^{\tau_n}, \qquad \tau_n = \begin{pmatrix} 0 & r_n \ell_n^{-2iE} \\ r_n^* \ell_n^{2iE} & 0 \end{pmatrix}, \qquad \forall E \in \mathbb{R}, \tag{146}$$

where taking $|\varrho_n| < 1$,

$$r_n = \frac{\varrho_n}{|\varrho_n|} \log \frac{1 + |\varrho_n|}{1 - |\varrho_n|}, \qquad \varrho_n = \frac{r_n}{|r_n|} \tanh \frac{|r_n|}{2}. \tag{147}$$

To derive Equation (146) we used the relation

$$\exp \begin{pmatrix} 0 & a \\ b & 0 \end{pmatrix} = \begin{pmatrix} \cosh(\sqrt{ab}) & \frac{a}{\sqrt{ab}} \sinh(\sqrt{ab}) \\ \frac{b}{\sqrt{ab}} \sinh(\sqrt{ab}) & \cosh(\sqrt{ab}) \end{pmatrix}, \qquad \forall a, b \in \mathbb{C} - \{0\}. \tag{148}$$

If $|\varrho_n| \ll 1$ one gets $r_n \simeq 2\varrho_n$, hence in that limit both parameters give the same result. Using Equation (146), the recursion relation (132) reads

$$|\mathbf{A}_k\rangle = e^{-\tau_k} e^{-\tau_{k-1}} \dots e^{-\tau_2} |\mathbf{A}_1\rangle, \qquad k \geq 2. \tag{149}$$

12.2. The Magnus Expansion

It is rather difficult to find an analytic expression of the product of matrices of Equation (149). However, we can estimate it replacing r_n by εr_n, and taking the limit $\varepsilon \to 0$. Under this replacement Equation (149) becomes

$$|\mathbf{A}_k\rangle = e^{-\varepsilon\tau_k}e^{-\varepsilon\tau_{k-1}}\ldots e^{-\varepsilon\tau_2}|\mathbf{A}_1\rangle \quad (k \geq 2). \tag{150}$$

The product of exponentials of matrices can be expressed as the exponential of a matrix given by the Magnus expansion [75]

$$e^{-\varepsilon\tau_k}e^{-\varepsilon\tau_{k-1}}\ldots e^{-\varepsilon\tau_2} = \exp\left(-\varepsilon \sum_{n=2}^{n} \tau_n - \frac{\varepsilon^2}{2}\sum_{n_1 > n_2 = 2}^{k}[\tau_{n_1}, \tau_{n_2}] + O(\varepsilon^3)\right) \quad (k \geq 2). \tag{151}$$

In the limit $\varepsilon \to 0$ we truncate this expression to the term of order ε,

$$e^{-\varepsilon\tau_k}e^{-\varepsilon\tau_{k-1}}\ldots e^{-\varepsilon\tau_2} \simeq \exp\begin{pmatrix} 0 & -\varepsilon\sum_{n=2}^{k}r_n\ell_n^{-2iE} \\ -\varepsilon\sum_{n=2}^{k}r_n^*\ell_n^{2iE} & 0 \end{pmatrix} \simeq \exp\begin{pmatrix} 0 & -\varepsilon M_z(k) \\ -\varepsilon M_z^*(k) & 0 \end{pmatrix}, \tag{152}$$

which using (136)

$$r_n = \frac{\mu(n)}{n^{1/2}} \tag{153}$$

gives

$$M_z(k) = 1 + \sum_{n=2}^{k} r_n \ell_n^{-2iE} = \sum_{n=1}^{k} \frac{\mu(n)}{n^z}, \quad z = \frac{1}{2} + iE. \tag{154}$$

We have added the constant 1 to $M_z(k)$, which does not affect the results in the limit $\varepsilon \to 0$. Using Equations (148), (150) and (152) we obtain

$$|\mathbf{A}_n\rangle \simeq \exp\begin{pmatrix} 0 & -\varepsilon M_z(n) \\ -\varepsilon M_z^*(n) & 0 \end{pmatrix}\begin{pmatrix} 1 \\ e^{i\vartheta} \end{pmatrix} \tag{155}$$

$$= \begin{pmatrix} \cosh(|\varepsilon M_z(n)|) & -\frac{\varepsilon M_z(n)}{|\varepsilon M_z(n)|}\sinh(|\varepsilon M_z(n)|) \\ -\frac{\varepsilon M_z^*(n)}{|\varepsilon M_z(n)|}\sinh(|\varepsilon M_z(n)|) & \cosh(|\varepsilon M_z(n)|) \end{pmatrix}\begin{pmatrix} 1 \\ e^{i\vartheta} \end{pmatrix}$$

$$= \begin{pmatrix} \cosh(|\varepsilon M_z(n)|) & -e^{-i\Phi_z(n)}\sinh(|\varepsilon M_z(n)|) \\ -e^{i\Phi_z(n)}\sinh(|\varepsilon M_z(n)|) & \cosh(|\varepsilon M_z(n)|) \end{pmatrix}\begin{pmatrix} 1 \\ e^{i\vartheta} \end{pmatrix}$$

$$\simeq \begin{pmatrix} e^{\frac{i}{2}(\vartheta-\Phi_z(n))}\left[e^{-|\varepsilon M_z(n)|}\cos(\frac{1}{2}(\vartheta - \Phi_z(n))) - ie^{|\varepsilon M_z(n)|}\sin(\frac{1}{2}(\vartheta - \Phi_z(n)))\right] \\ e^{\frac{i}{2}(\vartheta+\Phi_z(n))}\left[e^{-|\varepsilon M_z(n)|}\cos(\frac{1}{2}(\vartheta - \Phi_z(n))) + ie^{|\varepsilon M_z(n)|}\sin(\frac{1}{2}(\vartheta - \Phi_z(n)))\right] \end{pmatrix} \quad (n \geq 2),$$

where $\Phi_z(n)$ is the phase

$$e^{-i\Phi_z(n)} = \frac{M_z(n)}{|M_z(n)|}. \tag{156}$$

From (155) follows an estimate of the norm (145)

$$||\chi||_c^2 \simeq \mathcal{N}_z(\varepsilon) \equiv \sum_{n=1}^{\infty} \frac{1}{n}\left[e^{-2|\varepsilon M_z(n)|}(1 + \cos(\vartheta - \Phi_z(n))) + e^{2|\varepsilon M_z(n)|}(1 - \cos(\vartheta - \Phi_z(n)))\right], \tag{157}$$

whose convergence depends on the asymptotic behavior of $M_z(n)$ and $\Phi_z(n)$. $\mathcal{N}_z(\varepsilon)$ has the lower bound

$$\mathcal{N}_z(\varepsilon) \geq \sum_{n=1}^{\infty} \frac{2}{n} e^{-2|\varepsilon M_z(n)|}, \tag{158}$$

that follows from the inequality

$$a(1+b) + \frac{1}{a}(1-b) \geq 2a, \quad a \in (0,1], \quad b \in [-1,1]. \tag{159}$$

If $|M_z(n)|$ is bounded then the norm is infinite,

$$\text{if } |M_z(n)| < C, \; \forall n \implies \mathcal{N}_z(\varepsilon) \geq \sum_{n=1}^{\infty} \frac{2}{n} e^{-2|\varepsilon|C} = \infty. \tag{160}$$

This case corresponds in general to eigenstates belonging to the continuum. Eigenstates with finite norm require $|M_z(n)|$ to be unbounded. Notice that $\mathcal{N}_z(\varepsilon)$ is the sum of two series with non-negative terms. The convergence of the first summand in (157) is guaranteed if

$$\sum_{n=1}^{\infty} \frac{1}{n} e^{-2|\varepsilon M_z(n)|} < \infty, \tag{161}$$

which occurs if $|M_z(n)|$ diverges sufficiently fast with n. The convergence of the second summand in (157) requires $\Phi_z(n)$ to have a limit when $n \to \infty$, and to choose the parameter ϑ such that

$$\lim_{n \to \infty} \Phi_z(n) = \vartheta. \tag{162}$$

Moreover, $1 - \cos(\vartheta - \Phi_z(n))$ must approach 0 sufficiently fast in order to compensate the factor $\frac{1}{n} e^{2\varepsilon|M_z(n)|}$. We now pass to analyze the latter conditions in detail.

12.3. Perron Formula

Let us define the function

$$M'_z(x) \equiv \sum_{1 \leq n \leq x}{}' \frac{\mu(n)}{n^z}, \quad z = \frac{1}{2} + iE, \quad E \in \mathbb{R}, \tag{163}$$

where $\sum'_{1 \leq n \leq x}$ means that the last term in the sum is multiplied by $1/2$ when x is an integer. Figure 14 shows $|M'_z(n)|$ as a function of E for several values of n. Observe that $|M'_z(n)|$ increases with n when E is a *zero*. We shall derive below this behavior.

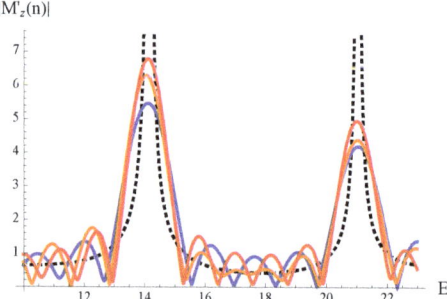

Figure 14. Plot of $|M'_z(n)|$ defined in Equation (163), for $E \in (10, 23)$ and $n = 50, 100, 150$ (blue, orange, red curves) and $1/|\zeta(1/2 + iE)|$ (black dotted line). Observe the increase with n at $E = 14.13\ldots$ and $E = 21.02\ldots$ which are the first two *zeros* of ζ.

To compute $M'_z(x)$ we use Perron's formula [76]

$$M'_z(x) = \lim_{T \to \infty} \int_{c-iT}^{c+iT} \frac{ds}{2\pi i} \frac{1}{\zeta(s+z)} \frac{x^s}{s}, \quad c > \frac{1}{2}, \tag{164}$$

where we have used that $\operatorname{Re} z = 1/2$. The integral (164) can be done by residue calculus [42]

$$\lim_{T\to\infty}\int_{c-iT}^{c+iT}\frac{ds}{2\pi i}F(s) = \sum_{\operatorname{Re} s_j<c}\operatorname{Res}_{s_j} F(s), \quad F(s) = \frac{1}{\zeta(s+z)}\frac{x^s}{s}, \qquad (165)$$

where the sum runs over the poles s_j of $F(s)$ located to the left of the line of integration $\operatorname{Re} s = c$, which is $\operatorname{Re} s_j < c$. The poles of $F(s)$ come from the zeros of $s\zeta(s+z)$. The pole at $s=0$ can be simple, or multiple, depending on the values of $\zeta(z)$ and its derivatives. The remaining poles of $F(s)$ come from the zeros of $\zeta(s+z)$, say $s_j + z = \rho_j$, and they lie to the left of the integration line, because the trivial and non-trivial zeros of ζ, satisfy $\operatorname{Re} \rho_j < 1$, which is

$$\operatorname{Re} s_j = \operatorname{Re}(\rho_j - z) = \operatorname{Re} \rho_j - \frac{1}{2} < \frac{1}{2} < c. \qquad (166)$$

To compute the residues of Equation (165) we consider the cases: $s = 0$, $s_j + z$ a trivial zero of ζ and $s_j + z$ a non-trivial zero of ζ:

- $s=0$. Let $m \geq 0$ be the lowest integer such that $\zeta^{(m)}(z) = d^m \zeta(z)/dz^m \neq 0$. Then $F(s)$ has a pole of order $m+1$ at $s=0$ with residue (The expression for $\operatorname{Res}_{s=0} F(s)$ corresponding to the case $m = 1$ contains the term $-\frac{1}{2}\zeta''(z)/(\zeta'(z))^2$ which was omitted in the reference [42].)

$$\operatorname{Res}_{s=0} F(s) = \begin{cases} 1/\zeta(z) & \text{if } \zeta(z) \neq 0, \\ \log x/\zeta'(z) - \frac{1}{2}\zeta''(z)/(\zeta'(z))^2 & \text{if } \zeta(z) = 0, \zeta'(z) \neq 0, \\ \vdots & \vdots \\ (\log x)^m/\zeta^{(m)}(z) + O((\log x)^{m-1}) & \text{if } \zeta(z) = \cdots = \zeta^{(m-1)}(z) = 0, \zeta^{(m)}(z) \neq 0. \end{cases} \qquad (167)$$

- $s_n = -2n - z$ ($n = 1, 2, \ldots$), where $F(s)$ has a simple pole due to the trivial zeros $-2n$ of ζ.

$$\operatorname{Res}_{s=-2n-z} F(s) = \frac{x^{-2n-z}}{-(2n+z)\zeta'(-2n)}, \quad n = 1, 2, \ldots, \infty. \qquad (168)$$

- $s_j = \rho_j - z \neq 0$, then $F(s)$ has a pole due to the non-trivial zero ρ_j of ζ

$$\operatorname{Res}_{s=s_j} F(s) = \begin{cases} \frac{x^{\rho_j - z}}{(\rho_j - z)\zeta'(\rho_j)}, & \text{if } \zeta(\rho_j) = 0, \zeta'(\rho_j) \neq 0 \\ \frac{m(\ln x)^{m-1} x^{\rho_j - z}}{(\rho_j - z)\zeta^{(m)}(\rho_j)} + O((\ln x)^{m-2}), & \text{if } \zeta(\rho_j) = \cdots = \zeta^{(m-1)}(\rho_j) = 0, \zeta^{(m)}(\rho_j) \neq 0, m \geq 2 \end{cases} \qquad (169)$$

To make further progress we shall assume that all the Riemann zeros are simple, a statement which is not known to hold. The eventual case where there is a *zero* with double multiplicity will be considered elsewhere. In the former situation we are led to consider only two cases depending on whether z is, or is not, a simple *zero* of ζ. Collecting terms, we get

$$M_z(x) = \frac{1}{\zeta(z)} + \sum_{\rho_j} \frac{x^{\rho_j - z}}{(\rho_j - z)\zeta'(\rho_j)} + \sum_{n=1}^{\infty} \frac{x^{-2n-z}}{-(2n+z)\zeta'(-2n)}, \quad \text{if } \zeta(z) \neq 0, \qquad (170)$$

$$M_z(x) = \frac{\log x}{\zeta'(z)} - \frac{\zeta''(z)}{2(\zeta'(z))^2} + \sum_{\rho_j \neq z} \frac{x^{\rho_j - z}}{(\rho_j - z)\zeta'(\rho_j)} + \sum_{n=1}^{\infty} \frac{x^{-2n-z}}{-(2n+z)\zeta'(-2n)}, \quad \text{if } \zeta(z) = 0, \zeta'(z) \neq 0. \qquad (171)$$

where the sum $\sum_{\rho_j}$ runs over the non-trivial zeros of ζ. These equations are verified numerically in Figure 15. The last term in these equations, which comes from the trivial *zeros*, converges quickly and is finite for all x due to the exponential increase of $\zeta'(-2n)$ [77]

$$\zeta'(-2n) = \frac{(-1)^n \zeta(2n+1)(2n)!}{2^{2n+1}\pi^{2n}} \xrightarrow{n\to\infty} (-1)^n \sqrt{\pi n}\left(\frac{n}{e\pi}\right)^{2n}. \qquad (172)$$

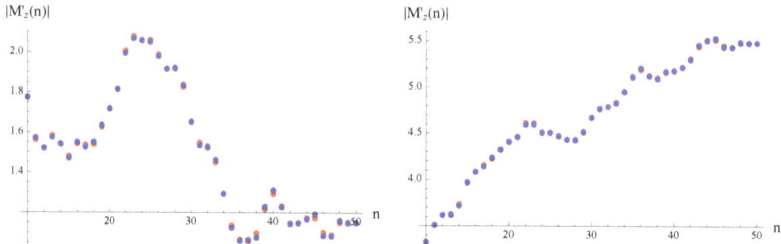

Figure 15. Plot of $|M'_z(n)|$ for $n = 10, \ldots, 50$ and $E = 20$ (**left**) and $E = 14.13$ (**right**). In red the values obtained doing the sum in Equation (163). In blue the sum of Equation (170) for $E = 20$ and Equation (171) for $E = 14.13$, including the first 100 Riemann zeros, and 20 trivial zeros. Observe the accuracy of the approximation. The slow increase in the latter plot is due to the factor $\log n$ in Equation (171).

We do not know an estimation of the term depending on the sum over the non-trivial zeros. If the Riemann hypothesis is true the term $x^{\rho_j - z}$ will oscillate as a function of x. We expect that for $\zeta(z) \neq 0$, $|M_z(x)|$ will not yield a finite norm such that the corresponding eigenstate will not belong to the discrete spectrum. When $\zeta(z)) = 0, \zeta'(z) \neq 0$, we shall make the approximation

$$M_z(x) \to \frac{\log x}{\zeta'(z)} \qquad x \to \infty, \tag{173}$$

where we neglect the finite part $\frac{\zeta''(z)}{2(\zeta'(z))^2}$; and the possible contribution of the sum over the Riemann zeros. Using that $\zeta(1/2 + iE) = e^{-i\theta(E)} Z(E)$ we find

$$M_x(z) \to \frac{i e^{i\theta(E)} \log x}{Z'(E)} \quad \text{as } x \to \infty, \tag{174}$$

hence $\Phi_z(n)$, given in Equation (156), behaves as

$$e^{-i\Phi_z(n)} \to i e^{i\theta(E)} \operatorname{sign} Z'(E) \quad \text{as } n \to \infty, \tag{175}$$

which has a well-defined asymptotic limit. We shall then choose ϑ according to Equation (162) namely

$$\vartheta = \lim_{n \to \infty} \Phi_z(n) = -\left(\theta(E) + \frac{\pi}{2} \operatorname{sign} Z'(E)\right), \tag{176}$$

that provides a necessary condition for the convergence of the norm. It remains to show that Equation (176) is also sufficient but this requires the knowledge of the next to leading correction to (174). Notice that ϑ depends on $\theta(E)$ and the sign of $Z'(E)$, a feature that is not left fixed in Equation (142). The norm (157) then becomes

$$\|\chi\|_c^2 \simeq \sum_{n=1}^{\infty} \frac{2}{n} e^{-2\varepsilon \log n / |Z'(E)|} = 2\zeta\left(1 + \frac{2\varepsilon}{|Z'(E)|}\right) < \infty, \tag{177}$$

that is finite for all $\varepsilon > 0$. This result indicates that a *zero* of the zeta function gives a normalizable state, in agreement with heuristic derivation proposed in the previous section, but there are some differences. First, the eigenvalue E does not need to be expanded in series of ε. It is taken to be a *zero* of ζ from the beginning. This choice generates the $\log x$ term in Equation (171) and is responsible for the finiteness of the norm after the appropriate choice of the phase (176) that also differs from the heuristic value (142). On the other hand, if ϑ does not satisfy Equation (176), then the norm of the state will diverge badly and so the *zero* E will be missing in the spectrum. Finally, if E is not a *zero*, we expect that the state will belong generically to the continuum. Figure 16 shows the expected spectrum of the model, which recalls Connes's scenario of missing spectral lines, except that in our case, one can pick up a zero at a time by tuning ϑ.

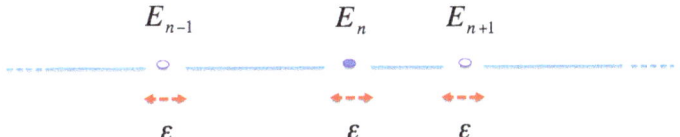

Figure 16. Graphical representation of the spectrum of the model. It is expected to consist of an infinite number of bands separated by forbidden regions of width proportional to ε. The latter regions may contain a *zero* E_n if the phase ϑ is chosen according to Equation (176). Otherwise, the *zeros* will be missing in the spectrum that is represented by the points E_{n-1} and E_{n+1}.

If the RH is false there will be at least four *zeros* outside the critical line, say $\rho_c = \sigma_c + iE_c, \bar{\rho}_c = \sigma_c - iE_c, 1 - \rho_c$ and $1 - \bar{\rho}_c$, with $\sigma_c > \frac{1}{2}, E_c \in \mathbb{R}_+$. We shall choose the highest value of σ_c. The asymptotic behavior of $M_z(x)$ will be given by the *zeros* located to the right of the critical line,

$$M_z(x) \to \frac{x^{\rho_c - z}}{(\rho_c - z)\zeta'(\rho_c)} + \frac{x^{\bar{\rho}_c - z}}{(\bar{\rho}_c - z)\zeta'(\bar{\rho}_c)} \quad \text{as } x \to \infty. \tag{178}$$

To simplify the discussion let us choose $E \gg E_c$, which yields the approximation

$$M_z(x) \to \frac{2i \, x^{\sigma_c - 1/2 - iE}}{E|\zeta'(\rho_c)|} \cos(E_c \log x - \phi_c) \quad \text{as } x \to \infty, \tag{179}$$

where $e^{i\phi_c} = \zeta'(\rho_c)/|\zeta'(\rho_c)|$. The phase $\Phi_z(n)$ is given by Equation (156)

$$\Phi_z(n) \to E \log n - \frac{\pi}{2}\text{sign}\left(\cos(E_c \log n - \phi_c)\right) \quad \text{as } n \to \infty. \tag{180}$$

Correspondingly, the norm (157) diverges so badly, $\propto \sum_n \frac{1}{n}\exp(Cn^{\sigma_c - 1/2})\ldots$, for any value of ϑ that the state will not be normalizable even using Dirac delta functions. This result occurs for all eigenenergies E. Therefore, the Hamiltonian will not admit a spectral decomposition, but this is impossible because it is a well-defined self-adjoint operator. We conclude that a *zero* outside the critical line does not exist which provides an argument likely to be persuasive to physicists for the truth of the Riemann hypothesis.

13. The Riemann Interferometer

The model considered in the previous sections looks at first glance quite difficult to simulate. We shall next show that this model is equivalent to another one that can be implemented in the Lab. We shall call this system the Riemann interferometer. The basic idea can be illustrated with the mapping between the quantum xp Hamiltonian and the momentum operator $\hat{p}$. Let us make the change of coordinates $x = \log \rho$ and relate the wave functions in both coordinates, $\phi(x)$ and $\psi(\rho)$, as follows

$$\phi(x) = \left(\frac{d\rho}{dx}\right)^{1/2} \psi(\rho) = e^{x/2}\psi(e^x). \tag{181}$$

An eigenstate of the Hamiltonian $(\rho\, \hat{p}_\rho + \hat{p}_\rho\, \rho)/2$, with eigenvalue E, is mapped by Equation (181) into an eigenstate of the momentum operator $\hat{p}_x = -i\partial_x$ with the same eigenvalue,

$$\psi(\rho) = \frac{1}{\rho^{1/2 - iE}} \implies \phi(x) = e^{iEx}. \tag{182}$$

This shows that the energy E can be seen as momentum. For a relativistic massless fermion, this is always the case. The measure that defines the scalar product of the corresponding Hilbert spaces are one-to-one related

$$\int_\ell^\infty d\rho\, \psi_1^*(\rho)\psi_2(\rho) = \int_{\log\ell}^\infty dx\, \phi_1^*(x)\phi_2(x). \tag{183}$$

The operator $(\rho\hat{p}_\rho + \hat{p}_\rho\rho)/2$ is self-adjoint in the interval $(0,\infty)$ but not in the interval $(1,\infty)$, just like $\hat{p}_x$ is self-adjoint in the real line $(-\infty,\infty)$ but not in the half-line $(0,\infty)$ [23,66]. The former case corresponds to the value $\ell = 0$ and the latter one to $\ell = 1$ in Equation (183). Let us now consider the Dirac Hamiltonian in the Rindler variable ρ, given in Equation (115). It becomes in the x variable

$$H = \begin{pmatrix} -i\partial_x & 0 \\ 0 & i\partial_x \end{pmatrix}. \tag{184}$$

Unlike $\hat{p}_x$, this Hamiltonian is self-adjoint in the interval $x \in (\log \ell_1, \infty)$. We choose for convenience $\ell_1 = 1$. The moving mirrors located at $\rho = \ell_n$ are now placed at the positions $x = d_n$, with $d_n = \log \ell_n$, so for $\ell_n = \sqrt{n}$, we get

$$d_n = \frac{1}{2}\log n, \tag{185}$$

where n are square free integers and the reflection coefficients are given by $r_n = \mu(n)/\sqrt{n}$. Figure 17 shows the array of mirrors satisfying Equation (185). One can easily generalize this interferometer to provide a spectral realization of the *zeros* of Dirichlet L-functions, by changing the reflection coefficients r_n,

$$L_\chi(s) = \sum_{n=1}^\infty \frac{\chi(n)}{n^s} \longrightarrow r_n = \frac{\mu(n)\chi(n)}{n^{1/2}}, \tag{186}$$

where $\chi(n)$ is the Dirichlet character associated with the L-function. It would be interesting to replace the massless fermions by massless bosons, say photons and study what kind of Riemann interferometer arise.

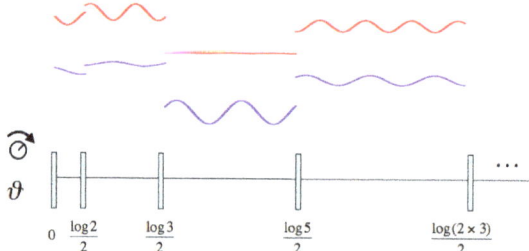

Figure 17. Graphical representation of the array of mirrors in Minkowski space that reproduce the Riemann zeros. The phase at the boundary ϑ must be chosen according to Equation (176) in order that E is an eigenvalue of the Hamiltonian. Recall Figure 13. Between the mirrors the wave functions are plane waves.

14. Dirac Models for a Class of Modified ζ and L Functions

Grosswald and Schnitzer proved in 1978 two very surprising theorems that we shall use below to generalize the construction done in the previous sections. Let us first consider a set on integers q_n satisfying the conditions

$$p_n \leq q_n \leq p_{n+1}, \quad n = 1,\ldots,\infty, \tag{187}$$

where p_n is the nth prime number. With these numbers define the infinite product

$$\zeta^*(s) = \prod_{n=1}^{\infty}(1-q_n^{-s})^{-1}. \tag{188}$$

One then has [78]:

Theorem 1. *This function is holomorphic for $\sigma = \operatorname{Re} s > 1$ and has the following properties:*

(i) $\zeta^*(s) \neq 0$, for $\sigma > 1$,
(ii) $\zeta^*(s)$ has a meromorphic extension to $\sigma > 0$,
(iii) in $\sigma > 0$, $\zeta^*(s)$ has a simple pole at $s = 1$ with residue r, $1/2 \leq r \leq 1$,
(iv) in $\sigma > 0$, $\zeta^*(s)$ has the same zeros as $\zeta(s)$ with the same multiplicity.

This theorem means that the relation between prime numbers and Riemann zeros via the zeta function is less rigid that one may have though. We shall use this freedom to associate a Hamiltonian to every series satisfying (187). Let us first write the inverse of (188) as

$$\frac{1}{\zeta^*(s)} = \sum_{n=1}^{\infty} \frac{\mu^*(n)}{n^s}, \qquad \mu^*(n) = n_{\text{even}} - n_{\text{odd}}, \tag{189}$$

where $n_{\text{even}}(n_{\text{odd}})$ is the number of times n can be written as the product of an even (odd) number of q_i numbers in the series (187). An example of a series satisfying (187) is

$$2,4,6,8,12,\ldots q_n = p_n + 1, \ldots \tag{190}$$

for which we have

$$\frac{1}{\zeta^*(s)} = 1 - \frac{1}{2^s} - \frac{2}{(2^6 \cdot 3)^s} + \frac{2}{(2^3 \cdot 3)^s} + \frac{1}{(2^8 \cdot 3)^s} - \frac{1}{2^{2s}} - \frac{1}{(2 \cdot 3)^s} + \ldots. \tag{191}$$

Notice that $\mu^*(2^6 \cdot 3) = -2$ because $2^6 \cdot 3 = 4 \cdot 6 \cdot 8 = 2 \cdot 8 \cdot 12$. Obviously $\mu^*(n) = \mu(n)$ if $q_n = p_n$, $\forall n$. Using Equation (189) we define a massless Dirac model with reflection coefficients (recall Equation (153))

$$r_n = \frac{\mu^*(n)}{n^{1/2}}, \quad n > 1. \tag{192}$$

Hence, by the arguments given in Section 12 and theorem 1, we shall find the Riemann zeros in the spectrum of the Hamiltonian H_ϑ by tuning the parameter ϑ in the limit $\varepsilon \to 0$.

The second theorem in reference [78] is an extension of theorem 1 to Dirichlet L-functions $L(s) = \prod_n (1-\chi(n)n^{-s})^{-1}$, where χ is a character modulo k. The series (187) is replaced by

$$p_n \leq q_n \leq p_n + K, \qquad p_n = q_n \bmod k \tag{193}$$

where $K \geq k$. The modified L-function is defined as

$$L^*(s) = \prod_{n=1}^{\infty}(1-\chi(q_n)q_n^{-s})^{-1}, \tag{194}$$

that can be extended to the region $\sigma > 0$, with the same zeros (and multiplicities) as $L(s)$. In this case, too, we can construct a Dirac model with reflection coefficients (recall Equation (186))

$$r_n = \frac{\chi(n)\mu^*(n)}{n^{1/2}}, \quad n > 1. \tag{195}$$

whose associated Hamiltonian H_ϑ contains the zeros of $L(s)$ by varying ϑ. Theorem 2 of [78] was mentioned by LeClair and Mussardo in [63] as a support to their approach to the Generalized Riemann hypothesis based on random walks and the Lemke Oliver-Soundararajan conjecture on the distribution of pairs of residues on consecutive primes [79] (for other statistical properties of the prime numbers see [80,81]). It will be worth to investigate if there is a relation between our approach and the one proposed in [62,63].

15. Conclusions

In this paper, we have reviewed the spectral approach to the RH that started with the Berry–Keating–Connes xp model and continued with several works aimed to provide a physical realization of the Riemann zeros. The main steps in this approach are: (i) spectral realization of Connes's xp model using the Landau model of an electron in a magnetic field and electrostatic potential, (ii) construction of modified quantum xp models whose spectra reproduce, on average, the behavior of the *zeros*, (iii) reformulation of the $x(p+1/p)$ model as a relativistic theory of a massive Dirac fermion in a region of Rindler spacetime, (iv) inclusion of prime numbers into the massless Dirac equation by means of delta function potentials acting as moving mirrors that, in the limit where they become semitransparent, leads to a spectral realization of the *zeros*, (v) a route for proving the Riemann Hypothesis, and (vi) proposal of an interferometer that may provide an experimental observation of the zeros of the Riemann zeta function and other Dirichlet L-functions.

The Pólya-Hilbert (PH) conjecture was proposed as a physical explanation of the RH based on the spectral properties of self-adjoint operators: there exists a *single* quantum Hamiltonian containing *all* the Riemann zeros in its spectrum which are therefore real numbers. This statement can be called the *global* version of the PH conjecture. Instead of this, we have found a *local* version according to which a Riemann zero E_n becomes an eigenvalue of the Hamiltonian H_ϑ provided the parameter ϑ, which characterizes the self-adjoint extension, is fine-tuned to the combination $\theta(E_n) + \frac{\pi}{2}\mathrm{sign}Z'(E_n)$. In this sense the Hamiltonian provides a physical realization of $\zeta(\frac{1}{2} + it)$, and not only of the Riemann-Siegel Z function. We have given arguments for a proof of the RH by contradiction: the existence of a *zero* off the critical line implies that the eigenstates of H_ϑ are non-normalizable in the discrete or continuum sense, which is impossible since H_ϑ is a self-adjoint operator. These results are obtained in the limit where the mirrors become semitransparent and assumes the convergence of some mathematical series that need to be analyzed more thoroughly. Finally, we have proposed an interferometer made of fermions propagating in an array of mirrors that may yield an experimental observation of the Riemann zeros in the Lab.

Funding: Grants FIS2012-33642, FIS2015-69167-C2-1-P, QUITEMAD+ S2013/ICE-2801; and SEV-2012-0249, and SEV-2016-0597 of the "Centro de Excelencia Severo Ochoa" Program.

Acknowledgments: I am grateful for fruitful discussions and comments to Julio Andrade, Manuel Asorey, Michael Berry, Ignacio Cirac, Charles Creffield, Jon Keating, José Ignacio Latorre, Giuseppe Mussardo, André LeClair, Miguel Angel Martín-Delgado, Javier Molina-Vilaplana, Javier Rodríguez-Laguna, Mark Srednicki and Paul Townsend. I thank Denis Bernard for pointing out an error in the first version of this manuscript.

Conflicts of Interest: The authors declare no conflict of interest.

References

1. Riemann, B. On the Number of Primes Less Than a Given Quantity. Available online: https://www.claymath.org/sites/default/files/ezeta.pdf (accessed on 30 December 2018).
2. Edwards, H.M. *Riemann's Zeta Function*; Academic Press: New York, NY, USA, 1974.
3. Titchmarsh, E.C. *The Theory of the Riemann Zeta Function*; Oxford University Press: Oxford, UK, 1986.
4. Davenport, H. *Multiplicative Number Theory*; Grad. Texts in Math.; Springer: New York, NY, USA, 2000; Volume 74.

5. Bombieri, E. Problems of the Millennium: The Riemann Hypothesis. Available online: https://www.researchgate.net/publication/247265052_Problems_of_the_Millennium_the_Riemann_Hypothesis (accessed on 30 December 2018).
6. Sarnak, P. Problems of the Millennium: The Riemann Hypothesis. Available online: http://www.claymath.org/library/annual_report/ar2004/04report_sarnak.pdf (accessed on 30 December 2018).
7. Conrey, B. The Riemann Hypothesis, Notices Amer. Math. Available online: https://www.ams.org/notices/200303/fea-conrey-web.pdf (accessed on 30 December 2018).
8. Pólya, G. See A. Odlyzko, Correspondence about the Origins of the Hilbert-Pólya Conjecture. unpublished (c. 1914). 1981–1982. Available online: http://www.dtc.umn.edu/~odlyzko/polya/index.html (accessed on 30 December 2018).
9. Montgomery, H.L. The pair correlation of the zeta function. *Proc. Symp. Pure Math.* **1973**, *24*, 181–193.
10. Odlyzko, A.M. Supercomputers and the Riemann zeta function. In *Conf. on Supercomputing*; International Supercomputing Institute: St. Petersburg, FL, USA, 1989; Volume 348.
11. Bohigas, O.; Gianonni, M.J.; Schmit, C. Characterization of chaotic quantum spectra and universality of level fluctuation. *Phys. Rev. Lett.* **1984**, *52*, 1–4. [CrossRef]
12. Berry, M.V. Riemann's zeta function: A model for quantum chaos? In *Quantum Chaos and Statistical Nuclear Physics*; Seligman, T.H., Nishioka, H., Eds.; Springer Lecture Notes in Physics; Springer: New York, NY, USA, 1986; Volume 263, p. 1.
13. Bogomolny, E.B.; Keating, J.P. Random matrix theory and the Riemann zeros I; three- and four-point correlations. *Nonlinearity* **1995**, *8*, 1115–1131. [CrossRef]
14. Keating, J.P. Periodic orbits, spectral statistics and the Riemann zeros. In *Supersymmetry and Trace Formulae: Chaos and Disorder*; Lerner, I.V., Keating, J.P., Khmelnitskii, D.E., Eds.; Kluwer Academic/Plenum Publishers: New York, NY, USA, 1999; pp. 1–15.
15. Keating, J.P.; Snaith, N.C. Random matrix theory and $\zeta(1/2+it)$. *Commun. Math. Phys.* **2000**, *214*, 57. [CrossRef]
16. Leboeuf, P.; Monastra, A.G.; Bohigas, O. The Riemannium. *Reg. Chaot. Dyn.* **2001**, *6*, 205. [CrossRef]
17. Hejhal, D. The Selberg trace formula and the Riemann zeta function. *Duke Math. J.* **1976**, *43*, 441–482. [CrossRef]
18. Berry, M.V.; Keating, J.P. $H = xp$ and the Riemann zeros. In *Supersymmetry and Trace Formulae: Chaos and Disorder*; Lerner, I.V., Keating, J.P., Khmelnitskii, D.E., Eds.; Kluwer Academic/Plenum Publishers: New York, NY, USA, 1999; pp. 355–367.
19. Berry, M.V.; Keating, J.P. The Riemann zeros and eigenvalue asymptotics. *SIAM Rev.* **1999**, *41*, 236–266. [CrossRef]
20. Connes, A. Trace formula in noncommutative geometry and the zeros of the Riemann zeta function. *Sel. Math. New Ser.* **1999**, *5*, 29. [CrossRef]
21. Aneva, B. Symmetry of the Riemann operator. *Phys. Lett. B* **1999**, *450*, 388–396. [CrossRef]
22. Sierra, G. The Riemann zeros and the cyclic renormalization group. *J. Stat. Mech. Theor. Exp.* **2005**, *2005*, P12006. [CrossRef]
23. Sierra, G. $H = xp$ with interaction and the Riemann zeros. *Nucl. Phys. B* **2007**, *776*, 327–364. [CrossRef]
24. Twamley, J.; Milburn, G.J. The quantum Mellin transform. *New J. Phys.* **2006**, *8*, 328. [CrossRef]
25. Sierra, G. Quantum reconstruction of the Riemann zeta function. *J. Phys. A Math. Theor.* **2007**, *40*, 1.
26. Sierra, G. A quantum mechanical model of the Riemann zeros. *New J. Phys.* **2008**, *10*, 033016. [CrossRef]
27. Lagarias, J.C. The Schroëdinger operator with Morse potential on the right half line. *Commun. Number Theory Phys.* **2009**, *3*, 323–361. [CrossRef]
28. Burnol, J.-F. On some bound and scattering states associated with the cosine kernel. *arXiv* **2008**, arXiv:0801.0530.
29. Sierra, G.; Townsend, P.K. The Landau model and the Riemann zeros. *Phys. Rev. Lett.* **2008**, *101*, 110201. [CrossRef] [PubMed]
30. Endres, S.; Steiner, F. The Berry-Keating operator on $L^2(R_>, dx)$ and on compact quantum graphs with general self-adjoint realizations. *J. Phys. A: Math. Theor.* **2010**, *43*, 095204. [CrossRef]
31. Regniers, G.; van der Jeugt, J. The Hamiltonian $H = xp$ and classification of $osp(1|2)$ representations. *AIP Conf. Proc.* **2010**, *1243*, 138.

32. Sierra, G.; Rodríguez-Laguna, J. The $H = xp$ model revisited and the Riemann zeros. *Phys. Rev. Lett.* **2011**, *106*, 200201. [CrossRef]
33. Srednicki, M. The Berry-Keating Hamiltonian and the Local Riemann Hypothesis. *J. Phys. A Math. Theor.* **2011**, *44*, 305202. [CrossRef]
34. Srednicki, M. Nonclasssical Degrees of Freedom in the Riemann Hamiltonian. *Phys. Rev. Lett.* **2011**, *107*, 100201. [CrossRef] [PubMed]
35. Sierra, G. General covariant xp models and the Riemann zeros. *J. Phys. A Math. Theor.* **2012**, *45*, 055209. [CrossRef]
36. Berry, M.V.; Keating, J.P. A compact hamiltonian with the same asymptotic mean spectral density as the Riemann zeros. *J. Phys. A Math. Theor.* **2011**, *44*, 285203. [CrossRef]
37. Gupta, K.S.; Harikumar, E.; de Queiroz, A.R. A Dirac type xp-Model and the Riemann Zeros. *Eur. Phys. Lett.* **2013**, *102* 10006. [CrossRef]
38. Molina-Vilaplana, J.; Sierra, G. An xp model on AdS_2 spacetime. *Nucl. Phys. B* **2013**, *877*, 107. [CrossRef]
39. Nucci, M.C. Spectral realization of the Riemann zeros by quantizing $H = w(x)(p + \ell_p^2/p)$: The Lie-Noether symmetry approach. *J. Phys. Conf. Ser.* **2014**, *482*, 012032. [CrossRef]
40. Andrade, J.C. Hilbert-Pólya conjecture, zeta-functions and bosonic quantum field theories. *Int. J. Mod. Phys. A* **2013**, *28*, 1350072. [CrossRef]
41. Kuipers, J.; Hummel, Q.; Richter, K. Quantum graphs whose spectra mimic the zeros of the Riemann zeta function. *Phys. Rev. Lett* **2014**, *112*, 070406. [CrossRef]
42. Sierra, G. The Riemann zeros as energy levels of a Dirac fermion in a potential built from the prime numbers in Rindler spacetime. *J. Phys. A Math. Theor.* **2014**, *47*, 325204. [CrossRef]
43. Bender, C.M.; Brody, D.C.; Müller, M.P. Hamiltonian for the zeros of the Riemann zeta function. *Phys. Rev. Lett.* **2017**, *118*, 130201. [CrossRef] [PubMed]
44. Bellissard, J.V. Comment on "Hamiltonian for the zeros of the Riemann zeta function". *arXiv* **2017**, arXiv:1704.02644.
45. Bender, C.M.; Brody, D.C.; Müller, M.P. Comment on 'Comment on "Hamiltonian for the zeros of the Riemann zeta function"'. *arXiv* **2017**, arXiv:1705.06767.
46. Schumayer, D.; Hutchinson, D.A.W. Physics of the Riemann Hypothesis. *Rev. Mod. Phys.* **2011**, *83*, 307–330. [CrossRef]
47. Pavlov, B.S.; Fadeev, L.D. Scattering theory and authomorphic functions. *Sov. Math.* **1975**, *3*, 522–548. [CrossRef]
48. Lax, P.D.; Phillips, R.S. *Scattering Theory for Automorphic Functions*; Princeton University Press: Princeton, NJ, USA, 1976.
49. Bhaduri, R.K.; Khare, A.; Law, J. Phase of the Riemann zeta function and the inverted harmonic oscillator. *Phys. Rev. E* **1995**, *52*, 486. [CrossRef]
50. LeClair, A. Interacting Bose and Fermi gases in low dimensions and the Riemann hypothesis. *Int. J. Mod. Phys. A* **2008**, *23*, 1371–1391. [CrossRef]
51. He, Y.-H.; Jejjala, V.; Minic, D. Eigenvalue Density, Li's Positivity, and the Critical Strip. *arXiv* **2009**, arXiv:0903.4321.
52. Berry, M.V. Riemann zeros in radiation patterns. *J. Phys. A Math. Theor.* **2012**, *45*, 302001. [CrossRef]
53. Latorre, J.I.; Sierra, G. Quantum Computation of Prime Number Functions. *Quant. Inf. Comp.* **2014**, *14*, 0577.
54. Menezes, G.; Svaiter, B.F.; Svaiter, N.F. Riemann zeta zeros and prime number spectra in quantum field theory. *Int. J. Mod. Phys. A* **2013**, *28*, 1350128. [CrossRef]
55. Ramos, R.V.; Mendes, F.V. Riemannian Quantum Circuit. *Phys. Lett. A* **2014**, *378*, 1346. [CrossRef]
56. Dueñas, J.G.; Svaiter, N.F. Riemann zeta zeros and zero-point energy. *Int. J. Mod. Phys. A* **2014**, *29*, 1450051. [CrossRef]
57. Feiler, C.; Schleich, W.P. Entanglement and analytical continuation: An intimate relation told by the Riemann zeta function. *New J. Phys* **2013**, *15*, 063009. [CrossRef]
58. Creffield, C.E.; Sierra, G. Finding zeros of the Riemann zeta function by periodic driving of cold atoms. *Phys. Rev. A* **2015**, *91*, 063608. [CrossRef]
59. França, G.; A. LeClair, A. Transcendental equations satisfied by the individual zeros of Riemann, Dirichlet and modular L-functions. *arXiv* **2015**, arXiv:1502.06003.
60. LeClair, A. Riemann Hypothesis and Random Walks: The Zeta case. *arXiv* **2016**, arXiv:1601.00914.

61. França, G.; LeClair, A. Some Riemann Hypotheses from Random Walks over Primes. *Commun. Cont. Math.* **2017**, *20*, 1750085. [CrossRef]
62. Mussardo, G.; LeClair, A. Generalized Riemann Hypothesis and Stochastic Time Series. *J. Stat. Mech.* **2018**, *2018*, 063205. [CrossRef]
63. LeClair, A.; Mussardo, G. Generalized Riemann Hypothesis, Time Series and Normal Distributions. *J. Stat. Mech.* **2019**, *2019*, 023203. [CrossRef]
64. Abramowitz, M.; Stegun, I.A. *Handbook of Mathematical Functions*; Dover: New York, NY, USA, 1974.
65. von Neumann, J. Allgemeine Eigenwerttheorie Hermitescher Funktionaloperatoren. *Math. Ann.* **1929**, *102*, 49–131. [CrossRef]
66. Galindo, A.; Pascual, P. *Quantum Mechanics I*; Springer: Berlin, Germany, 1991.
67. Rindler, W. Kruskal space and the uniformly accelerated frame. *Am. J. Phys.* **1966**, *34*, 1174. [CrossRef]
68. Unruh, W.G. Notes on black-hole evaporation. *Phys. Rev. D* **1976**, *14*, 870. [CrossRef]
69. Pólya, G. Bemerkung uber die integraldarstellung der Riemannschen zeta-funktion. *Acta Math.* **1926**, *48*, 305. [CrossRef]
70. Hejhal, D.A. On a result of G. Pólya concerning the Riemann ζ-function. *J. d' Analyse Mathématique* **1990**, *55*, 59. [CrossRef]
71. Asorey, M.; Ibort, A.; Marmo, G. Global Theory of Quantum Boundary Conditions and Topology Change. *Int. J. Mod. Phys.* **2005**, *A20*, 1001. [CrossRef]
72. Julia, B. *Statistical Theory of Numbers, in Number Theory and Physics*; Luck, J.M., Moussa, P., Waldschmidt, M., Eds.; Springer Proceedings in Physics; Springer: Berlin, Germany, 1990; Volume 47, p. 276.
73. Spector, D. Supersymmetry and the Moebius Inversion Function. *Commun. Math. Phys.* **1990**, *127*, 239. [CrossRef]
74. Mussardo, G. The quantum mechanical potential for the prime numbers. *arXiv* **1997**, arXiv:cond-mat.9712010.
75. Blanes, S.; Casas, F.; Oteo, J.A.; Ros, J. The Magnus expansion and some of its applications. *Phys. Rep.* **2008**, *470*, 151–238. [CrossRef]
76. Apostol, T.M. *Introduction to Analytic Number Theory*; Springer: New York, NY, USA, 1976.
77. Borwein, P.; Choi, S.; Rooney, B.; Weirathmueller, A. (Eds.) *The Riemann Hypothesis. A Resource for the Afficionado and Virtuoso Alike*; CMS Books in Mathematics; Springer: Berlin, Germany, 2008.
78. Grosswald, E.; Schnitzer, F.J. A class of modified ζ and L-functions. *Pacific. Jour. Math.* **1978**, *74*, 357–364. [CrossRef]
79. Oliver, R.J.L.; Soundararajan, K. Unexpected biases in the distribution of consecutive primes. *Proc. Nat. Acad. Sci. USA* **2016**, *113*, E4446–E4454 [CrossRef] [PubMed]
80. Kristyan, S. On the statistical distribution of prime numbers: A view from where the distribution of prime numbers are not erratic. *AIP Conf. Proc.* **2017**, *1863*, 560013.
81. Kristyan, S. Note on the cardinality difference between primes and twin primes and its impact on function $x/\ln(x)$ in prime number theorem. *AIP Conf. Proc.* **2018**, *1978*, 470064.

© 2019 by the authors. Licensee MDPI, Basel, Switzerland. This article is an open access article distributed under the terms and conditions of the Creative Commons Attribution (CC BY) license (http://creativecommons.org/licenses/by/4.0/).

Article

Universal Quantum Computing and Three-Manifolds

Michel Planat [1,*], Raymond Aschheim [2], Marcelo M. Amaral [2] and Klee Irwin [2]

1. Institut FEMTO-ST CNRS UMR 6174, Université de Bourgogne/Franche-Comté, 15 B Avenue des Montboucons, F-25044 Besançon, France
2. Quantum Gravity Research, Los Angeles, CA 90290, USA; raymond@QuantumGravityResearch.org (R.A.); Marcelo@quantumgravityresearch.org (M.M.A.); Klee@quantumgravityresearch.org (K.I.)
* Correspondence: michel.planat@femto-st.fr

Received: 23 November 2018; Accepted: 14 December 2018; Published: 19 December 2018

Abstract: A single qubit may be represented on the Bloch sphere or similarly on the 3-sphere S^3. Our goal is to dress this correspondence by converting the language of universal quantum computing (UQC) to that of 3-manifolds. A magic state and the Pauli group acting on it define a model of UQC as a positive operator-valued measure (POVM) that one recognizes to be a 3-manifold M^3. More precisely, the d-dimensional POVMs defined from subgroups of finite index of the modular group $PSL(2, \mathbb{Z})$ correspond to d-fold M^3- coverings over the trefoil knot. In this paper, we also investigate quantum information on a few 'universal' knots and links such as the figure-of-eight knot, the Whitehead link and Borromean rings, making use of the catalog of platonic manifolds available on the software SnapPy. Further connections between POVMs based UQC and M^3's obtained from Dehn fillings are explored.

Keywords: quantum computation; IC-POVMs; knot theory; three-manifolds; branch coverings; Dehn surgeries

PACS: 03.67.Lx; 03.65.Wj; 03.65.Aa; 02.20.-a; 02.10.Kn; 02.40.Pc; 02.40.Sf

MSC: 81P68; 81P50; 57M25; 57R65; 14H30; 20E05; 57M12

Manifolds are around us in many guises.

As observers in a three-dimensional world, we are most familiar with two-manifolds: the surface of a ball or a doughnut or a pretzel, the surface of a house or a tree or a volleyball net...

Three-manifolds may be harder to understand at first. But as actors and movers in a three-dimensional world, we can learn to imagine them as alternate universes.

(William Thurston [1]).

1. Introduction

Mathematical concepts pave the way for improvements in technology. As far as topological quantum computation is concerned, non-abelian anyons have been proposed as an attractive (fault-tolerant) alternative to standard quantum computing which is based on a universal set of quantum gates [2–5]. Anyons are two-dimensional quasiparticles with world lines forming braids in space-time. Whether non-abelian anyons do exist in the real world and/or would be easy to create artificially, is still open to discussion. In this paper, we propose an alternative to anyon-based universal quantum computation (UQC) thanks to three-dimensional topology. Our proposal relies on appropriate 3-manifolds whose fundamental group is used for building the magic states for UQC. Three-dimensional topological quantum computing would federate the foundations of quantum mechanics and cosmology, a recurrent dream of many physicists. Three-dimensional topology was already investigated by several groups in the context of quantum information [6,7], high energy physics [8,9], biology [10] and consciousness studies [11].

Recall the context of our work and clarify its motivation. Bravyi & Kitaev introduced the principle of 'magic state distillation': universal quantum computation, the possibility to implement an arbitrary quantum gate, may be realized thanks to the stabilizer formalism (Clifford group unitaries, preparations and measurements) and the ability to prepare an appropriate single qubit non-stabilizer state, called a 'magic state' [12]. Then, irrespectively of the dimension of the Hilbert space where the quantum states live, a non-stabilizer pure state was called a magic state [13]. An improvement of this concept was carried out in [14,15] showing that a magic state could be at the same time a fiducial state for the construction of an informationally complete positive operator-valued measure, or IC-POVM, under the action on it of the Pauli group of the corresponding dimension. Thus UQC in this view happens to be relevant both to such magic states and to IC-POVMs. In [14,15], a d-dimensional magic state follows from the permutation group that organizes the cosets of a subgroup H of index d of a two-generator free group G. This is due to the fact that a permutation may be seen as a permutation matrix/gate and that mutually commuting matrices share eigenstates—they are either of the stabilizer type (as elements of the Pauli group) or of the magic type. In the calculation, it is enough to keep magic states that are simultaneously fiducial states for an IC-POVM and we are done. Remarkably, a rich catalog of the magic states relevant to UQC and IC-POVMs can be obtained by selecting G as the two-letter representation of the modular group $\Gamma = PSL(2, \mathbb{Z})$ [16]. The next step, developed in this paper, is to relate the choice of the starting group G to three-dimensional topology. More precisely, G is taken as the fundamental group $\pi_1(S^3 \setminus K)$ of a 3-manifold M^3 defined as the complement of a knot or link K in the 3-sphere S^3. A branched covering of degree d over the selected M^3 has a fundamental group corresponding to a subgroup of index d of π_1 and may be identified as a sub-manifold of M^3, the one leading to an IC-POVM is a model of UQC. In the specific case of Γ, the knot involved is the left-handed trefoil knot T_1, as shown in Section 2.

While Γ serves as a motivation for investigating the trefoil knot manifold in relation to UQC and the corresponding ICs, it is important to put the UQC problem in the wider frame of Poincaré conjecture, the Thurston's geometrization conjecture and the related 3-manifolds [1]. For example, ICs may also follow from hyperbolic or Seifert 3-manifolds as shown in Tables of this paper.

More details are provided at the next subsections.

1.1. From Poincaré Conjecture to UQC

The Poincaré conjecture is the elementary (but deep) statement that every simply connected, closed 3-manifold is homeomorphic to the 3-sphere S^3 [17]. Having in mind the correspondence between S^3 and the Bloch sphere that houses the qubits $\psi = a\left|0\right\rangle + b\left|1\right\rangle$, $a, b \in \mathbb{C}$, $|a|^2 + |b|^2 = 1$, one would desire a quantum translation of this statement. For doing this, one may use the picture of the Riemann sphere $\mathbb{C} \cup \infty$ in parallel to that of the Bloch sphere and follow F. Klein lectures on the icosahedron to perceive the platonic solids within the landscape [18]. This picture fits well the Hopf fibrations [19], their entanglements described in [20,21] and quasicrystals [22,23]. However, we can be more ambitious and dress S^3 in an alternative way that reproduces the historic thread of the proof of Poincaré conjecture. Thurston's geometrization conjecture, from which Poincaré conjecture follows, dresses S^3 as a 3-manifold not homeomorphic to S^3. The wardrobe of 3-manifolds M^3 is huge but almost every dress is hyperbolic and W. Thurston found the recipes for them [1]. Every dress is identified thanks to a signature in terms of invariants. For our purpose, the fundamental group π_1 of M^3 does the job.

The three-dimensional space surrounding a knot K—the knot complement $S^3 \setminus K$—is an example of a three-manifold [1,24]. We will be especially interested by the trefoil knot that underlies work of the first author [16] as well as the figure-of-eight knot, the Whitehead link and the Borromean rings because they are universal (in a sense described below), hyperbolic and allow to build 3-manifolds from platonic manifolds [25]. Such manifolds carry a quantum geometry corresponding to quantum computing and (possibly informationally complete) POVMs identified in our earlier work [14–16].

According to [26], the knot K and the fundamental group $G = \pi_1(S^3 \setminus K)$ are universal if every closed and oriented 3-manifold M^3 is homeomorphic to a quotient $\mathbb{H}/G$ of the hyperbolic 3-space $\mathbb{H}$ by a subgroup H of finite index d of G. As just announced, the figure-of-eight knot, the Whitehead link and Borromean rings are universal. The catalog of the finite index subgroups of their fundamental group G and of the corresponding 3-manifolds defined from the d-fold coverings [27] can easily be established up to degree 8, using the software SnapPy [28].

In paper [16] of the first author, it has been found that minimal d-dimensional IC-POVMs (sometimes called MICs) may be built from finite index subgroups of the modular group $\Gamma = PSL(2, \mathbb{Z})$. To such an IC (or MIC) is associated a subgroup of index d of Γ, a fundamental domain in the Poincaré upper-half plane and a signature in terms of genus, elliptic points and cusps as summarized in ([16] Figure 1). There exists a relationship between the modular group Γ and the trefoil knot T_1 since the fundamental group $\pi_1(S^3 \setminus T_1)$ of the knot complement is the braid group B_3, the central extension of Γ. However, the trefoil knot and the corresponding braid group B_3 are not universal [29] which forbids the relation of the finite index subgroups of B_3 to all three-manifolds.

It is known that two coverings of a manifold M with fundamental group $G = \pi_1(M)$ are equivalent if there exists a homeomorphism between them. Besides, a d-fold covering is uniquely determined by a subgroup of index d of the group G and the inequivalent d-fold coverings of M correspond to conjugacy classes of subgroups of G [27]. In this paper we will fuse the concepts of a three-manifold M^3 attached to a subgroup H of index d and the POVM, possibly informationally complete (IC), found from H (thanks to the appropriate magic state and related Pauli group factory).

1.2. Minimal Informationally Complete POVMs and UQC

In our approach [15,16], minimal informationally complete (IC) POVMs are derived from appropriate fiducial states under the action of the (generalized) Pauli group. The fiducial states also allow to perform universal quantum computation [14].

A POVM is a collection of positive semi-definite operators $\{E_1, \ldots, E_m\}$ that sum to the identity. In the measurement of a state ρ, the i-th outcome is obtained with a probability given by the Born rule $p(i) = \text{tr}(\rho E_i)$. For a minimal IC-POVM (or MIC), one needs d^2 one-dimensional projectors $\Pi_i = |\psi_i\rangle \langle \psi_i|$, with $\Pi_i = dE_i$, such that the rank of the Gram matrix with elements $\text{tr}(\Pi_i \Pi_j)$, is precisely d^2. A SIC-POVM (the S means symmetric) obeys the relation $|\langle \psi_i | \psi_j \rangle|^2 = \text{tr}(\Pi_i \Pi_j) = \frac{d\delta_{ij}+1}{d+1}$, that allows the explicit recovery of the density matrix as in ([30] Equation (29)).

New minimal IC-POVMs (i.e., whose rank of the Gram matrix is d^2) and with Hermitian angles $|\langle \psi_i | \psi_j \rangle||_{i \neq j} \in A = \{a_1, \ldots, a_l\}$ have been discovered [16]. A SIC (i.e., a SIC-POVM) is equiangular with $|A| = 1$ and $a_1 = \frac{1}{\sqrt{d+1}}$. The states encountered are considered to live in a cyclotomic field $\mathbb{F} = \mathbb{Q}[\exp(\frac{2i\pi}{n})]$, with $n = \text{GCD}(d, r)$, the greatest common divisor of d and r, for some r. The Hermitian angle is defined as $|\langle \psi_i | \psi_j \rangle||_{i \neq j} = \|(\psi_i, \psi_j)\|^{\frac{1}{\deg}}$, where $\|.\|$ means the field norm of the pair (ψ_i, ψ_j) in $\mathbb{F}$ and deg is the degree of the extension $\mathbb{F}$ over the rational field $\mathbb{Q}$ [15].

The fiducial states for SIC-POVMs are quite difficult to derive and seem to follow from algebraic number theory [31].

Except for $d = 3$, the IC-POVMs derived from permutation groups are not symmetric and most of them can be recovered thanks to subgroups of index d of the modular group Γ ([16] Table 2). The geometry of the qutrit Hesse SIC is shown in Figure 1a. It follows from the action of the qutrit Pauli group on magic/fiducial states of type $(0, 1, \pm 1)$. For $d = 4$, the action of the two-qubit Pauli group on the magic/fiducial state of type $(0, 1, -\omega_6, \omega_6 - 1)$ with $\omega_6 = \exp(\frac{2i\pi}{6})$ results into a minimal IC-POVM whose geometry of triple products of projectors Π_i turns out to correspond to the commutation graph of Pauli operators, see Figure 1b and ([16] Figure 2). For $d = 5$, the geometry of an IC consists of copies of the Petersen graph reproduced in Figure 1c. For $d=6$, the geometry consists of components looking like Borromean rings (see [16] Figure 2 and Table 1 below).

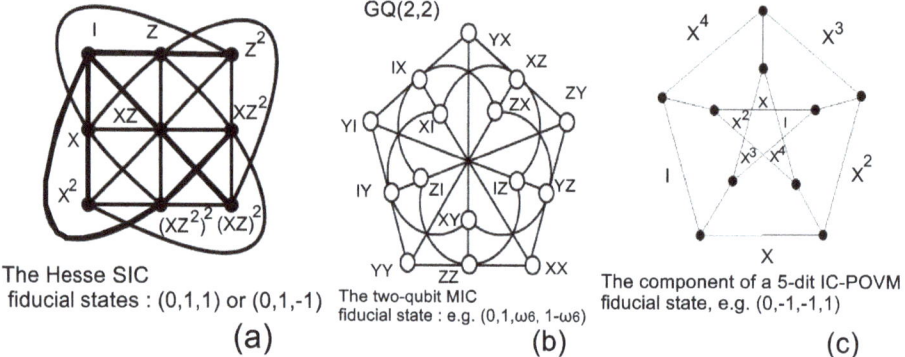

Figure 1. Geometrical structure of low dimensional MICs: (a) the qutrit Hesse SIC, (b) the two-qubit MIC that is the generalized quadrangle of order two $GQ(2,2)$, (c) the basic component of the 5-dit MIC that is the Petersen graph. The coordinates on each diagram are the d-dimensional Pauli operators that act on the fiducial state, as shown.

1.3. Organization of the Paper

The paper is organized as follows. Section 2 deals with the relationship between quantum information seen from the modular group Γ and from the trefoil knot 3-manifold. Section 3 deals with the (platonic) 3-manifolds related to coverings over the figure-of-eight knot, Whitehead link and Borromean rings, see Figure 2, and how they relate to minimal IC-POVMs. Section 4 describes the important role played by Dehn fillings for describing the many types of 3-manifolds that may relate to topological quantum computing.

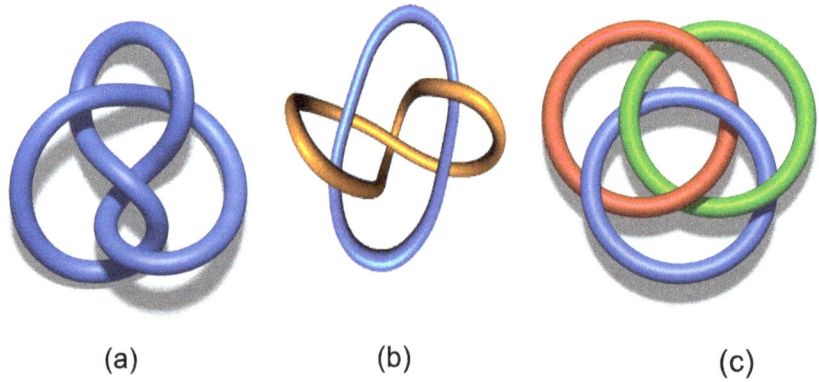

Figure 2. (a) The figure-of-eight knot: $K4a1 = otet02_{00001} = m004$, (b) the Whitehead link $L5a1 = ooct01_{00001} = m129$, (c) Borromean rings $L6a4 = ooct02_{00005} = t12067$.

2. Quantum Information from the Modular Group Γ and the Related Trefoil Knot T_1

In this section, we describe the results established in [16] in terms of the 3-manifolds corresponding to coverings of the trefoil knot complement $S^3 \setminus T_1$.

Let us introduce to the group representation of a knot complement $\pi_1(S^3 \setminus K)$. A Wirtinger representation is a finite representation of π_1 where the relations are of the form $wg_iw^{-1} = g_j$ where

w is a word in the k generators $\{g_1, \cdots, g_k\}$. For the trefoil knot $T_1 = K3a1 = 3_1$ shown in Figure 3a, a Wirtinger representation is [32]

$$\pi_1(S^3 \setminus T_1) = \langle x, y | yxy = xyx \rangle \text{ or equivalently } \pi_1 = \left\langle x, y | y^2 = x^3 \right\rangle.$$

In the rest of the paper, the number of d-fold coverings of the manifold M^3 corresponding to the knot T will be displayed as the ordered list $\eta_d(T), d \in \{1..10\ldots\}$. For T_1 it is

$$\eta_d(T_1) = \{1, 1, 2, 3, 2, \ 8, 7, 10, 18, 28, \ldots\}.$$

Details about the corresponding d-fold coverings are in Table 1. As expected, the coverings correspond to subgroups of index d of the fundamental group associated to the trefoil knot T_1.

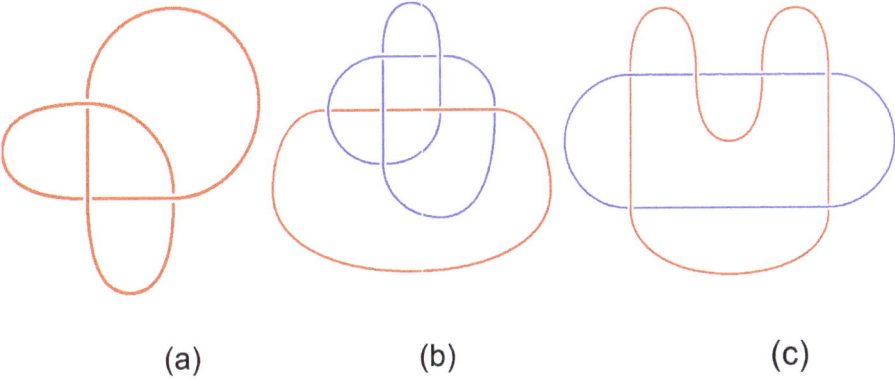

Figure 3. (a) The trefoil knot $T_1 = K3a1 = 3_1$, (b) the link $L7n1$ associated to the Hesse SIC, (c) the link $L6a3$ associated to the two-qubit IC.

2.1. Cyclic Branched Coverings over the Trefoil Knot

Let p, q, r be three positive integers (with $p \leq q \leq r$), the Brieskorn 3-manifold $\Sigma(p, q, r)$ is the intersection in $\mathbb{C}^3$ of the 5-sphere S^5 with the surface of equation $z_1^p + z_2^q + z_3^r = 1$. In [33], it is shown that a r-fold cyclic covering over S^3 branched along a torus knot or link of type (p, q) is a Brieskorn 3-manifold $\Sigma(p, q, r)$ (see also Section 4.1). For the spherical case $p^{-1} + q^{-1} + r^{-1} > 1$, the group associated to a Brieskorn manifold is either dihedral [that is the group D_r for the triples $(2, 2, r)$], tetrahedral [that is A_4 for $(2, 3, 3)$], octahedral [that is S_4 for $(2, 3, 4)$] or icosahedral [that is A_5 for $(2, 3, 5)$]. The Euclidean case $p^{-1} + q^{-1} + r^{-1} = 1$ corresponds to $(2, 3, 6)$, $(2, 4, 4)$ or $(3, 3, 3)$. The remaining cases are hyperbolic.

The cyclic branched coverings with spherical groups for the trefoil knot (which is of type $(2, 3)$) are identified in the right hand side column of Table 1.

2.2. Irregular branched coverings over the trefoil knot

The right hand side column of Table 1 shows the subgroups of Γ identified in ([16] Table 1) as corresponding to a minimal IC-POVM. Let us give a few more details on how to attach a MIC to some coverings/subgroups of the trefoil knot fundamental group $\pi_1(T_1)$. Columns 1 to 6 in Table 1 contain information available in SnapPy [28], with d, ty, hom, cp, $gens$ and CS the degree, the type, the first homology group, the number of cusps, the number of generators and the Chern-Simons invariant of the relevant covering, respectively. In column 7, a link is possibly identified by SnapPy when the fundamental group and other invariants attached to the covering correspond to those of the link. For our purpose, we are also interested in the possible recognition of a MIC behind some manifolds in the table.

Table 1. Coverings of degree *d* over the trefoil knot found from SnapPy [28]. The related subgroup of modular group Γ and the corresponding IC-POVM [16] (when applicable) is in the right column. The covering is characterized by its type ty, homology group hom (where 1 means ℤ), the number of cusps cp, the number of generators gens of the fundamental group, the Chern-Simons invariant CS and the type of link it represents (as identified in SnapPy). The links L7n1 (shown in Figure 3b) and L6a3 (shown in Figure 3c) correspond to the Hesse SIC and the two-qubit IC, respectively. The case of cyclic coverings corresponds to Brieskorn 3-manifolds as explained in the text: the spherical groups for these manifolds is given at the right hand side column.

d	ty	hom	cp	Gens	CS	Link	Type in [16]
2	cyc	$\frac{1}{3}+1$	1	2	$-1/6$		
3	irr	$1+1$	2	2	$1/4$	L7n1	$\Gamma_0(2)$, Hesse SIC
.	cyc	$\frac{1}{2}+\frac{1}{2}+1$	1	3	.		A_4
4	irr	$1+1$	2	2	$1/6$	L6a3	$\Gamma_0(3)$, 2QB IC
.	irr	$\frac{1}{2}+1$	1	3	.		$4A^0$, 2QB-IC
.	cyc	$\frac{1}{3}+1$	1	2	.		S_4
5	cyc	1	1	2	$5/6$		A_5
.	irr	$\frac{1}{3}+1$	1	3	.		$5A^0$, 5-dit IC
6	reg	$1+1+1$	3	3	0	L8n3	$\Gamma(2)$, 6-dit IC
.	cyc	$1+1+1$	1	3	.		Γ', 6-dit IC
.	irr	$1+1+1$	3	3	.		
.	irr	$\frac{1}{2}+1+1$	2	3	.		$3C^0$, 6-dit IC
.	irr	$\frac{1}{2}+1+1$	2	3	.		$\Gamma_0(4)$, 6-dit IC
.	irr	$\frac{1}{2}+1+1$	2	3	.		$\Gamma_0(5)$, 6-dit IC
.	irr	$\frac{1}{2}+\frac{1}{2}+\frac{1}{2}+1$	1	4	.		
.	irr	$\frac{1}{3}+\frac{1}{3}+1$	1	3	.		
7	cyc	1	1	2	$-5/6$		
.	irr	$1+1$	2	3	.		NC 7-dit IC
.	irr	$\frac{1}{2}+\frac{1}{2}+1$	1	4	.		$7A^0$ 7-dit IC
8	irr	$1+1$	2	2	$-1/6$		
.	cyc	$\frac{1}{3}+1$	2	2	.		
.	cyc	$\frac{1}{3}+1+1$	2	3	.		
.	cyc	$\frac{1}{6}+1$	1	4	.		$8A^0$, ∼8-dit IC

For the irregular covering of degree 3 and first homology ℤ + ℤ, the fundamental group provided by SnapPy is $\pi_1(M^3) = \langle a,b | ab^{-2}a^{-1}b^2 \rangle$ that, of course, corresponds to a representative H of one of the two conjugacy classes of subgroups of index 3 of the modular group Γ, following the theory of [27]. The organization of cosets of H in the two-generator group $G = \langle a,b | a^2, y^3 \rangle \cong \Gamma$ thanks to the Coxeter-Todd algorithm (implemented in the software Magma [34]) results in the permutation group $P = \langle 3 | (1,2,3), (2,3) \rangle$, as in ([16] Section 3.1). This permutation group is also the one obtained from the congruence subgroup $\Gamma_0(2) \cong S_3$ of Γ (where S_3 is the three-letter symmetric group) whose fundamental domain is in ([16] Figure 1b). Then, the eigenstates of the permutation matrix in S_3 of type $(0,1,\pm 1)$ serve as magic/fiducial state for the Hesse SIC [15,16].

A similar reasoning applied to the irregular coverings of degree 4, and first homology ℤ + ℤ and $\frac{\mathbb{Z}}{2}+\mathbb{Z}$ leads to the recognition of congruence subgroups $\Gamma_0(3)$ and $4A^0$, respectively, behind the corresponding manifolds. It is known from ([16] Section 3.2) that they allow the construction of two-qubit minimal IC-POVMs. For degree 5, the equiangular 5-dit MIC corresponds to the irregular covering of homology $\frac{\mathbb{Z}}{3}+\mathbb{Z}$ and to the congruence subgroup $5A^0$ in Γ (as in [16] Section 3.3).

Five coverings of degree 6 allow the construction of the (two-valued) 6-dit IC-POVM whose geometry contain the picture of Borromean rings ([16] Figure 2c). The corresponding congruence subgroups of Γ are identified in Table 1. The first, viz $\Gamma(2)$, define a 3-manifold whose fundamental group is the same as the one of the link L8n3. The other three coverings leading to the 6-dit IC are the congruence subgroups γ', $3C^0$, $\Gamma_0(4)$ and $\Gamma_0(5)$.

3. Quantum Information from Universal Knots and Links

In the previous section, we found the opportunity to rewrite results about the existence and construction of d-dimensional MICs in terms of the three-manifolds corresponding to some degree d coverings of the trefoil knot T_1. However, neither T_1 nor the manifolds corresponding to its covering are hyperbolic. In the present section, we proceed with hyperbolic (and universal) knots and links and display the three-manifolds behind the low dimensional MICs. The method is as above in Section 2 in the sense that the fundamental group of a 3-manifold M^3 attached to a degree d-covering is the one of a representative of the conjugacy class of subgroups of the corresponding index in the relevant knot or link.

3.1. Three-Manifolds Pertaining to the Figure-of-Eight Knot

The fundamental group for the figure-of-eight knot K_0 is

$$\pi_1(S^3 \setminus K_0) = \left\langle x, y | y * x * y^{-1} xy = xyx^{-1}yx \right\rangle.$$

and the number of d-fold coverings is in the list

$$\eta_d(K_0) = \{1, 1, 1, 2, 4, 11, 9, 10, 11, 38, \ldots\}.$$

Table 2 establishes the list of 3-manifolds corresponding to subgroups of index $d \leq 7$ of the universal group $G = \pi_1(S^3 \setminus K_0)$. The manifolds are labeled otetN_n in [25] because they are oriented and built from $N = 2d$ tetrahedra, with n an index in the table. The identification of 3-manifolds of finite index subgroups of G was first obtained by comparing the cardinality list $\eta_d(H)$ of the corresponding subgroup H to that of a fundamental group of a tetrahedral manifold in SnapPy table [28]. However, there is a more straightforward way to perform this task by identifying a subgroup H to a degree d covering of K_0 [27]. The full list of d-branched coverings over the figure eight knot up to degree 8 is available in SnapPy. Extra invariants of the corresponding M^3 may be found there. In addition, the lattice of branched coverings over K_0 was investigated in [35].

Table 2. Table of 3-manifolds M^3 found from subgroups of finite index d of the fundamental group $\pi_1(S^3 \setminus K_0)$ (alias the d-fold coverings of K_0). The terminology in column 3 is that of Snappy [28]. The identified M^3 is made of $2d$ tetrahedra and has cp cusps. When the rank rk of the POVM Gram matrix is d^2 the corresponding IC-POVM shows pp distinct values of pairwise products as shown.

d	ty	M^3	cp	rk	pp	Comment
2	cyc	otet04$_{00002}$, $m206$	1	2		
3	cyc	otet06$_{00003}$, $s961$	1	3		
4	irr	otet08$_{00002}$, $L10n46$, t_{12840}	2	4		Mom-4s [36]
	cyc	otet08$_{00007}$, $t12839$	1	16	1	2-qubit IC
5	cyc	otet10$_{00019}$	1	21		
	irr	otet10$_{00006}$, $L8a20$	3	15, 21		
	irr	otet10$_{00026}$	2	25	1	5-dit IC
6	cyc	otet12$_{00013}$	1	28		
	irr	otet12$_{00041}$	2	36	2	6-dit IC
	irr	otet12$_{00039}$, otet12$_{00038}$	1	31		
	irr	otet12$_{00017}$	2	33		
	irr	otet12$_{00000}$	2	36	2	6-dit IC
7	cyc	otet14$_{00019}$	1	43		
	irr	otet14$_{00002}$, $L14n55217$	3	49	2	7-dit IC
	irr	otet14$_{00035}$	1	49	2	7-dit IC

Let us give more details about the results summarized in Table 2. Using Magma, the conjugacy class of subgroups of index 2 in the fundamental group G is represented by the subgroup on three generators and two relations as follows $H = \left\langle x, y, z | y^{-1}zx^{-1}zy^{-1}x^{-2}, z^{-1}yxz^{-1}yz^{-1}xy \right\rangle$, from which the sequence of subgroups of finite index can be found as $\eta_d(M^3) = \{1, 1, 5, 6, 8, 33, 21, 32, \cdots \}$. The manifold M^3 corresponding to this sequence is found in Snappy as otet04$_{00002}$, alias $m206$.

The conjugacy class of subgroups of index 3 in G is represented as

$$H = \left\langle x, y, z | x^{-2}zx^{-1}yz^2x^{-1}zy^{-1}, z^{-1}xz^{-2}xz^{-2}y^{-1}x^{-2}zy \right\rangle,$$

with $\eta_d(M^3) = \{1, 7, 4, 47, 19, 66, 42, 484, \cdots \}$ corresponding to the manifold otet06$_{00003}$, alias $s961$.

As shown in Table 2, there are two conjugacy classes of subgroups of index 4 in G corresponding to tetrahedral manifolds otet08$_{00002}$ (the permutation group P organizing the cosets is $\mathbb{Z}_4$) and otet08$_{00007}$ (the permutation group organizing the cosets is the alternating group A_4). The latter group/manifold has fundamental group

$$H = \left\langle x, y, z | yx^{-1}y^{-1}z^{-1}xy^{-2}xyzx^{-1}y, zx^{-1}yx^{-1}yx^{-1}zyx^{-1}y^{-1}z^{-1}xy^{-1} \right\rangle,$$

with cardinality sequences of subgroups as $\eta_d(M^3) = \{1, 3, 8, 25, 36, 229, 435 \cdots \}$. To H is associated an IC-POVM [15,16] which follows from the action of the two-qubit Pauli group on a magic/fiducial state of type $(0, 1, -\omega_6, \omega_6 - 1)$, with $\omega_6 = \exp(2i\pi/6)$ a six-root of unity.

For index 5, there are three types of 3-manifolds corresponding to the subgroups H. The tetrahedral manifold otet10$_{00026}$ of cardinality sequence $\eta_d(M^3) = \{1, 7, 15, 88, 123, 802, 1328 \cdots \}$, is associated to a 5-dit equiangular IC-POVM, as in ([15] Table 5).

For index 6, the 11 coverings define six classes of 3-manifolds and two of them: otet12$_{00041}$ and otet12$_{00000}$ are related to the construction of ICs. For index 7, one finds three classes of 3-manifolds with two of them: otet14$_{00002}$ (alias L14n55217) and otet14$_{00035}$ are related to ICs. Finally, for index 7, 3 types of 3-manifolds exist, two of them relying on the construction of the 7-dit (two-valued) IC. For index 8, there exists 6 distinct 3-manifolds (not shown) none of them leading to an IC.

A Two-Qubit Tetrahedral Manifold

The tetrahedral three-manifold otet08$_{00007}$ is remarkable in the sense that it corresponds to the subgroup of index 4 of G that allows the construction of the two-qubit IC-POVM. The corresponding hyperbolic polyhedron taken from SnapPy is shown in Figure 4a. Of the 29 orientable tetrahedral manifolds with at most 8 tetrahedra, 20 are two-colorable and each of those has at most 2 cusps. The 4 three-manifolds (with at most 8 tetrahedra) identified in Table 2 belong to the 20's and the two-qubit tetrahedral manifold otet08$_{00007}$ is one with just one cusp ([37] Table 1).

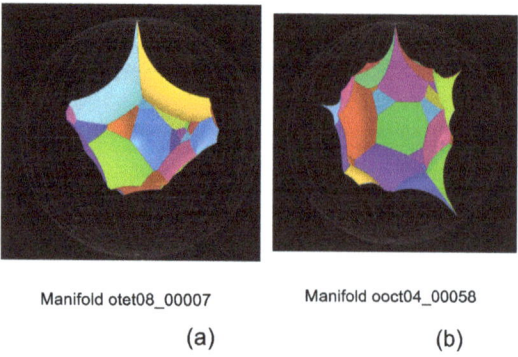

Manifold otet08_00007   Manifold ooct04_00058

(a)   (b)

Figure 4. Two platonic three-manifolds leading to the construction of the two-qubit MIC. Details are given in Tables 2 and 3.

Table 3. A few 3-manifolds M^3 found from subgroups of the fundamental group associated to the Whitehead link. For $d \geq 4$, only the M^3's leading to an IC are listed.

d	ty	M^3	cp	rk	pp	Comment
2	cyc	ooct02$_{00003}$, t12066, L8n5	3	2		Mom-4s [36]
	cyc	ooct02$_{00018}$, t12048	2	2		Mom-4s [36]
3	cyc	ooct03$_{00011}$, L10n100	4	3		
	cyc	ooct03$_{00018}$	2	3		
	irr	ooct03$_{00014}$, L12n1741	3	9	1	qutrit Hesse SIC
4	irr	ooct04$_{00058}$	4	16	2	2-qubit IC
	irr	ooct04$_{00061}$	3	16	2	2-qubit IC
5	irr	ooct05$_{00092}$	3	25	1	5-dit IC
	irr	ooct05$_{00285}$	2	25	1	5-dit IC
	irr	ooct05$_{00098}$, L13n11257	4	25	1	5-dit IC
6	cyc	ooct06$_{06328}$	5	36	2	6-dit IC
	irr	ooct06$_{01972}$	3	36	2	6-dit IC
	irr	ooct06$_{00471}$	4	36	2	6-dit IC

3.2. Three-Manifolds Pertaining to the Whitehead Link

One could also identify the 3-manifold substructure of another universal object, viz the Whitehead link L_0 [38].

The cardinality list corresponding to the Whitehead link group $\pi_1(L_0)$ is

$$\eta_d(L_0) = \{1, 3, 6, 17, 22, 79, 94, 412, 616, 1659 \ldots\},$$

Table 3 shows that the identified 3-manifolds for index d subgroups of $\pi_1(L_0)$ are aggregates of d octahedra. In particular, one finds that the qutrit Hesse SIC can be built from ooct03$_{00014}$ and that the two-qubit IC-POVM may be built from ooct04$_{00058}$. The hyperbolic polyhedron for the latter octahedral manifold taken from SnapPy is shown in Figure 4b. The former octahedral manifold follows from the link L12n1741 shown in Figure 5a and the corresponding polyhedron taken from SnapPy is shown in Figure 5b.

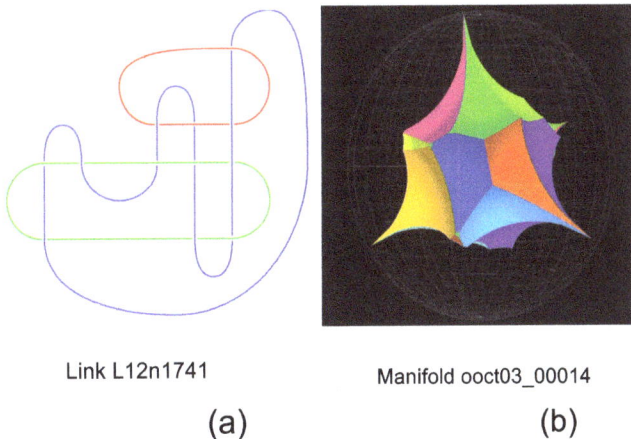

Link L12n1741   Manifold ooct03_00014

(a)   (b)

Figure 5. (a) The link $L12n1741$ associated to the qutrit Hesse SIC, (b) The octahedral manifold ooct03$_{00014}$ associated to the 2-qubit IC.

3.3. A Few Three-Manifolds Pertaining to Borromean Rings

Three-manifolds corresponding to coverings of degree 2 and 3 of the 3-manifold branched along the Borromean rings L6a4 (that is a not a (3,3)-torus link but an hyperbolic link) (see Figure 1c) are given in Table 4. The identified manifolds are hyperbolic octahedral manifolds of volume 14.655 (for the degree 2) and 21.983 (for the degree 3).

Table 4. Coverings of degrees 2 to 4 branched over the Borromean rings. The identification of the corresponding hyperbolic 3-manifold M^3 is at the 5th column. Only two types of 3-manifolds allow the building of the Hesse SIC. The two 3-manifolds of degree 4 allow the construction of the two-qubit MIC to be identified by the cardinality structure of their subgroups/coverings.

d	ty	hom	cp	M^3	Comment
2	cyc	$\frac{1}{2}+\frac{1}{2}+1+1+1$	3	ooct04$_{00259}$	
.	.	$\frac{1}{2}+1+1+1+1$	4	ooct04$_{00055}$	
.	.	$1+1+1+1+1$	5	ooct04$_{00048}$, L12n2226	
3	cyc	$\frac{1}{3}+\frac{1}{3}+1+1+1$	3	ooct06$_{07427}$	
.	.	$\frac{1}{3}+1+1+1+1+1+1$	5	ooct06$_{00463}$	
.	.	$1+1+1+1+1+1+1+1+1$	7	ooct06$_{00411}$	
.	irr	$1+1+1+1$	4	ooct06$_{00466}$	Hesse SIC
.	.	$1+1+1+1+1+1$	4	ooct06$_{00398}$	Hesse SIC
.	.	$1+1+1+1+1+1$	5	ooct06$_{00407}$, L14n63856	
4	irr	$\frac{1}{2}+\frac{1}{2}+1+1+1+1$	4	$\{63,300,10747\cdots\}$	2QB MIC
.	.	$\frac{1}{2}+1+1+1+1+1+1$	4	$\{127,2871,478956,\cdots\}$	2QB MIC

4. A Few Dehn Fillings and Their POVMs

To summarize our findings of the previous section, we started from a building block, a knot (viz the trefoil knot T_1) or a link (viz the figure-of-eight knot K_0) whose complement in S^3 is a 3-manifold M^3. Then a d-fold covering of M^3 was used to build a d-dimensional POVM, possibly an IC. Now we apply a kind of 'phase surgery' on the knot or link that transforms M^3 and the related coverings while preserving some of the POVMs in a way to be determined. We will start with our friend T_1 and arrive at a few standard 3-manifolds of historic importance, the Poincaré homology sphere [alias the Brieskorn sphere $\Sigma(2,3,5)$], the Brieskorn sphere $\Sigma(2,3,7)$ and a Seifert fibered toroidal manifold Σ'. Then we introduce the 3-manifold Σ_Y resulting from 0-surgery on the figure-of-eight knot [39]. Later in this section, we will show how to use the $\{3,5,3\}$ Coxeter lattice and surgery to arrive at a hyperbolic 3-manifold Σ_{120e} of maximal symmetry whose several coverings (and related POVMs) are close to the ones of the trefoil knot [40].

Let us start with a Lens space $L(p,q)$ that is 3-manifold obtained by gluing the boundaries of two solid tori together, so that the meridian of the first solid torus goes to a (p,q)-curve on the second solid torus [where a (p,q)-curve wraps around the longitude p times and around the meridian q times]. Then we generalize this concept to a knot exterior, i.e., the complement of an open solid torus knotted like the knot. One glues a solid torus so that its meridian curve goes to a (p,q)-curve on the torus boundary of the knot exterior, an operation called Dehn surgery ([1] (p. 275), [24] (p. 259), [41]). According to Lickorish's theorem, every closed, orientable, connected 3-manifold is obtained by performing Dehn surgery on a link in the 3-sphere. For example, surgeries on the trefoil knot allow to build the most important spherical 3-manifolds—the ones with a finite fundamental group—that are the basis of ADE correspondence. The acronym ADE refers to simply laced Dynkin diagrams that connect apparently different objects such as Lie algebras, binary polyhedral groups, Arnold's theory of catastophes, Brieskorn spheres and quasicrystals, to mention a few [42].

4.1. A Few Surgeries on the Trefoil Knot

The Poincaré Homology Sphere

The Poincaré dodecahedral space (alias the Poincaré homology sphere) was the first example of a 3-manifold not the 3-sphere. It can be obtained from $(-1,1)$ surgery on the left-handed trefoil knot T_1 [43].

Let p, q, r be three positive integers and mutually coprime, the Brieskorn sphere $\Sigma(p,q,r)$ is the intersection in $\mathbb{C}^3$ of the 5-sphere S^5 with the surface of equation $z_1^p + z_2^q + z_3^r = 1$. The homology of a Brieskorn sphere is that of the sphere S^3. A Brieskorn sphere is homeomorphic but not diffeomorphic to S^3. The sphere $\Sigma(2,3,5)$ may be identified to the Poincaré homology sphere. The sphere $\Sigma(2,3,7)$ [39] may be obtained from $(1,1)$ surgery on T_1. Table 5 provides the sequences η_d for the corresponding surgeries $(\pm 1, 1)$ on T_1. Plain digits in these sequences point out the possibility of building ICs of the corresponding degree. This corresponds to a considerable filtering of the ICs coming from T_1.

Table 5. A few surgeries (column 1), their name (column 2) and the cardinality list of d-coverings (alias conjugacy classes of subgroups). Plain characters are used to point out the possible construction of an IC-POVM in at least one the corresponding three-manifolds (see [16] and Section 2 for the ICs corresponding to T_1).

T	Name	$\eta_d(T)$
T_1	trefoil	{1,1,**2**,**3**,**2**, **8**,**7**,10,**10**,**28**, **27**,**88**,**134**,**171**,**354**}
$T_1(-1,1)$	$\Sigma(2,3,5)$	{1,0,0,0,**1**, **1**,0,0,0,**1**, 0,**1**,0,0,**1**}
$T_1(1,1)$	$\Sigma(2,3,7)$	{1,0,0,0,0, 0,**2**,**1**,**1**,0, 0,0,0,**9**,**3**}
$T_1(0,1)$	Σ'	{1,1,**2**,**2**,1, **5**,**3**,**2**,**4**,1, 1,12,**3**,**3**,**4**}
$K_0(0,1)$	Σ_Y	{1,1,1,**2**,**2**, **5**,1,**2**,**2**,**4**, **3**,**17**,1,1,**2**}
$v_{2413}(-3,2)$	Σ_{120e}	{1,1,**1**,**4**,1, 7,**2**,25,**3**,10, **10**,**62**,1,30,23}

For instance, the smallest IC from $\Sigma(2,3,5)$ has dimension five and is precisely the one coming from the congruence subgroup $5A^0$ in Table 1. However, it is built from a non modular (fundamental) group whose permutation representation of the cosets is the alternating group $A_5 \cong \langle (1,2,3,4,5), (2,4,3) \rangle$ (compare [15] Section 3.3).

The smallest dimensional IC derived from $\Sigma(2,3,7)$ is 7-dimensional and two-valued, the same as the one arising from the congruence subgroup $7A^0$ given in Table 1. However, it arises from a non modular (fundamental) group with the permutation representation of cosets as $PSL(2,7) \cong \langle (1,2,4,6,7,5,3), (2,5,3)(4,6,7) \rangle$.

4.2. The Seifert Fibered Toroidal Manifold Σ'

An hyperbolic knot (or link) in S^3 is one whose complement is 3-manifold M^3 endowed with a complete Riemannian metric of constant negative curvature, i.e., it has a hyperbolic geometry and finite volume. A Dehn surgery on a hyperbolic knot is exceptional if it is reducible, toroidal or Seifert fibered (comprising a closed 3-manifold together with a decomposition into a disjoint union of circles called fibers). All other surgeries are hyperbolic. These categories are exclusive for a hyperbolic knot. In contrast, a non-hyperbolic knot such as the trefoil knot admits a toroidal Seifert fiber surgery Σ' obtained by $(0,1)$ Dehn filling on T_1 [44].

The smallest dimensional ICs built from Σ' are the Hesse SIC that is obtained from the congruence subgroup $\Gamma_0(2)$ (as for the trefoil knot) and the two-qubit IC that comes from a non modular fundamental group [with cosets organized as the alternating group $A_4 \cong \langle (2,4,3), (1,2,3) \rangle$].

4.3. Akbulut's Manifold Σ_Y

Exceptional Dehn surgery at slope $(0, 1)$ on the figure-of-eight knot K_0 leads to a remarkable manifold Σ_Y found in [39] in the context of 3-dimensional integral homology spheres smoothly bounding integral homology balls. Apart from its topological importance, we find that some of its coverings are associated to already discovered ICs and those coverings have the same fundamental group $\pi_1(\Sigma_Y)$.

The smallest IC-related covering (of degree 4) occurs with integral homology $\mathbb{Z}$ and the congruence subgroup $\Gamma_0(3)$ also found from the trefoil knot (see Table 1). Next, the covering of degree 6 and homology $\frac{\mathbb{Z}}{5} + \mathbb{Z}$ leads to the 6-dit IC of type $3C^0$ (also found from the trefoil knot). The next case corresponds to the (non-modular) 11-dimensional (3-valued) IC.

4.4. The Hyperbolic Manifold Σ_{120e}

The hyperbolic manifold closest to the trefoil knot manifold known to us was found in [40]. The goal in [40] is the search of—maximally symmetric—fundamental groups of 3-manifolds. In two dimensions, maximal symmetry groups are called Hurwitz groups and arise as quotients of the $(2,3,7)$-triangle groups. In three dimensions, the quotients of the minimal co-volume lattice Γ_{min} of hyperbolic isometries, and of its orientation preserving subgroup Γ_{min}^+, play the role of Hurwitz groups. Let C be the $\{3,5,3\}$ Coxeter group, Γ_{min} the split extension $C \rtimes \mathbb{Z}_2$ and Γ_{min}^+ one of the index two subgroups of Γ_{min} of presentation

$$\Gamma_{min}^+ = \left\langle x, y, z | x^3, y^2, z^2, (xyz)^2, (xzyz)^2, (xy)^5 \right\rangle.$$

According to ([40] Corollary 5), all torsion-free subgroups of finite index in Γ_{min}^+ have index divisible by 60. There are two of them of index 60, called Σ_{60a} and Σ_{60b}, obtained as fundamental groups of surgeries $m017(-4,3)$ and $m016(-4,3)$. Torsion-free subgroups of index 120 in Γ_{min}^+ are given in Table 6. It is remarkable that these groups are fundamental groups of oriented three-manifolds built with a single icosahedron except for Σ_{120e} and Σ_{120g}.

Table 6. The index 120 torsion-free subgroups of Γ_{min}^+ and their relation to the single isosahedron 3-manifolds [40]. The icosahedral symmetry is broken for Σ_{120e} (see the text for details).

Manifold T	Subgroup	$\eta_d(T)$
oicocld01$_{00001}$ = s897$(-3,2)$	Σ_{120a}	$\{1,0,0,0,0, 8,2,1,1,8\}$
oicocld01$_{00000}$ = s900$(-3,2)$	Σ_{120b}	$\{1,0,0,0,5, 8,10,15,5,24\}$
oicocld01$_{00003}$ = v2051$(-3,2)$	Σ_{120c}	$\{1,0,0,0,0, 4,8,12,6,6\}$
oicocld01$_{00002}$ = s890$(3,2)$	Σ_{120d}	$\{1,0,1,5,0, 9,0,35,9,2\}$
v2413$(-3,2) \neq$ oicocld01$_{00004}$	Σ_{120e}	$\{1,1,1,4,1, 7,2,25,3,10\}$
oicocld01$_{00005}$ = v3215$(1,2)$	Σ_{120f}	$\{1,0,0,0,0, 14,10,5,10,17\}$
v3318$(-1,2)$	Σ_{120g}	$\{1,3,1,2,0, 11,0,23,12,14\}$

The group Σ_{120e} is special in the sense that many small dimensional ICs may be built from it in contrast to the other groups in Table 6. The smallest ICs that may be built from Σ_{120e} are the Hesse SIC coming from the congruence subgroup $\Gamma_0(2)$, the two-qubit IC coming the congruence subgroup $4A^0$ and the 6-dit ICs coming from the congruence subgroups $\Gamma(2)$, $3C^0$ or $\Gamma_0(4)$ (see [16] Section 3 and Table 1). Higher dimensional ICs found from Σ_{120e} do not come from congruence subgroups.

5. Conclusions

The relationship between 3-manifolds and universality in quantum computing has been explored in this work. Earlier work of the first author already pointed out the importance of hyperbolic geometry and the modular group Γ for deriving the basic small dimensional IC-POVMs. In Section 2, the move from Γ to the trefoil knot T_1 (and the braid group B_3) to non-hyperbolic 3-manifolds could be investigated by making use of the *d*-fold coverings of T_1 that correspond to *d*-dimensional POVMs (some of them being IC). Then, in Section 3, we went on to universal links (such as the figure-of-eight knot, Whitehead link and Borromean rings) and the related hyperbolic platonic manifolds as new models for quantum computing based POVMs. Finally, in Section 4, Dehn fillings on T_1 were used to explore the connection of quantum computing to important exotic 3-manifolds (i.e., $\Sigma(2,3,5)$ and $\Sigma(2,3,7)$), to the toroidal Seifert fibered Σ', to Akbulut's manifold Σ_Y and to a maximum symmetry hyperbolic manifold Σ_{120e} slightly breaking the icosahedral symmetry. It is expected that our work will have importance for new ways of implementing quantum computing and for the understanding of the link between quantum information and cosmology [45–47]. A subsequent paper of ours develops the field of 3-manifold based UQC with its relationship to Bianchi groups [48].

Author Contributions: All authors contributed significantly to the content of the paper. M.P. wrote the manuscript and the co-authors reviewed it.

Funding: The first author acknowledges the support by the French "Investissements d'Avenir" program, project ISITE-BFC (contract ANR-15-IDEX-03). The other resources came from Quantum Gravity Research.

Conflicts of Interest: The authors declare no competing interests.

References

1. Thurston, W.P. *Three-Dimensional Geometry and Topology*; Princeton University Press: Princeton, NJ, USA, 1997; Volume 1.
2. Yu Kitaev, A. Fault-tolerant quantum computation by anyons. *Ann. Phys.* **2003**, *303*, 2–30. [CrossRef]
3. Nayak, C.; Simon, S.; Stern, A.; Freedman, M.; Sarma, S.D. Non-Abelian anyons and topological quantum computation. *Rev. Mod. Phys.* **2008**, *80*, 1083. [CrossRef]
4. Wang, Z. *Topological Quantum Computation*; American Mathematical Soc.: Providence, RI, USA, 2010.
5. Pachos, J.K. *Introduction to Topological Quantum Computation*; Cambridge University Press: Cambridge, UK, 2012.
6. Kauffman, L.H.; Baadhio, R.L. *Quantum Topology*; Series on Knots and Everything; World Scientific: Singapore, 1993.
7. Kauffman, L.H. Knot logic and topological quantum computing with Majorana fermions. In *Linear and Algebraic Structures in Quantum Computing*; Lecture Notes in Logic 45; Chubb, J., Eskandarian, A., Harizanov, V., Eds.; Cambridge Univ. Press: Cambridge, UK, 2016.
8. Seiberg, N.; Senthil, T.; Wang, C.; Witten, E. A duality web in 2 + 1 dimensions and condensed matter physics. *Ann. Phys.* **2016**, *374*, 395–433. [CrossRef]
9. Gang, D.; Tachikawa, Y.; Yonekura, K. Smallest 3*d* hyperbolic manifolds via simple 3*d* theories. *Phys. Rev. D* **2017**, *96*, 061701(R). [CrossRef]
10. Lim, N.C.; Jackson, S.E. Molecular knots in biology and chemistry. *J. Phys. Condens. Matter* **2015**, *27*, 354101. [CrossRef] [PubMed]
11. Irwin, K. Toward a Unification of Physics and Number Theory. Available online: https://www.researchgate.net/publication/314209738 (accessed on 1 January 2018).
12. Bravyi, S.; Kitaev, A. Universal quantum computation with ideal Clifford gates and noisy ancillas. *Phys. Rev.* **2005**, *A71*, 022316. [CrossRef]
13. Veitch, V.; Mousavian, S.A.; Gottesman, D.; Emerson, J. The resource theory of stabilizer quantum computation. *New J. Phys.* **2014**, *16*, 013009. [CrossRef]
14. Planat, M.; Haq, R.U. The magic of universal quantum computing with permutations. *Adv. Math. Phys.* **2017**, *217*, 5287862. [CrossRef]
15. Planat, M.; Gedik, Z. Magic informationally complete POVMs with permutations. *R. Soc. Open Sci.* **2017**, *4*, 170387. [CrossRef]
16. Planat, M. The Poincaré half-plane for informationally complete POVMs. *Entropy* **2018**, *20*, 16. [CrossRef]
17. Milnor, J. The Poincaré Conjecture 99 Years Later: A Progress Report (The Clay Mathematics Institute 2002 Annual Report, 2003). Available online: http://www.math.sunysb.edu/S\sim$jack/PREPRINTS/poiproof.pdf (accessed on 1 January 2018).

18. Planat, M. On the geometry and invariants of qubits, quartits and octits. *Int. J. Geom. Methods Mod. Phys.* **2011**, *8*, 303–313. [CrossRef]
19. Manton, N.S. Connections on discrete fiber bundles. *Commun. Math. Phys.* **1987**, *113*, 341–351. [CrossRef]
20. Mosseri, R.; Dandoloff, R. Geometry of entangled states, Bloch spheres and Hopf fibrations. *Int. J. Phys. A Math. Gen.* **2001**, *34*, 10243. [CrossRef]
21. Nieto, J.A. Division-Algebras/Poincare-Conjecture Correspondence. *J. Mod. Phys.* **2013**, *4*, 32–36. [CrossRef]
22. Fang, F.; Hammock, D.; Irwin, K. Methods for calculating empires in quasicrystals. *Crystals* **1997**, *7*, 304. [CrossRef]
23. Sen, A.; Aschheim, R.; Irwin, K. Emergence of an aperiodic Dirichlet space from the tetrahedral units of an icosahedral internal space. *Mathematics* **2017**, *5*, 29.
24. Adams, C.C. *The Knot Book, An Elementary Introduction to the Mathematical Theory of Knots*; W. H. Freeman and Co.: New York, NY, USA, 1994.
25. Fominikh, E.; Garoufalidis, S.; Goerner, M.; Tarkaev, V.; Vesnin, A. A census of tethahedral hyperbolic manifolds. *Exp. Math.* **2016**, *25*, 466–481. [CrossRef]
26. Hilden, H.M.; Lozano, M.T.; Montesinos, J.M.; Whitten, W.C. On universal groups and three-manifolds. *Invent. Math.* **1987**, *87*, 441–445. [CrossRef]
27. Mednykh, A.D. A new method for counting coverings over manifold with finitely generated fundamental group. *Dokl. Math.* **2006**, *74*, 498–502. [CrossRef]
28. Culler, M.; Dunfield, N.M.; Goerner, M.; Weeks, J.R. SnapPy, a Computer Program for Studying the Geometry and Topology of 3-Manifolds. Available online: http://snappy.computop.org (accessed on 1 January 2018).
29. Hilden, H.M.; Lozano, M.T.; Montesinoos, J.M. On knots that are universal. *Topology* **1985**, *24*, 499–504. [CrossRef]
30. Fuchs, C.A. On the quantumness of a Hibert space. *Quant. Inf. Comp.* **2004**, *4*, 467–478.
31. Appleby, M.; Chien, T.Y.; Flammia, S.; Waldron, S. Constructing Exact Symmetric Informationally Complete Measurements from Numerical Solutions. *arXiv* **2018**, arXiv:1703.05981.
32. Rolfsen, D. *Knots and Links*; Mathematics Lecture Series 7; Publish of Perish: Houston, TX, USA, 1990.
33. Milnor, J. On the 3-dimensional Brieskorn manifolds $M(p,q,r)$. In *Knots, Groups and 3-Manifolds*; Neuwirth, L.P., Ed.; Princeton Univ. Press: Princeton, NJ, USA, 1975; pp. 175–225.
34. Bosma, W.; Cannon, J.J.; Fieker, C.; Steel, A. Eds. *Handbook of Magma Functions*; University of Sydney: Sydney, Australia, 2017.
35. Hempel, J. The lattice of branched covers over the Figure-eight knot. *Topol. Appl.* **1990**, *34*, 183–201. [CrossRef]
36. Haraway, R.C. Determining hyperbolicity of compact orientable 3-manifolds with torus boundary. *arXiv* **2014**, arXiv:1410.7115.
37. Ballas, S.A.; Danciger, J.; Lee, G.S. Convex projective structures on non-hyperbolic three-manifolds. *arXiv* **2018**, arXiv:1508.04794.
38. Gabai, D. The Whitehead manifold is a union of two Euclidean spaces. *J. Topol.* **2011**, *4*, 529–534. [CrossRef]
39. Akbulut, S.; Larson, K. Brieskorn spheres bounding rational balls. *arXiv* **2017**, arXiv:1704.07739.
40. Conder, M.; Martin, G.; Torstensson, A. Maximal symmetry groups of hyperbolic 3-manifolds. *N. Z. J. Math.* **2006**, *35*, 3762.
41. Gordon, C.M. *Dehn Filling: A survey, Knot Theory*; Banach Center Publ.: Warsaw, Poland, 1998; Volume 42, pp. 129–144.
42. Sirag, S.-P. *ADEX Theory, How the ADE Coxeter Graphs Unify Mathematics and Physics*; World Scientific: Singapore, 2016.
43. Kirby, R.C.; Scharlemann, M.G. Eight faces of the Poincaré homology 3-sphere. In *Geometric Topology*; Acad. Press: New York, NY, USA, 1979; pp. 113–146.
44. Wu, Y. Seifert fibered surgery on Montesinos knots. *arXiv* **2012**, arXiv:1207.0154.
45. Chan, K.T.; Zainuddin, H.; Atan, K.A.M.; Siddig, A.A. Computing Quantum Bound States on Triply Punctured Two-Sphere Surface. *Chin. Phys. Lett.* **2016**, *33*, 090301. [CrossRef]
46. Aurich, R.; Steiner, F.; Then, H. Numerical computation of Maass waveforms and an application to cosmology. In *Hyperbolic Geometry and Applications in Quantum Chaos and Cosmology*; Jens, B., Frank, S., Eds.; Cambridge Univ. Press: Cambridge, UK, 2012.

47. Asselmeyer-Maluga, T. Smooth quantum gravity: Exotic smoothness and Quantum gravity. In *At the Frontier of Spacetime Scalar-Tensor Theory, Bells Inequality, Machs Principle, Exotic Smoothness*; Fundamental Theories of Physics Book Series (FTP); Asselmeyer-Maluga, T., Ed.; Springer: Cham, Switzerland, 2016; pp. 247–308.
48. Planat, M.; Aschheim, R.; Amaral, M.M.; Irwin, K. Quantum computing with Bianchi groups. *arXiv* **2018**, arXiv:1808.06831.

© 2018 by the authors. Licensee MDPI, Basel, Switzerland. This article is an open access article distributed under the terms and conditions of the Creative Commons Attribution (CC BY) license (http://creativecommons.org/licenses/by/4.0/).

Article

Braids, 3-Manifolds, Elementary Particles: Number Theory and Symmetry in Particle Physics

Torsten Asselmeyer-Maluga

German Aeorspace Center, Rosa-Luxemburg-Str. 2, 10178 Berlin, Germany; torsten.asselmeyer-maluga@dlr.de; Tel.: +49-30-67055-725

Received: 25 July 2019; Accepted: 2 October 2019; Published: 15 October 2019

Abstract: In this paper, we will describe a topological model for elementary particles based on 3-manifolds. Here, we will use Thurston's geometrization theorem to get a simple picture: fermions as hyperbolic knot complements (a complement $C(K) = S^3 \setminus (K \times D^2)$ of a knot K carrying a hyperbolic geometry) and bosons as torus bundles. In particular, hyperbolic 3-manifolds have a close connection to number theory (Bloch group, algebraic K-theory, quaternionic trace fields), which will be used in the description of fermions. Here, we choose the description of 3-manifolds by branched covers. Every 3-manifold can be described by a 3-fold branched cover of S^3 branched along a knot. In case of knot complements, one will obtain a 3-fold branched cover of the 3-disk D^3 branched along a 3-braid or 3-braids describing fermions. The whole approach will uncover new symmetries as induced by quantum and discrete groups. Using the Drinfeld–Turaev quantization, we will also construct a quantization so that quantum states correspond to knots. Particle properties like the electric charge must be expressed by topology, and we will obtain the right spectrum of possible values. Finally, we will get a connection to recent models of Furey, Stoica and Gresnigt using octonionic and quaternionic algebras with relations to 3-braids (Bilson–Thompson model).

Keywords: standard model of elementary particles; 4-manifold topology; particles as 3-Braids; branched coverings; knots and links; charge as Hirzebruch defect; umbral moonshine; number of generations

1. Introduction

General relativity (GR) deepens our view on space-time. In parallel, the appearance of quantum field theory (QFT) gives us a different view of particles, fields and the measurement process. One approach for the unification of QFT and GR, to a quantum gravity, starts with a proposal to quantize GR and its underlying structure, space-time. Here, there is a unique opinion in the community about the relation between geometry and quantum theory: the geometry as used in GR is classical and should emerge from a quantum gravity in the usual limit that Planck's constant tends to zero. Most theories went a step further and try to get the spacetime directly from quantum theory. As a consequence, the used model of a smooth manifold cannot be used to describe quantum gravity. However, currently, there is no real sign for a discrete spacetime structure or higher dimensions in current experiments [1]. Therefore, in this work, we conjecture that the model of spacetime as a smooth 4-manifold can be also used in the quantum gravitational regime. As a consequence, one has to find geometrical/topological representations for quantities in QFT (submanifolds for particles or fields, etc.) as well in order to quantize GR. In this paper, we will tackle this problem to get a geometrical/topological description of the standard model of elementary particle physics. Recently, there were efforts by Furey [2–5], Gresnigt [6–8] and Stoica [9] to use octonions and Clifford algebras to get a coherent model to describe the particle generations in the standard model. In the past, the stability of matter was related to topology like in the approach of Lord Kelvin [10] with knotted aether vortices. The proposal to derive matter from space was considered by Clifford as well by Einstein, Eddington, Schrödinger and Wheeler with only partial success (see [11,12]). Giulini [13] discussed the status of geometrodynamics in establishing particle properties like spin from spacetime by using special solutions of general relativity. The usage of knots and links to model particles, like the electron, neutrinos, etc., was firstly observed by Jehle [14]. The phenomenological description of particle properties by using the quantum group $SU_q(n)$ is given in the work of Gavrilik [15]. Here, the (deformation) parameter q of $SU_q(n)$ was linked with the flavor mixing angle (Cabibbo angle). Furthermore, torus knots as given by 2-braids were associated with vector mesons (vector quarkonia) of different flavors. Later, Finkelstein [16] used the representation of knots by quantum groups for his particle model. Similar ideas are discussed in the model of Bilson–Thompson [17] in its loop theoretic extension [18]. Even for the Bilson–Thompson model, Gresnigt found a link between Furey's approach and this model. However, some properties of the Bilson–Thompson model remained mysterious, like the definition of the charge. Open is also the meaning of the braiding. If there is a connection between spacetime and matter, then one has to construct the known fermions and bosons directly. Here, it seems that the main problem is the determination of the underlying spacetime. In this paper, we will follow this way with an heuristic argument for the spacetime to be the K3 surface in Section 3. Then, we will analyze this spacetime by using branched covers to find two suitable substructures, a knot complement representing the fermions and a link complement to represent the bosons. Here, we profit from ideas by Duston [19–21] as well from the work of Denicola, Marcolli and al-Yasry [22]. As a byproduct, we also found interesting relations to octonions. The representation of the knot and link complements by branched covers gives the link to the original Bilson–Thompson model [17] but also to Gresnigt's work [6]. In Section 5, we will discuss the electric charge and construct the corresponding operator by using the underlying $U(1)$ gauge theory by using the Hirzebruch defect. Finally, we will obtain the correct charge spectrum: fermions carry the charges $0, \pm\frac{1}{3}, \pm\frac{2}{3}, \pm 1$ in units of the unit charge e. Here, the factor $\frac{1}{3}$ is related to 4-dimensional topology (Hirzebruch signature theorem). In Section 6, we will discuss the independence of the particular braid from the particle. The braid is connected with the state of the particle as shown by using the Drinfeld–Turaev quantization. Finally, we finish the paper with some speculations about the number of generations, given by the number of $S^2 \times S^2$ parts in the spacetime (K3 surface), and a global symmetry, the group $PGL(3,4)$ of 3×3 matrices of the 4-element field F_4, induced from the K3 surface by using umbral moonshine.

Before we start with the description of the model, we will discuss the key arguments and scope of the model. The model is based on a smooth spacetime that is described by a smooth 4-manifold. First, it is argued that this spacetime is the K3 surface where the evolution of the cosmos is submanifold. By using this model, the calculation of cosmic parameters (cosmological constant, inflation parameters, etc.) matches with the experimental results. For the following, we use the characterization of 4-manifold (i.e., the spacetime) by surfaces that are connected to complements of links and knots (3-manifolds with boundary). Interestingly, both 3-manifolds have a meaning by our previous work: knot complements are fermions and link complements are bosons. Here, we find a representation of both 3-manifolds by braids of three strands (3-braids), which is the connection to the work of Bilson–Thompson and Gresnigt. In case of the K3 surface as spacetime, the surfaces mentioned above are arranged with a strong connection to octonions, which is the link to Furey's work. To connect fermions, one needs a special class of link complements, so-called torus bundles, which can be interpreted as gauge fields. Interestingly, there are only three classes of torus bundles and we were able to connect them with the gauge groups $U(1), SU(2), SU(3)$. The main result of the paper is definition and interpretation of the electric charge. The electric charge is a topological invariant (Hirzebruch defect). For the fermions (knot complements), we obtained the spectrum $\{0, \pm 1, \pm 2, \pm 3\}$ which agrees with the observation (normalized in $e/3$ units). The discussion about the quantization of the model and the number of generations closes the paper.

The model uses a classical spacetime. Quantum properties are not included in an obvious way. We obtained the correct charge spectrum, but we do not know the direct connection between knot complement and fermion (electron, neutrino, quark). There are ideas to use a quantization by a change of the smoothness structure, known as Smooth Quantum Gravity [23]. In principle, the number of generations is connected with the spacetime. We got the minimal value of three generations, but it is only a lower bound. The model produces only the particles of the standard model. No supersymmetry or other extensions of the standard model can be derived by this model. Currently, we also have no idea to calculate the coupling constants and masses. It seems that these parameters are connected with the topological property of the spacetime.

2. Preliminaries: Branched Coverings of 3- and 4-Manifolds

According to Alexander (see [24] for instance), every manifold M^n can be represented as p-fold branched covering $\pi : M^n \to S^n$ along an $n-2$ dimensional subcomplex $N^{n-2} \subset S^n$, the branching set. In detail, the map π is a covering except for the branching set, i.e., for every point $b \in S^n \setminus N^{n-2}$ the map $\pi^{-1}(b)$ consists of p points and the neighborhood $U(b)$ is homeomorphic to one component $\pi^{-1}U(b)$. Then, $M^n \setminus \pi^{-1}(N^{n-2}) \to S^n \setminus N^{n-2}$ is a p-fold covering. Usually, N^{n-2} has the structure of a simplicial subcomplex. The p-fold covering is completely determined by the representation $\pi_1(S^n \setminus N^{n-2}) \to S_p$ in the permutation group S_p of p symbols. Before going into the details, we will look at some examples. First, there are no branched coverings for 1-manifolds except for a trivial one. The first interesting example is given by a compact 2-manifold, i.e., by a surface of genus g. Here, the branching set consists of a finite number of points (0-dimensional branching set). The branching set of 3-manifolds is a one-dimensional, complex and we will later see that knots and links are the appropriated structures. In case of 4-manifolds, one has a two-dimensional complex as a branching set that was shown to be a surface. These facts are easy to understand, but one parameter of a branched covering is open: how many folds are minimally needed to represent every manifold in a fixed dimension, or, how large is p minimally for a manifold of dimension n?

2.1. As Warmup: Branched Coverings of 2-Manifolds

Let us start with the simplest case, the surface. By results of Riemann and Hurwitz, every surface can be represented by a 2-fold covering of S^2. As example, let us take the torus T^2 with the 2-fold covering $T^2 \to S^2$ branched along four points. The idea of the construction is simple: choose a symmetry axis so that the genus g surface can be generated by a rotation (see Figure 1).

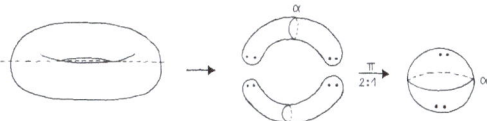

Figure 1. 2-fold covering of torus, α is the equator and the four-point branching set.

This axis meets the surface in $4g$ points which are the branching points of the covering.

2.2. Branched Coverings of 3-Manifolds

For a 3-manifold, one has a one-dimensional, branching set and a result of Alexander states that this branching set is a link with a finite number of components. Later, Hirsch, Hilden and Montesinos [25–27] obtained independently the result that every closed, compact 3-manifold can be represented as a 3-fold branched covering of the 3-sphere branched along a knot. As an example, consider the Poincare sphere $\Sigma(2,3,5)$ that can be represented by a 3-fold covering $\Sigma(2,3,5) \to S^3$ branched along the $(2,5)$ or $(5,2)$ torus knot. Now, let us consider a closed, compact 3-manifold N^3 with a 3-fold branched covering $N^3 \to S^3$ branched along a knot K. It means that the map $N^3 \setminus K \to S^3 \setminus K$ is a real $3 : 1$ map. Interestingly, there is a diffeomorphism between $N^3 \setminus K$ and $S^3 \setminus K$ so that the $3 : 1$ covering map is now given by $S^3 \setminus K \to S^3 \setminus K$. Furthermore, the 3-fold covering is completely determined by the map $\pi_1(S^3 \setminus K) \to S_3$, the representation of the fundamental group into the permutation group S_3

of three letters. A simple extension of the S_3 by considering the order of the permutations gives the braid group B_3 of three strands. In principle, the minimal number of folds $p = 3$ is the root for the description of particles by 3-braids, as shown later on.

2.3. Branched Covering of 4-Manifolds

In a similar manner, one would expect that every 4-manifold M^4 can be represented as 4-fold branched covering of S^4 branched along a surface. Piergallini [28] was able to show something similar, but the surface is only immersed and admits a finite number of singularities (cusps and nodes). If one adds an additional sheet, getting a 5-fold branched covering, then one can omit these singularities (thus getting a locally flat embedded surface) [29]. For a better understanding, we will discuss the way to this result shortly. In [28], Piergallini considered the possible transformations or changes of the branching set for a 3-manifold N^3, i.e., the knot. Amazingly, he found two possible changes (see Figure 2).

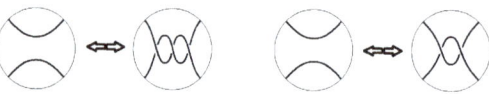

Figure 2. Branching set changes, Left: Move 1, Right: Move 2 (so-called Montesino moves).

All changes do not affect the underlying 3-manifold N^3, i.e., he found different knots and links representing the same 3-manifold as a 3-fold branched cover. Then, he used this result to find a branched covering for a 4-manifold. For that purpose, he introduced the concept of an additional leaf or fold in the covering, i.e., if a 3-manifold N^3 is represented by a 3-fold covering, then it is also represented by a 4-fold covering. Then, he extended this 4-fold covering of N^3 to a 4-fold covering of $N^3 \times [0,1]$ (i.e., a trivial cobordism). At the same time, the knot K_1 as branching set of N^3 at one side of $N^3 \times [0,1]$ (i.e., $N^3 \times \{0\}$) is related to the changed knot K_2 as branching set of N^3 on the other side of $N^3 \times [0,1]$ (i.e., $N^3 \times \{1\}$). For the covering of $N^3 \times [0,1]$, one will get a surface with two boundaries, the disjoint union $K_1 \sqcup K_2$ of the two branching knots. This procedure can be done for the two possible changes [30]. For the first change, Piergallini got the trefoil knot at the boundary, whereas, for the second case, he obtained the Hopf link (see the Figures 3).

Figure 3. Boundary of the branching set change—Left: Move 1 leading to the Trefoil knot, Right: Move 2 leading to the Hopf link.

In dimension 4, one will get the cone over the trefoil, also known as cusp, and the cone over the Hopf link, also known as node, as singularities of the surface as a branching set of the 4-manifold. Expressed differently, the trefoil knot is the link of the cusp singularity $z^2 + w^3$; the Hopf link (oriented correctly) is the link of the node singularity $z^2 + w^2$. As explained in the previous subsection, we have to consider the two fundamental groups $G_P = \pi_1(S^3 \setminus \{trefoil\})$ and $G_{WW} = \pi_1(S^3 \setminus \{Hopf\})$ for the corresponding branched cover. Both groups are known and can be simply calculated to be

$$G_P = \langle a, b \,|\, aba = bab \rangle = B_3 \qquad G_{WW} = \langle a, b, c \,|\, ab^{-1}a^{-1}b \rangle = \mathbb{Z} \oplus \mathbb{Z}, \tag{1}$$

which is a surprising result. From the point of 4-manifolds we will get two possible 3-manifolds as boundary of the singularities: a 3-manifold (knot complement) with one boundary and a 3-manifold (link complement) with two boundaries. These two 3-manifolds can be simply interpreted: the knot complement has one boundary component and can be seen as a fermion (see also [31]) and the link complement has two boundary components and can be interpreted as interaction (see also [32]).

2.4. Branched Coverings of Knot Complements

Now, we will describe the case of a 3-manifold (the knot complement) with boundary a torus T^2. First, we will change the 2-fold covering of the torus to a 3-fold covering by adding a trivial sheet. For that purpose, we consider the 2-fold branched covering $T^2 \to S^2$ and add a 2-sphere $T^2 \# S^2 \to S^2 \# S^2$ which changes nothing ($S \# S^2$ is diffeomorphic to S for every surface S). However, at the same time, we will obtain a 3-fold covering (see Figure 4).

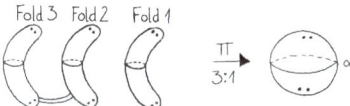

Figure 4. 3-fold covering torus.

For a 3-manifold Σ with boundary T^2, one has to consider a branched covering $\Sigma \to S^3 \setminus D^3$ (3-sphere with one puncture or p punctures for p boundary components). For the construction of the covering, one needs another representation of a 3-manifold, the Heegard decomposition. There, one considers two handle bodies H_g, H'_g of genus g, i.e., the sum of g copies of the solid torus $D^2 \times S^1$. The gluing $H_g \cup_\phi H'_g$ of these handle bodies along the boundary using a diffeomorphism $\phi : \partial H_g \to \partial H'_g$ (to be precise: ϕ is an element of the mapping class group) produces every compact, closed 3-manifold. For the 3-sphere, one obtains a decomposition $H_1 \cup H'_1$ using two solid tori where the meridian of H_1 is mapped to the longitude of H'_1 and vice versa. H_g can be obtained by a branched covering $H_g \to D^3$ with a branching set $g + 2$ arcs (see Figure 5).

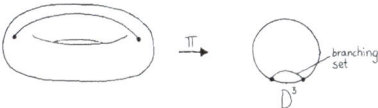

Figure 5. Covering of the handle body.

The diffeomorphism ϕ is represented by a braid connecting the two handle bodies where the braid closes above and below to get a link. In case of a 3-manifold with a boundary, the braid closes only on one side and we obtain a braid again (or, more generally, a tangle), see Figure 6.

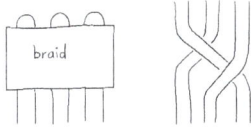

Figure 6. Left: branching set of knot complement (6-plat), **Right**: 3-Braid as 6-Braid.

The underlying braid must be also a 3-braid but represented as a 6-braid.

3. Reconstructing a Spacetime: The K3 Surface and Particle Physics

After so many years of experimental research, the standard model of particle physics and of cosmology as well general relativity are confirmed with high precision. There is no real contradicting result which shows the necessity to introduce new physics. An exception may be cosmology with the unknown components of dark matter and dark energy. However, this situation is by no means satisfying. Both standard models have a bunch of free parameters (19 parameters in particle physics, for instance). If there is no sign for new physics, how did we get these parameters? Here, we will argue that these parameters can be determined by topology. However, at first glance, this idea seems hopeless. There are infinitely many suitable topologies for the spacetime, seen as 4-manifold, and, for the space, seen as 3-manifold. Here, we will go a different way. Why not try to determine the space $\mathcal{M}$ of all possible spacetime-events? Thus, let me start with a definition: let $\mathcal{M}$ be the space of all possible spacetime events, i.e., the set of all spacetime events carrying a manifold structure. In principle, $\mathcal{M}$ can be identified with the spacetime. Then, a specific physical situation is an embedding of a 3-manifold into $\mathcal{M}$, a dynamics is an embedding of a cobordism between 3-manifolds into $\mathcal{M}$. Here, we assume implicitly that everything can be geometrically/topologically expressed as submanifolds (see [31,32]). In the following, we will try to discuss this approach and how far one can go. Some heuristic arguments are rather obvious:

1. $\mathcal{M}$ is a smooth 4-manifold,
2. any sequence of spacetime event has to converge to a spacetime event and
3. any loop (time-like or not) must be contracted.

A dynamics is a mapping of a spacetime event to a new spacetime event. It is usually smooth (differential equations), motivating the first argument. The second argument expresses the fact that any initial spacetime event must converge to a final spacetime event, or the limit of any sequence of spacetime events must converge to a spacetime event. Then, $\mathcal{M}$ is a compact, smooth 4-manifold. The usual spacetime is an open subset of $\mathcal{M}$. The third argument above is motivated to neglect time-like loops. The spacetime is an open subset of $\mathcal{M}$ or the spacetime is embedded in $\mathcal{M}$. Now, consider a loop in the spacetime. By changing the embedding via diffeomorphisms (this procedure is called isotopy), every loop is contractable. Therefore, this argument implies that there is no time-like loops (implying causality). Finally, $\mathcal{M}$ is a compact, simply connected, smooth 4-manifold.

Now, we will restrict $\mathcal{M}$ in a manner so that we are able to determine it. For the following, we implicitly assume that the equations of general relativity are valid without any restrictions. Then, the vacuum equations are equivalent to

$$R_{\mu\nu} = 0,$$

demanding Ricci flatness. However, as shown in [33,34] and in recent years in [23,31,32], the coupling to matter can be described by a change of the smoothness structure. Therefore, the modification of the smoothness structure will produce matter (or sources of gravity). However, at the same time, we need a smoothness structure that can be interpreted as a vacuum given by a Ricci-flat metric. Therefore, we will demand that

1. $\mathcal{M}$ has to admit a smoothness structure with Ricci-flat metric representing the vacuum.

Interestingly, these four demands are restrictive enough to determine the topology of $\mathcal{M}$. With the help of Yau's seminal work [35], we will obtain that $\mathcal{M}$ is homeomorphic to the K3 surface, using Yaus's work that there is only one compact, simply connected Ricci-flat 4-manifold. However, it is known by the work of LeBrun [36] that there are non-Ricci-flat smoothness structures. In the next step, we will determine the smoothness structure of $\mathcal{M}$. For that purpose, we will present some deep results in differential topology of 4-manifolds:

- there is a compact, contactable submanifold $A \subset \mathcal{M}$ (called Akbulut cork) so that cutting out A and reglue it (by an involution) will produce a new smoothness structure,
- $\mathcal{M}$ splits topologically into

$$|E_8 \oplus E_8| \# \underbrace{\left(S^2 \times S^2\right) \# \left(S^2 \times S^2\right) \# \left(S^2 \times S^2\right)}_{3(S^2 \times S^2)} = 2|E_8| \# 3(S^2 \times S^2) \qquad (2)$$

- two copies of the E_8 manifold and three copies of $S^2 \times S^2$ and
- the 3-sphere S^3 is a submanifold of A.

In [37], we already discussed this case. Interestingly, there is always a topological 4-manifold for all combinations of E_8 and $S^2 \times S^2$, but not all topological 4-manifolds are smooth manifolds. Let us consider the 4-manifold that splits topologically into p copies of the $|E_8|$ manifold and q copies of $S^2 \times S^2$ or

$$p|E_8|\#q\left(S^2 \times S^2\right).$$

Then, this 4-manifold is smoothable for every q but $p = 0$ and the first combination for $p \neq 0$ is the pair of numbers $p = 2, q = 3$ (which is the K3 surface). Any other combination ($p = 2, q < 3$ or every q and $p = 1$) is forbidden as shown by Donaldson [38].

Now, we consider the smooth K3 surface that is Ricci-flat, simply connected, smooth. The main part in the following discussion will be the use of the smoothness condition. As discussed above, the smoothness structure is determined by the Akbulut cork A. Furthermore, as argued above, the smoothness structure is strongly related to the appearance of matter and this process is strongly connected to the evolution of our cosmos. It is known as reheating after the inflationary phase. Therefore, the Akbulut cork (including its embedding) should represent the inflationary phase with reheating.

The Akbulut cork is built from a homology 3-sphere that will become the boundary ∂A. The difference to a usual 3-sphere S^3 is given by the so-called fundamental group, the equivalence class of closed loops up to deformation (homotopy) with concatenation as group operation. In principle, one constructs a cobordism between S^3 and the homology 3-sphere ∂A. All elements of the fundamental group will be killed by adding appropriate disks. At the end, one can add a 4-disk to get the full contractable cork A. After this short discussion, we are able to identify the first topological transition: if the cosmos starts as small 3-sphere (conjectural of Planck size), then the space changes to ∂A, or

$$S^3 \to \partial A.$$

The topology of ∂A depends strongly on the topology of $\mathcal{M}$. In case of the K3 surface, ∂A is known to be a Brieskorn spheres, precisely the 3-manifold

$$\Sigma(2,5,7) = \left\{x, y, z \in \mathbb{C} \mid x^2 + y^5 + z^7 = 0 \, |x|^2 + |y|^2 + |z|^2 = 1\right\}.$$

The embedding of the Akbulut cork is essential for the following results. In [39], it was shown that the embedded cork admits a hyperbolic geometry if the underlying K3 surface has an exotic smoothness structure. This simple property has far-reaching consequences. Hyperbolic manifolds of dimension three or higher are rigid, i.e., geometric properties like volume or curvature are topological invariants (Mostow-Prasad rigidity). If we assume that the cork A represents the cosmic evolution, then geometric properties like the curvature of ∂A or the change of the size after the transition $S^3 \to \partial A$ are connected with topological properties of the embedded cork A and of the underlying K3 surface by using Mostow–Prasad rigidity. This simple idea opens the door to explicit calculations. In case of the transition $S^3 \to \partial A = \Sigma(2,5,7)$, the corresponding results can be found in [39]. If one assumes a Planck-size (L_P) 3-sphere at the Big Bang, then the scale a of $\Sigma(2,5,7)$ changes like

$$a = L_P \cdot \exp\left(\frac{3}{2 \cdot CS(\Sigma(2,5,7))}\right)$$

with the Chern–Simons invariant

$$CS(\Sigma(2,5,7)) = \frac{9}{4 \cdot (2 \cdot 5 \cdot 7)} = \frac{9}{280}$$

and the Planck scale of order $10^{-34}m$ changes to $10^{-15}m$. Obviously, this transition has an exponential or inflationary behavior. Surprisingly, the number of e-folds can be explicitly calculated (see [40]) to be

$$N = \frac{3}{2 \cdot CS(\Sigma(2,5,7))} + \ln(8\pi^2) \approx 51, \tag{3}$$

and we also obtain the energy and time scale of this transition (see [40,41])

$$E_{GUT} = \frac{E_P}{1 + N + \frac{N^2}{2} + \frac{N^3}{6}} \approx 10^{15} GeV \quad t = t_P \left(1 + N + \frac{N^2}{2} + \frac{N^3}{6}\right) \approx 10^{-39} s \tag{4}$$

right at the conjectured scale of the Grand Unified Theory (GUT) (E_P, t_P Planck energy and time, respectively). In our recent work [41], this transition was analyzed in a detailed manner. There, it was shown that the transition can be described by a scalar field model which conformally agrees (as shown in [42]) with the Starobinsky-R^2 theory [43]. However, then, the dimension-less free parameter $\alpha \cdot M_P^{-2}$ as well as spectral tilt n_s and the tensor-scalar ratio r can be determined to be

$$\alpha \cdot M_P^{-2} = 1 + N + \frac{N^2}{2} + \frac{N^3}{6} \approx 10^{-5}, \; n_s \approx 0.961 \; r \approx 0.0046,$$

using equation (3), which is in good agreement with current measurements. The embedding of the cork A is based on the topological structure of the K3 surface $\mathcal{M}$. As discussed above, $\mathcal{M}$ splits topologically into a 4-manifold $|E_8 \oplus E_8|$ and the sum of three copies of $S^2 \times S^2$ (see [44]). In the topological splitting (2), the 4-manifold $|E_8 \oplus E_8|$ has a boundary that is the sum of two Poincaré spheres $P\#P$. Here, we used the fact that a smooth 4-manifold of type $|E_8|$ must have a boundary (which is the Poincaré sphere P); otherwise, it would contradict the Donaldson's theorem [38]. Then, any closed version of $|E_8 \oplus E_8|$ does not exist and this fact is the reason for the existence of an exotic $\mathbb{R}^4$. To express it differently, the neighborhood of the embedded cork lies between the 3-manifold $\Sigma(2,5,7)$ (boundary of the cork) and the sum of two Poincaré spheres $P\#P$. Therefore, we have two topological transitions resulting from the embedding

$$S^3 \xrightarrow{cork} \Sigma(2,5,7) \xrightarrow{gluing} P\#P. \tag{5}$$

The transition $\Sigma(2,5,7) \to P\#P$ has a different character as discussed in [39]. A direct consequence is the appearance of a cosmological constant as a direct consequence of the topological invariance of the curvature of a hyperbolic manifold. With respect to the critical density, the final formula for normalized cosmological constant, denoted by Ω_Λ, reads

$$\Omega_\Lambda = \frac{c^5}{24\pi^2 \hbar G H_0^2} \cdot \exp\left(-\frac{3}{CS(\Sigma(2,5,7))} - \frac{3}{CS(P\#P)} - \frac{\chi(A)}{4}\right).$$

The Chern–Simons invariants $CS(P\#P) = \frac{1}{60}$, $CS(\Sigma(2,5,7)) = \frac{9}{280}$ and the Euler characteristics of the cork $\chi(A) = 1$ together with the Hubble constant (see [45,46] combined with [47])

$$(H_0)_{Planck+Hubble} = 69,2 \; \frac{km}{s \cdot Mpc}$$

gives the value

$$\Omega_\Lambda \approx 0.7029,$$

which is in excellent agreement with the measurements. The numerical results above illustrate the power of the main idea to use topology to fit the parameter in the standard model of cosmology. Interestingly, these parameters are also important for particle physics. The existence of two transitions (5) implies in the formalism above the existence of two different energy scales, the

GUT scale of the first transition and a scale of Higgs mass order $126 GeV$ for the second transition. These two scales are the right input for the see-saw mechanism to generate a tiny neutrino mass (see [40]). Secondly, the formalism also provides a favor regarding the existence of a right-handed neutrino. The energy scale of the two transitions $S^3 \to \Sigma(2,5,7) \to P\#P$ can be expressed as a mass (via Mc^2), and we obtain

$$M = \sqrt{\frac{4\hbar c}{G}} \left(\frac{\exp\left(-\frac{1}{2\cdot CS(P\#P)}\right)}{1+N+\frac{N^2}{2}+\frac{N^3}{6}} \right) \approx 126.4 GeV, \tag{6}$$

which agrees with the mass of the Higgs boson (see [40]). Then, the Higgs boson can be expressed as the result of a topological transition (see [48]). Now, we will use the two energy scales to generate the neutrino mass. For that purpose, we start with the non-diagonal mass matrix

$$\begin{pmatrix} 0 & M \\ M & B \end{pmatrix}$$

with two mass scales B and M fulfilling $M \ll B$. This matrix has eigenvalues

$$\lambda_1 \approx B \qquad \lambda_2 \approx -\frac{M^2}{B}$$

so that λ_1 is the mass of the right-handed neutrino, and λ_2 represents the mass of the left-handed neutrino. Now, we will use the two scales (4) and (6)

$$B \approx 0.67 \cdot 10^{15} GeV, \quad M \approx 126.4 GeV,$$

and we will obtain for the neutrino mass

$$m = \frac{M^2}{B} \approx 0.024 eV,$$

which is in good agreement with the constraints from the PLANCK mission.

These results seem to support our view that the K3 surface can be the underlying spacetime (as seen as the set of all possible spacetime events). The evolution of the cosmos is a suitable subset of this space.

4. From K3 Surfaces to Octonions, 3-Braids and Particles

The results of the previous section illustrated the power of the approach and its relation to particle physics. In this section, we will discuss the topological reasons and the relation to the models of Furey, Gresnigt and Bilson–Thompson. In these models, 3-braids, octonions and quaternions play a key role. Therefore, we have to understand how these structures will naturally appear in the K3 surface.

4.1. K3 Surfaces and Octonions

The starting point for the description of any K3 surface is the topological splitting (2)

$$2|E_8|\#3(S^2 \times S^2).$$

The K3 surface is a closed, compact, simply connected 4-manifold. According to Freedman [49], the topology is uniquely given by the intersection form, a quadratic form on the second homology characterizing the intersections of the generators as given by surfaces. The K3 surface has the intersection form

$$Q_{K3} = E_8 \oplus E_8 \oplus \left(\oplus_3 \begin{pmatrix} 0 & 1 \\ 1 & 0 \end{pmatrix}\right) := 2E_8 \oplus 3H, \tag{7}$$

with the the matrix E_8:

$$E_8 = \begin{bmatrix} 2 & 1 & 0 & 0 & 0 & 0 & 0 & 0 \\ 1 & 2 & 1 & 0 & 0 & 0 & 0 & 0 \\ 0 & 1 & 2 & 1 & 0 & 0 & 0 & 0 \\ 0 & 0 & 1 & 2 & 1 & 0 & 0 & 0 \\ 0 & 0 & 0 & 1 & 2 & 1 & 0 & 1 \\ 0 & 0 & 0 & 0 & 1 & 2 & 1 & 0 \\ 0 & 0 & 0 & 0 & 0 & 1 & 2 & 0 \\ 0 & 0 & 0 & 0 & 1 & 0 & 0 & 2 \end{bmatrix}. \qquad (8)$$

This matrix is also the Cartan matrix of the Lie algebra E_8 and here is where the connection to the octonions starts. For this purpose, we have to deal with the root system of the Lie algebra E_8. Consider a semi-simple Lie algebra G and its Cartan subalgebra $H = (H_1, \ldots, H_r)$, where r is the rank of G. This subalgebra is usually considered in the Cartan basis with the non-Hermitian generators E_k and their conjugates E_{-k}. E_k is associated with the root vector $r_m^{(k)}$ such that

$$[H_m, E_{\pm k}] = \pm r_m^{(k)} E_{\pm m}.$$

When the rank r is 1; 2; 4 or 8, we can combine the operators H_m and the vectors $r_m^{(k)}$ (or eigenvalues) into elements of a division algebra with imaginary units e_i:

$$H = H_0 + e_i H_i \qquad r^{(k)} = r_0^{(k)} + e_i r_i^{(k)}$$

so that

$$[H, E_{\pm k}] = \pm r^{(k)} E_{\pm k}.$$

For the Lie algebras of the groups $SU(2), O(4), O(8)$, one can construct the real numbers $\mathbb{R}$, the complex numbers $\mathbb{C}$ and the quaternions $\mathbb{H}$, respectively. Interestingly, the triality of the $O(8)$ group reflects the symmetry of the three quaternionic units $e_i = -i\sigma_i$, where σ_i are the Pauli matrices. The case of E_8 was worked out by Coxeter [50] in connection with 8-dimensional regular solids and corresponds to the octonions $\mathbb{O}$. There are 240 rational points on the unit sphere S^7 represented by integer octonions that correspond to the 240 roots of E_8. We first introduce octonionic imaginary units e_α ($\alpha = 1, \ldots, 7$) with the multiplication rule

$$e_\alpha e_\beta = -\delta_{\alpha\beta} + \psi_{\alpha\beta\gamma} e_\gamma,$$

with $\psi_{\alpha\beta\gamma}$ as a third rank antisymmetric tensor that is non-zero and equal to one for the index triples $123, 246, 435, 367, 651, 572, 714$. Now, we define the special element

$$h = \frac{1}{2}(e_1 + e_2 + e_3 + e_7).$$

Then, $1; e_7; e_2; e_6$, together with $h; eh; e_2h; e_7h$ correspond to one possible set of principal positive roots. These elements are also forming the Dynkin diagram of the root system of the E_8. The matrix E_8 in Equation (8) above is the Cartan matrix with entry (i, j) defined by

$$2 \frac{\langle r_i, r_j \rangle}{\langle r_i, r_i \rangle}$$

the scalar products between the root vectors r_i. Then, the whole approach showed that simple combinations of the octonionic imaginary units are corresponding to generators of the second homology groups for a 4-manifold having the matrix E_8 as an intersection form. In case of the K3 surface, one has the intersection form containing the matrix $E_8 \oplus E_8$ which corresponds to two copies of octonions $\mathbb{O} \times \mathbb{O}$. Here, there is a link to the recent work [8] of complex sedions. Now, every

4.2. From Immersed Surfaces in K3 Surfaces to Fermions and Knot Complements

In the previous subsection, we described a relation between the octonions and a system of eight intersecting surfaces in the K3 surface. In this system of surfaces, every surface has two self-intersections (the diagonal of the matrix (8)). Therefore, every surface is not embedded but immersed in the K3 surface. For immersed surfaces, there is a whole theory, called Weierstrass representation, with a close connection between immersed surfaces and spinors. The following discussion is borrowed from [32], and we will present it here again for completeness. First, we start with the immersion $I: \Sigma \to \mathbb{R}^3$ of a surface Σ into $\mathbb{R}^3$. This immersion I can be defined by a spinor φ on Σ fulfilling the Dirac equation

$$D\varphi = H\varphi, \tag{9}$$

with $|\varphi|^2 = 1$ (or an arbitrary constant) (see Theorem 1 of [51]). A spinor bundle over a surface splits into two sub-bundles $S = S^+ \oplus S^-$, representing spinors of different helicity, with the corresponding splitting of the spinor φ in components

$$\varphi = \begin{pmatrix} \varphi^+ \\ \varphi^- \end{pmatrix},$$

and we have the Dirac equation

$$D\varphi = \begin{pmatrix} 0 & \partial_z \\ \partial_{\bar{z}} & 0 \end{pmatrix} \begin{pmatrix} \varphi^+ \\ \varphi^- \end{pmatrix} = H \begin{pmatrix} \varphi^+ \\ \varphi^- \end{pmatrix}$$

with respect to the coordinates $(z, \bar{z})$ on Σ. In dimension 3, the spinor bundle has the same fiber dimension as the spinor bundle S (but without a splitting $S = S^+ \oplus S^-$ into two sub-bundles). Now, we define the extended spinor ϕ over the 3-torus $\Sigma \times [0,1]$ via the restriction $\phi|_{T^2} = \varphi$. The spinor ϕ is constant along the normal vector $\partial_N \phi = 0$ fulfilling the three-dimensional Dirac equation

$$D^{3D}\phi = \begin{pmatrix} \partial_N & \partial_z \\ \partial_{\bar{z}} & -\partial_N \end{pmatrix} \phi = H\phi \tag{10}$$

induced from the Dirac equation (9) via restriction and where $|\phi|^2 = const.$ In this picture, we shift the description from surfaces to 3-manifolds. The description above showed that the essential information is contained in the surface, but fermions are at least three-dimensional objects: fermions and bosons appear beginning with dimension 3 (irreducible representation of the group $SO(3)$ as given by the lift to $SU(2)$). In dimension 2, we have anyons with a spin of any rational number. However, how did we get the corresponding 3-manifold representing the fermion?

To answer this question, we consider the branched covering of the K3 surface M. As explained above, it must be a 4-fold covering $M \to S^4$ branched along a surface with singularities of two types cusp and fold. The cusp can be described as a cone over the trefoil knot, whereas the fold is the cone over the Hopf link (see Figure 9 in [30]). Now, we consider a 4-manifold with boundary, for instance by cutting out a 4-disk D^4 form M to get 4-manifold $M \setminus D^4$ with boundary $\partial(M \setminus D^4) = S^3$, the 3-sphere. Then, the branched covering of M induced a branched covering of the boundary ∂M, so that the branching set of M, a surface, induces a branching set of ∂M, a knot or link. In our case, the singularities of the surface (cusp and fold) given as cones over the trefoil knot and Hopf link will correspond to the trefoil knot and Hopf link in the 3-sphere. Then, the branched covering is given by the mappings of the complements $S^3 \setminus \{trefoil\}$ and $S^3 \setminus \{Hopf - link\}$ to the permutation group S_3. We see the appearance of these two complements as a sign to use these structures as particles and interactions. The complement of the knot is a 3-manifold with one boundary component. In contrast, the complement of the link looks like a cylinder $T^2 \times [0,1]$ which can connect two knot complements. Therefore, we have the conjecture that knot complements are fermions and link complements are bosons.

4.3. Fermions as Knot Complements

In this section, we will discuss the topological reasons for the identification of knot complements with fermions. In our paper [31], we obtained a relation between an embedded 3-manifold and a spinor in the spacetime. The main idea can be simply described by the following line of argumentation. Let $\iota : \Sigma \hookrightarrow M$ be an embedding of the 3-manifold Σ into the 4-manifold M with the normal vector $\vec{N}$ so that a small neighborhood U_ϵ of $\iota(\Sigma) \subset M$ looks like $U_\epsilon = \iota(\Sigma) \times [0,\epsilon]$. Every 3-manifold admits a spin structure with a SPIN BUNDLE, i.e., a principal $Spin(3) = SU(2)$ bundle (spin bundle) as a lift of the frame bundle (principal $SO(3)$ bundle associated with the tangent bundle). Furthermore, there is a (complex) vector bundle associated with the spin bundle (by a representation of the spin group), called SPINOR BUNDLE S_Σ. Now, we meet the usual definition in physics: a section in the spinor bundle is called a spinor field (or a spinor). In general, the unitary representation of the spin group in D dimensions is $2^{\lfloor D/2 \rfloor}$-dimensional. From the representational point of view, a spinor in four dimensions is a pair of spinors in dimension 3. Therefore, the spinor bundle S_M of the 4-manifold splits into two sub-bundles $S_M^\pm$ where one sub-bundle, say S_M^+, can be related to the spinor bundle S_Σ of the 3-manifold. Then, the spinor bundles are related by $S_\Sigma = \iota^* S_M^+$ with the same relation $\phi = \iota_* \Phi$ for the spinors ($\phi \in \Gamma(S_\Sigma)$ and $\Phi \in \Gamma(S_M^+)$). Let $\nabla_X^M, \nabla_X^\Sigma$ be the covariant derivatives in the spinor bundles along a vector field X as section of the bundle $T\Sigma$. Then, we have the formula

$$\nabla_X^M(\Phi) = \nabla_X^\Sigma \phi - \frac{1}{2}(\nabla_X \vec{N}) \cdot \vec{N} \cdot \phi \qquad (11)$$

with the embedding $\phi \mapsto \begin{pmatrix} 0 \\ \phi \end{pmatrix} = \Phi$ of the spinor spaces from the relation $\phi = \iota_* \Phi$. Here, we remark that, of course, there are two possible embeddings. For later use, we will use the left-handed version. The expression $\nabla_X \vec{N}$ is the second fundamental form of the embedding where the trace $tr(\nabla_X \vec{N}) = 2H$ is related to the mean curvature H. Then, from (11), one obtains the following relation between the corresponding Dirac operators

$$D^M \Phi = D^\Sigma \phi - H\phi \qquad (12)$$

with the Dirac operator D^Σ on the 3-manifold Σ. In [32], we extend the spinor representation of an immersed surface into the 3-space to the immersion of a 3-manifold into a 4-manifold according to the work in [51]. Then, the spinor ϕ defines directly the embedding (via an integral representation) of the 3-manifold. Then, the restricted spinor $\Phi|_\Sigma = \phi$ is parallel transported along the normal vector and Φ is constant along the normal direction (reflecting the product structure of U_ϵ). However, then the spinor Φ has to fulfill

$$D^M \Phi = 0 \qquad (13)$$

in U_ϵ i.e., Φ is a parallel spinor. Finally, we get

$$D^\Sigma \phi = H\phi \qquad (14)$$

with the extra condition $|\phi|^2 = const.$ (see [51] for the explicit construction of the spinor with $|\phi|^2 = const.$ from the restriction of Φ). The idea of the paper [31] was the usage of the Einstein–Hilbert action for a spacetime with boundary Σ. The boundary term is the integral of the mean curvature for the boundary; see [52,53]. Then by the relation (14) we will obtain

$$\int_\Sigma H \sqrt{h}\, d^3x = \int_\Sigma \bar\phi D^\Sigma \phi \sqrt{h}\, d^3x \tag{15}$$

using $|\phi|^2 = const$. As shown in [31], the extension of the spinor ϕ to the 4-dimensional spinor Φ by using the embedding

$$\Phi = \begin{pmatrix} 0 \\ \phi \end{pmatrix} \tag{16}$$

can be only seen as embedding, if (and only if) the 4-dimensional Dirac equation

$$D^M \Phi = 0 \tag{17}$$

on M is fulfilled (using relation (12)). This Dirac equation is obtained by varying the action

$$\delta \int_M \bar\Phi D^M \Phi \sqrt{g}\, d^4x = 0. \tag{18}$$

In [31], we went a step further and discussed the topology of the 3-manifold leading to a fermion. On general grounds, one can show that a fermion is given by a knot complement admitting a hyperbolic structure. However, for hyperbolic manifolds (of dimension greater than 2), one has the important property of Mostow rigidity where geometric expressions like the volume are topological invariants. This rigidity is a property which we should expect for fermions. The usual matter is seen as dust matter (incompressible $p = 0$). The scaling behavior of the energy density ρ for dust matter is determined by the time-dependent scaling parameter a to be $\rho \sim a^{-3}$. Thus, if one represents matter by very small regions in the space equipped with a geometric structure, then this scaling can be generated by an invariance of these small regions with respect to a rescaling. Mostow rigidity now singles out the hyperbolic geometry (and the hyperbolic 3-manifold as the corresponding small region) to have the correct behavior. All other geometries allow a scaling at least along one direction. Finally, *Fermions are represented by hyperbolic knot complements.*

4.4. Torus Bundle as Gauge Fields

Now, we have the following situation: two knot complements $C(K_1)$ and $C(K_2)$ can be connected by a so-called tube $T(K_1, K_2)$ along the boundary, a torus. This tube $T(K_1, K_2)$ can be described by the complement of a link with two components defined by the knots K_1, K_2. In the simplest case, it is the 3-manifold $T^2 \times [0,1]$. The knot complements are fermions. Therefore, both knot complements have to carry a hyperbolic structure, i.a. a space of constant negative curvature. The frame bundle of a 3-manifold is always trivial, so that we need a flat connection of this bundle to describe this space. Let $Isom(\mathbb{H}^3) = SO(3,1)$ be the isometry group of the three-dimensional hyperbolic space. There are suitable subgroups $G_1, G_2 \subset Isom(\mathbb{H}^3)$ so that (the interior of)$C(K_1)$ is diffeomorphic to $Isom(\mathbb{H}^3)/G_1$ (and similar with $C(K_2)$). As usual, the space of all flat $SO(3,1)$ connections of $C(K_1)$ is the space of all representations $\pi_1(C(K_1)) \to SO(3,1)$, where $SO(3,1)$ acts in the adjoint representation on this space (as gauge transformations). We note the fact that every $SO(3,1)$ connections lifts uniquely to a $SL(2,\mathbb{C})$ connection. Now, near the boundary, we have a flat $SL(2,\mathbb{C})$ connection in $C(K_1)$ which is connected to a flat $SL(2,\mathbb{C})$ connection in $C(K_2)$ by $T(K_1, K_2)$. The action for a flat connection A with values in the Lie algebra $\mathfrak{g}$ of the Lie group G as a subgroup of the $SL(2,\mathbb{C})$ in a 3-manifold Σ (with vanishing curvature $F = DA = 0$) is given by

$$\int_\Sigma A \wedge F$$

also known as background field model (BF model). By a small redefinition of the connection, one can also choose the Chern–Simons action:

$$CS(A, \Sigma) = \int_\Sigma \left(A \wedge dA + \frac{2}{3} A \wedge A \wedge A \right).$$

The variation of the Chern–Simons action $CS(A, \Sigma)$ gets flat connections $DA = 0$ as solutions. The flow of solutions $A(t)$ in $T(K_1, K_2) \times [0, 1]$ (parametrized by the variable t) between the flat connection $A(0)$ in $T(K_1, K_2) \times \{0\}$ to the flat connection $A(1)$ in $T(K_1, K_2) \times \{1\}$ will be given by the gradient flow equation (see [54] for instance)

$$\frac{d}{dt} A(t) = \pm * F(A) = \pm * DA, \tag{19}$$

where the coordinate t is normal to $T(K_1, K_2)$. Therefore, we are able to introduce a connection $\tilde{A}$ in $T(K_1, K_2) \times [0, 1]$ so that the covariant derivative in the t-direction agrees with $\partial/\partial t$. Then, we have for the curvature $\tilde{F} = D\tilde{A}$ where the fourth component is given by $\tilde{F}_{4\mu} = d\tilde{A}_\mu/dt$. Thus, we will get the instanton equation with (anti-)self-dual curvature

$$\tilde{F} = \pm * \tilde{F}.$$

However, now we have to extend the Chern–Simons action (of the 3-manifold) to the 4-manifold. It follows that

$$CS(A, T(K_1, K_2) \times \{1\}) - CS(A, T(K_1, K_2) \times \{0\}) = \int_{T(K_1, K_2) \times [0,1]} tr(\tilde{F} \wedge \tilde{F})$$

i.e., we obtain the second Chern class and finally

$$S_{EH}([0,1] \times T(K_1, K_2)) = \int_{T(K_1, K_2) \times [0,1]} tr(\tilde{F} \wedge \tilde{F}) = \pm \int_{T(K_1, K_2) \times [0,1]} tr(\tilde{F} \wedge *\tilde{F})$$

i.e., the action of the gauge field. The whole procedure remains true for an extension, i.e.,

$$S_{EH}(\mathbb{R} \times T(K_1, K_2)) = \pm \int_{T(K_1, K_2) \times \mathbb{R}} tr(\tilde{F} \wedge *\tilde{F}). \tag{20}$$

The gauge field action (20) is only defined along the tubes $T(K_1, K_2)$. For the extension of the action to the whole 4-manifold M, we need some non-trivial facts from the theory of 3-manifolds. We presented the ideas in [32]. Finally, we obtain the gauge field action

$$\int_M tr(\tilde{F} \wedge *\tilde{F}). \tag{21}$$

Now, we will discuss the possible gauge group. Again, for completeness, we will present the argumentation in [32] again. The gauge field in the action (21) has values in the Lie algebra of the maximal compact subgroup $SU(2)$ of $SL(2,\mathbb{C})$. However, in the derivation of the action, we used the connecting tube $T(K_1,K_2)$ between two tori, which is a cobordism. This cobordism $T(K_1,K_2)$ is also known as torus bundle (see [55] Theorem 1.15), which can be always decomposed into three elementary pieces—*finite order*, Dehn twist and the so-called Anosov map (named after the russian mathematician Dmitri Anosov).

The idea of this construction is very simple: one starts with two trivial cobordisms $T^2 \times [0,1]$ and glues them together by using a diffeomorphism $g: T^2 \to T^2$, which we call *gluing diffeomorphism*. From the geometrical point of view, we have to distinguish between three different types of torus bundles. The three types of torus bundles are distinguished by the splitting of the tangent bundle:

- finite order (orders 2, 3, 4, 6): the tangent bundle is three-dimensional,
- Dehn-twist (left/right twist): the tangent bundle is a sum of a two-dimensional and a one-dimensional bundle,
- Anosov: the tangent bundle is a sum of three one-dimensional bundles.

Following Thurston's geometrization program (see [56]), these three torus bundles are admitting a geometric structure, i.e., it has a metric of constant curvature. Apart from this geometric properties, all torus bundles are determined by the gluing diffeomorphism $g: T^2 \to T^2$, which also determines the fundamental group of the torus bundle. Therefore, this gluing diffeomorphism also has an influence on the structure of the diffeomorphism group of the torus bundle, which will be discussed now. From the physical point of view, we have two types of diffeomorphisms: local and global. Any coordinate transformation can be described by an infinitesimal or local diffeomorphism (coordinate transformation). In contrast, there are global diffeomorphisms like an orientation reversing diffeomorphism. Two diffeomorphisms not connected via a sequence of local diffeomorphism are part of different connecting components of the diffeomorphism group, i.e., the set of isotopy classes $\pi_0(Diff(M))$ (also called the mapping class group). Isotopy classes are important in order to understand the configuration space topology of general relativity (see Giulini [57]). In principle, the state space in geometrodynamics is the set of all isotopy classes, where every class represents one physical situation, or isotopy classes label two different physical situations. By definition, the two 3-manifolds in different isotopy classes cannot be connected by a sequence of local diffeomorphisms (local coordinate transformations). Again, these two different isotopy classes represent two different physical situations; see [13] for the relation of isotopy classes to particle properties like spin. In case of the torus bundle, we consider the isotopy classes $\pi_0(Diff(M,\partial M))$ relative to the boundary represented by the automorphisms of the fundamental group. Using the geometrization program, we obtain a relation between the isotopy classes $\pi_0(Diff(M,\partial M))$ and the isometry classes (connecting components of the isometry group) with respect to the geometric structure of the torus bundle (see, for instance, [58,59]). Then, the isotopy classes of the torus bundles are given by

- finite order: 2 isotopy classes (= no/even twist or odd twist),
- Dehn-twist: 2 isotopy classes (= left or right Dehn twists),
- Anosov: 8 isotopy classes (= all possible orientations of the three line bundles forming the tangent bundle).

From the geometrical point of view, we can rearrange the scheme above:

- torus bundle with no/even twists: one isotopy class,
- torus bundle with twist (Dehn twist or odd finite twist): three isotopy classes,
- torus bundle with Anosov map: eight isotopy classes.

This information creates a starting point for the discussion on how to derive the gauge group. Given a Lie group G with Lie algebra $\mathfrak{g}$, the rank of $\mathfrak{g}$ is the dimension of the maximal abelian subalgebra, also called Cartan algebra (see above for a definition). It is the same as the dimension of the maximal torus $T^n \subset G$. The curvature F of the gauge field takes values in the adjoint representation of the Lie algebra and the action $tr(F \wedge *F)$ forms an element of the Cartan subalgebra (the Casimir operator). However, each isotopy class contributes to the action and therefore we have to take the sum over all possible isotopy classes. Let t_a be the generator in the adjoint representation; then, we obtain for the Lie algebra part of the action $tr(F \wedge *F)$

- torus bundle with no twists: one isotopy class with t^2,
- torus bundle with twist: three isotopy classes with $t_1^2 + t_2^2 + t_3^2$,
- torus bundle with Anosov map: eight isotopy classes with $\sum_{a=1}^{8} t_a^2$.

The Lie algebra with one generator t corresponds uniquely to the Lie group $U(1)$ where the three generators t_1, t_2, t_3 form the Lie algebra of the $SU(2)$ group. Then, the last case with eight generators t_a has to correspond to the Lie algebra of the $SU(3)$ group. We remark the similarity with an idea from brane theory: n parallel branes (each decorated with an $U(1)$ theory) are described by an $U(n)$ gauge theory (see [60]). Finally, we obtain the maximal group $U(1) \times SU(2) \times SU(3)$ as a gauge group for all possible torus bundles (in the model: connecting tubes between the solid tori).

At the end, we will speculate about the identification of the isotopy classes for the torus bundle with the vector bosons in the gauge field theory. Obviously, the isotopy class of the torus bundle with no twist must be the photon. Then, the isotopy class of the other bundle of finite order should be identified with the Z^0 boson and the two isotopy classes of the Dehn twist bundles are the $W^\pm$ bosons. Here, we remark that this identification is consistent with the definition of the charge lateron. Furthermore, we remark that this scheme contains automatically the mixing between the photon and the Z^0 boson (the corresponding torus bundle are both of finite order). The isotopy classes of the Anosov map bundle have to correspond to the eight gluons. Later, we expect that these ideas will lead to an additional relation for the scattering amplitudes induced by the geometry of the torus bundles.

4.5. Fermions, Bosons and 3-Braids

In the previous subsection, we identified the hyperbolic knot complement with a fermion and the torus bundle between them as gauge bosons. The first natural question is then: which knot is related to a fermion like the electron? However, this question is meaningless in our approach. The knot/link complements are induced from the singularities of the branching set of the 4-fold branched covering of the K3 surface. However, these singularities are another expression of the change of the branching set of the 3-manifolds. With other words: the branching set, a knot or link, of a 3-manifold is not unique. There are transformations of the branching sets representing the same 3-manifold. Therefore, these complements are not uniquely connected to particles/fermions. Later, we will see that the knots represent mainly the state. However, what are the invariant properties? For that purpose, we will study the branched covering of the knot/link complement to understand the invariant properties.

Let $C(K) = S^3 \setminus (K \times D^2)$ be a knot complement which is a compact 3-manifold with boundary $\partial C(K) = T^2$ the 2-torus. $C(K)$ is given by a 3-fold branched covering $C(K) \to D^3$ inducing a 3-fold branched covering of the boundary $\partial C(K) \to \partial D^3$ or $T^2 \to S^2$. The 3-fold covering $T^2 \to S^2$ has six points as a branching set, the end points of a 6-plat or tangle (see Figure 7). In a similar manner, let $C(L) = C(K_1, K_2) = S^3 \setminus (L \times D^2)$ be a link complement of a link with two linked knots K_1, K_2 (the two linking components). $C(L)$ is a compact 3-manifold with two boundaries given by two tori. Then, the 3-fold branched covering $C(L) \to S^3 \setminus (D^3 \sqcup D^3)$ induces again 3-fold branched coverings $T^2 \to S^2$ of the two boundaries. Every covering $T^2 \to S^2$ has six points as a branching set again. The corresponding branching set of $C(L)$ is a braid of six strands but represented as a 3-braid (see Figure 7).

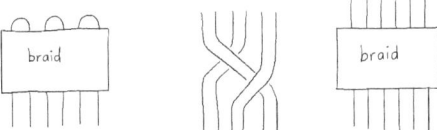

Figure 7. Branching sets—**Left**: 6-Plat for the knot cpomplement, **Center**: a 3-Braid as 6-Braid as example, **Right**: 6-Braid for the link complement

Finally, *bosons and fermions are represented as 3-braids*. Our model agrees with the Bilson–Thompson model but with the exception that we do not fix the braid. In particular, we do not believe that the difference between an electron and myon is given by a different braid.

5. Electric Charge and Quasimodularity

We argued above that all knot complements admitting a hyperbolic geometry (geometry of negative, constant scalar curvature) have the properties of fermions, i.e., spin $\frac{1}{2}$, are pressureless $p = 0$ in cosmology and fulfill the Dirac equation (see also the previous section for the action functional). However, a particle can carry charges (electrical or others like color, etc.).

5.1. Electric Charge as Dehn Twist of the Boundary

We described above the knot complement as a branched covering branched along a braid. What is the meaning of a charge in this description? Let us start with the case of an electric charge. Given a complex line bundle over $C(K) \times (0,1)$ with connection A and curvature $F = dA$, we then have the Maxwell equations:

$$dF = 0 \qquad d * F = *j,$$

with the Hodge operator $*$ and the 4-current 1-form j. The electric charge Q_e is given by

$$Q_e = \int_{\partial C(K) = T^2} *F$$

in the temporal gauge (normal to the boundary of $C(K)$) using $d * F = *j$ and Stokes theorem. The magnetic charge Q_m is defined in an analogous way

$$Q_m = \int_{\partial C(K) = T^2} F = 0,$$

but it is zero because of $dF = 0$. By the formulas above, we obtain a restriction of the complex line bundle to the boundary $\partial C(K) = T^2$. A complex line bundle over T^2 is determined by the twists of the fibers w.r.t. the lattice $\mathbb{Z}^2$ in the definition of the torus $T^2 = \mathbb{C}/\mathbb{Z}^2$. However, which twist is related to the electric charge? Consider a cylinder $S^1 \times [0,1]$ and identify the ends of the interval to get T^2 again. A complex line bundle over $S^1 \times [0,1]$ with curvature F gives the integral

$$\int_{S^1 \times [0,1]} F = \int_{S^1 \times \{1\}} A - \int_{S^1 \times \{0\}} A$$

using $F = dA$, which is only non-trivial if the two integrals differ. It can be realized by a twist of one side (say $S^1 \times \{1\}$), also called a Dehn twist. Dually by using $d * F = *j$, we get

$$Q_e = \int_{[0,1]} *j = \int_{[0,1]} d * F = *F|_1 - *F|_0,$$

which is non-zero by a Dehn twist along $[0,1]$. Therefore, charges can be detected by Dehn twists along the boundary. A Dehn twist along the meridian represents the magnetic charge, whereas a Dehn twist along the longitude is an electric charge. The number of twists is the charge, i.e., we obtain automatically a quantization of the electric and magnetic charge. Furthermore, there is a simple algebraic description of the twists, which agrees with the description of electromagnetic duality using $SL(2,\mathbb{Z})$. As noted above, the torus can be obtained by $T^2 = \mathbb{C}/\mathbb{Z}^2$ w.r.t. the lattice $\mathbb{Z}^2$. An automorphism of the torus is given by a the group $SL(2,\mathbb{Z})$ acting via rational transformations on $\mathbb{C}$, i.e.,

$$\begin{pmatrix} a & b \\ c & d \end{pmatrix} \in SL(2,\mathbb{Z}),\ ad - bc = 1 \to \begin{pmatrix} a & b \\ c & d \end{pmatrix} \cdot z = \frac{az+b}{cz+d}.$$

Then, the two possible Dehn twists are given by

$$z \mapsto z+1,$$
$$z \mapsto -\frac{1}{z},$$

which is known from electromagnetic duality. However, this group also has another meaning. Let $Diff(T^2)$ be the diffeomorphism group of the torus. All coordinate transformations, known as diffeomorphisms connected to the identity, are forming a (normal) subgroup $Diff_0(T^2) \subset Diff(T^2)$. Then, the factor space $MCG(T^2) = Diff(T^2)/Diff_0(T^2)$ is a group, known as a mapping classes group, and generated by Dehn twists, i.e., $MCG(T^2) = SL(2,\mathbb{Z})$ or the mapping class group is the modular group. An element of the mapping class group is a global diffeomorphism (also called *isotopy*) that cannot be described by coordinate transformations, i.e., full twists cannot be undone by a sequence of infinitesimal rotations. Then, different charges belong to different mapping (or isotopy) classes. Up to now, we have a full symmetry between electric and magnetic charges (geometrically expressed by the torus). Now, we will show that this behavior changed for the extension to the knot complement. Technically, it will be expressed by a change from modular to quasimodular functions.

5.2. Electric Charge as a Frame of the Knot Complement

However, what does change in the knot complement and in the branched covering? As a toy example, we consider the complement of the unknot $D^2 \times S^1$. Then, the Dehn twist along the meridian of the boundary torus will be trivialized. By a result of McCullough (see for instance [61]), every Dehn twist along the longitude induces a diffeomorphism of the solid torus. Then, the complement of the unknot can carry an electric charge (by a Dehn twist) but no magnetic charge. This result can be generalized to all knot complements (which are homologically equivalent to the solid torus). The effect on the branched covering can also be obtained by considering the boundary. The boundary is a torus written as two-fold branched covering branched along 4 points. A Dehn twist is given by a permutation of the branching points that leads to a twist of the braid as a branching set of the knot complement (see Figure 8).

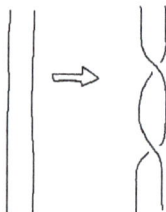

Figure 8. Dehn twist represented by a braid.

We obtain again the quantization of the electric charge as the number of Dehn twists.

However, more is true: the isotopy classes of the boundary determine the isotopy classes of the hyperbolic knot complement up to a finite subgroup [59]. The mapping class group $MCG(C(K))$ consists of a disjoint union of isotopy classes of framings, i.e., trivializations of the tangent bundle $TC(K)$ seen as sections of the frame bundle ($SO(3)$ principal bundle) up to homotopy. Therefore, the change of the number of Dehn twists at the boundary induces a change of the framing for the knot complement. However, there is also a direct way using obstruction theory. In [62], we described the quantization of the electric charge by using exotic smoothness as a substitute for a magnetic monopole. A magnetic monopole is a substitute for an element in the cohomology $H^2(S^2, \mathbb{Z})$ leading to the quantization of the electric charge

$$Q_m \cdot Q_e = \frac{c\hbar}{2} \cdot n, \, n \in \mathbb{Z}$$

for the magnetic charge Q_m and for the electric charge Q_e. Using the canonical isomorphism

$$H^2(S^2, \mathbb{Z}) \simeq H^3(S^3, \mathbb{Z}),$$

we can transform the monopole class (as first Chern class of a complex line bundle) into a class in $H^3(S^3, \mathbb{Z})$. Now, let P be a principal $SO(3)$ bundle over S^3, called the frame bundle. The obstruction for a section in this bundle lies at $H^4(S^3, \pi_3(SO(3))) = 0$, where the vanishing of the cocycle guarantees the existence. The number of sections is given by the elements in $H^3(S^3, \pi_3(SO(3))) = H^3(S^3, \mathbb{Z})$ using $\pi_3(SO(3)) = \mathbb{Z}$. By Hodge duality, we obtain the same line of argumentation for the class $*F$ getting the electric charge (using also the quantization condition). The class in $H^3(S^3, \mathbb{Z})$ can be related to a relative class in the 4-manifold $S^3 \times [0,1]$, i.e.,

$$H^3(S^3, \mathbb{Z}) \simeq H^4(S^3 \times [0,1], S^3 \times \partial[0,1], \mathbb{Z})$$

called the relative Pontrjagin class p_1. Now, we extend the whole discussion to an arbitrary 3-manifold Σ, which we identify with $\Sigma = C(K) \cup (D^2 \times S^1)$. Using this 4-dimensional interpretation, we obtain the framing as the Hirzebruch defect [63]. For that purpose, we consider the 4-manifold M with $\partial M = \Sigma$. Let $\sigma(M)$ be the signature of M, i.e., the number of positive minus negative eigenvalues of the intersection form of M. Furthermore, let $p_1(M, \Sigma)$ be the relative Pontrjagin class as an element of $H^4(M, \partial M = \Sigma, \mathbb{Z})$. Then, the Hirzebruch defect h is given by

$$h = 3\sigma(M) - p_1(M, \Sigma) = Q_e \quad (22)$$

and identified with the framing, i.e., with the charge. This definition is motivated by the Hirzebruchs signature formula for a closed 4-manifold relating the signature $\sigma(M)$ and the first Pontrjagin class $p_1(M)$ (of the tangent bundle TM) via $\sigma(M) = \frac{1}{3} p_1(M)$ (see [64]).

5.3. The Charge Spectrum

Now, we will discuss the expression (22) for the electric charge. By the argumentation above, the relative Pontrjagin class gives an integer expressing the framing of the knot complement for a fixed time. The appearance of the signature $\sigma(M)$ added a 4-dimensional element which describes more complex cases with many components. In [65], this case was also considered with a similar result: this formula is valid for links where $\sigma(M)$ is now the signature of the linking matrix and p_1 is the sum of framings for each component. The signature can be minimally changed by ± 1 leading to a change of the charge by ± 3. Therefore, the minimal change for one component can be

$$Q_e \bmod 3\mathbb{Z} = \{0, \pm 1, \pm 2, \pm 3\},$$

i.e., we obtain a spectrum containing four possible values. If we normalize the charge to be a multiple of $e/3$, then we have the charge spectrum

$$Q_e = \left\{0, \pm\frac{1}{3}e, \pm\frac{2}{3}e, \pm e\right\}$$

in agreement with the experiment. Then, we have the description of one particle generation: two leptons (neutrino of charge 0 and lepton of charge -1) and two quarks (quark of charge $-\frac{1}{3}$ and quark of charge $+\frac{2}{3}$).

5.4. Vanishing of the Magnetic Charge and Quasimodularity

One may wonder whether there is no magnetic charge anymore. Our argument is only partially satisfying because there are many incompressible surfaces inside of a hyperbolic knot complement serving as representatives for magnetic charges. Therefore, we need a stronger argument why the symmetry between electric and magnetic charge is broken. As explained above, the Dehn twists of the boundary torus are the generators of the mapping class (or isotopy) group. According to Atiyah [63], the framing can be used to define a central extension $\hat{\Gamma}$

$$1 \to \mathbb{Z} \to \hat{\Gamma} \to \Gamma \to 1$$

of the mapping class group Γ so that there is a section $s : \Gamma \to \hat{\Gamma}$ inducing a splitting of the sequence above. This section defines a canonical 2-cocycle c for the central extension that is given by the signature of the corresponding 4-manifold (see [63] for the details). However, in case of the torus, for the group $\Gamma = SL(2,\mathbb{Z})$, there is no non-zero homomorphism $\Gamma \to \mathbb{Z}$ and so the splitting $s_1 : \Gamma \to \hat{\Gamma}$ is unique. Therefore, the canonical section s is not a homomorphism and the framing (used in the definition of this section) leads to a breaking of the modular invariance i.e., the invariance w.r.t. Γ. This fact is simply expressed by considering the difference of the two sections $s(\gamma)$ and $s_1(\gamma)$ for $\gamma \in \Gamma$, which is given by the logarithm of the Dedekind η-function, related to quasimodular functions. Thus, our definition of the electric charge breaks explicitly the electro-magnetic duality and we get a vanishing magnetic charge.

6. Drinfeld–Turaev Quantization and Quantum States

In [66,67], we discussed the appearance of quantum states from knots known as Turaev–Drinfeld quantization. The idea for the following construction can be simply expressed. We start with two 3-manifolds and consider a cobordism between them. This cobordism is a 4-manifold with a branched covering branched over a surface with self-intersections. Here, it is enough to restrict to a special class of these surfaces, so-called ribbon surfaces (see [68]). The 3-manifolds are chosen to be hyperbolic knot complements, denoted by Y_1, Y_2. A hyperbolic structure is defined by a homomorphism $\pi_1(Y_i) \to SL(2,\mathbb{C})$ ($\in Hom(\pi_1(Y_i), SL(2,\mathbb{C}))$) up to conjugation. Now, we extend this structure to the entire cobordism, denoted by $Cob(Y_1, Y_2)$. The branching set of $Cob(Y_1, Y_2)$ is a surface S with non-trivial fundamental group $\pi_1(S)$. This surface can be changed without any change of $Cob(Y_1, Y_2)$. One change can be described as crossing change. Now, we have all ingredients for the Drinfeld–Turaev quantization:

- The surface S (branching set of $Cob(Y_1, Y_2)$) is inducing a representation $\pi_1(S) \to SL(2,\mathbb{C})$.
- The space of all representations $X(S, SL(2,\mathbb{C})) = Hom(\pi_1(S), SL(2,\mathbb{C}))/SL(2,\mathbb{C})$ has a natural Poisson structure (induced by the bilinear on the group) and the Poisson algebra $(X(S, SL(2,\mathbb{C})), \{\ \})$ of complex functions over them is the algebra of observables.
- The Skein module $K_{-1}(S \times [0,1])$ (i.e., $t = -1$) has the structure of an algebra isomorphic to the Poisson algebra $(X(S, SL(2,\mathbb{C})), \{\ \})$. (see also [69,70]).

- The skein algebra $K_t(S \times [0,1])$ is the quantization of the Poisson algebra $(X(S, SL(2,\mathbb{C})),\{\ \})$ with the deformation parameter $t = \exp(h/4)$ (see also [69]).

To understand these statements, we have to introduce the skein module $K_t(M)$ of a 3-manifold M (see [71]). For that purpose, we consider the set of links $\mathcal{L}(M)$ in M up to isotopy and construct the vector space $\mathbb{C}\mathcal{L}(M)$ with basis $\mathcal{L}(M)$. Then, one can define $\mathbb{C}\mathcal{L}[[t]]$ as ring of formal polynomials having coefficients in $\mathbb{C}\mathcal{L}(M)$. Now, we consider the link diagram of a link, i.e., the projection of the link to the $\mathbb{R}^2$ having the crossings in mind. Choose a disk in $\mathbb{R}^2$ so that one crossing is inside this disk. If the three links differ by the three crossings $L_\infty, L_0, L_{\infty}$ (see Figure 9) inside of the disk, then these links are skein related.

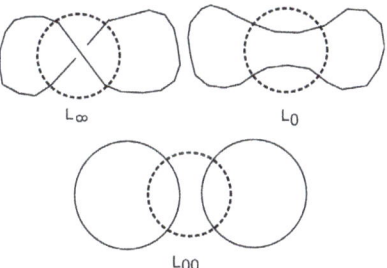

Figure 9. Crossings L_∞, L_0, L_{00}.

Then, in $\mathbb{C}\mathcal{L}[[t]]$, one writes the skein relation $L_\infty - tL_0 - t^{-1}L_{00}$ which depends only on the group $SL(2,\mathbb{C})$. Furthermore, let $L \sqcup O$ be the disjoint union of the link with a circle. Then, one writes the framing relation $L \sqcup O + (t^2 + t^{-2})L$. Let $S(M)$ be the smallest submodule of $\mathbb{C}\mathcal{L}[[t]]$ containing both relations. Then, we define the Kauffman bracket skein module by $K_t(M) = \mathbb{C}\mathcal{L}[[t]]/S(M)$. The modification of S by using the skein relations is one of the allowed changes of the branching set to keep $Cob(Y_1, Y_2)$.

Now, we list the following general results about this module:

- The module $K_{-1}(M)$ for $t = -1$ is a commutative algebra.
- Let S be a surface. Then, $K_t(S \times [0,1])$ carries the structure of an algebra.

The algebra structure of $K_t(S \times [0,1])$ can be simply seen by using the diffeomorphism between the sum $S \times [0,1] \cup_S S \times [0,1]$ along S and $S \times [0,1]$. Then, the product ab of two elements $a, b \in K_t(S \times [0,1])$ is a link in $S \times [0,1] \cup_S S \times [0,1]$ corresponding to a link in $S \times [0,1]$ via the diffeomorphism. The algebra $K_t(S \times [0,1])$ is in general non-commutative for $t \neq -1$. For the following, we will omit the interval $[0,1]$ and denote the skein algebra by $K_t(S)$.

As shown in [23,66,72], the skein algebra serves as the observable algebra of a quantum field theory. For this approach via branched coverings, the branching sets of knot complements (representing the fermions) are special braids (6-plats, see above). Any different braid is a different state or better than a different quantum state but not a different particle. As explained above, the charge spectrum is enough to describe one generation of particles (two leptons, two quarks). The appearance of different generations will be discussed below.

7. Fermions and Number Theory

In this section, we will present some ideas to uncover some explicit relations between fermions, given as hyperbolic knot complements, and number theory, notable quaternionic trace fields and algebraic K theory/Bloch group. However, we first need some definitions.

A quaternion algebra over a field F is a four-dimensional central simple F-algebra. A quaternion algebra has a basis $1, i, j, ij$ where $i^2, j^2 \in F^\times$ and $ij = -ji$. A subgroup of $PSL(2,\mathbb{C})$, the isometry group of the three-dimensional hyperbolic space isomorphic to the Lorentz group $SO(3,1)$, is said to be derived from a quaternion algebra if it can be obtained through the following construction. Let F be a number field that has exactly two embeddings into $\mathbb{C}$ whose image is not contained in $\mathbb{R}$. Let A be a quaternion algebra over F such that, for any embedding $\tau : F \to \mathbb{R}$, the algebra $A \otimes_\tau \mathbb{R}$ is isomorphic to the quaternions. Let $\mathcal{O}^1$ be the group of elements in the order of A with a 1. An order of a quaternionic algebra A is a finitely generated submodule $\mathcal{O}$ of A of reduced norm 1 and let Γ be its image in the 2×2 matrices $M_2(\mathbb{C})$ via $\phi : A \to M_2(\mathbb{C})$. We then consider the Kleinian group obtained as the image in $\text{PSL}(2,\mathbb{C})$ of $\phi(\mathcal{O}^1)$. This subgroup is called an arithmetic Kleinian group. An arithmetic hyperbolic three-manifold is the quotient of hyperbolic space $\mathbb{H}^3$ by an arithmetic Kleinian group. The complement of the figure 8 knot is one example of an arithmetic hyperbolic 3-manifold.

This class of 3-manifolds shows the strong relation between quaternions and 3-manifolds. We discussed above the relation between the K3 surface and the octonions. The starting point for the use of number theory in Kleinian groups is Mostow's rigidity theorem. A consequence of this theorem is that the matrix entries in $SL(2,\mathbb{C})$ of a finite covolume Kleinian group Γ may be taken to lie in a number field that is a finite extension of $\mathbb{Q}$. However, it is true that there is a strong relation between certain number theoretic functions (Bloch–Wigner function, dilogarithm) and the volume of the hyperbolic 3-manifolds: the volume is the sum over all Bloch–Wigner functions for the ideal tetrahedrons forming this 3-manifold. For more information about the relation between hyperbolic 3-manifolds and number theory, consult the book [73]. We hope to use this relation in the future to obtain more properties of fermions by using number theory.

8. The K3 Surface and the Number of Generations

In Section 4.1, we discussed a relation between the K3 surface and octonions by using the intersection form. Here, we use only the E_8 matrix, i.e., the Cartan matrix of the Lie algebra E_8. In this section, we will speculate about the other part

$$H = \begin{pmatrix} 0 & -1 \\ -1 & 0 \end{pmatrix}$$

of the intersection form (now with a different orientation). H is the intersection form of the 4-manifold $S^2 \times S^2$. To express it explicitly, there are homology classes $\alpha, \beta \in H_2(S^2 \times S^2)$ with $\alpha^2 = \beta^2 = 0$ and $\alpha \cdot \beta = -1$. Therefore, every S^2 of this manifold has no self-intersections. For the topology of the K3 surface with intersection form (7), this form has the desired form, but, as explained above, we will change the smoothness structure. The central idea is the usage of Casson handles CH for the 4-manifolds $S^2 \times S^2 \setminus pt$, the one-point complement of $S^2 \times S^2$. Here, the homology classes α, β are given (up to homotopy) by $\alpha^2, \beta^2 = 0 \bmod 2$ and $\alpha \cdot \beta = -1$; see [74]. However, then one has $\alpha^2 = 2n$. Interestingly, the existence of a spin structure is connected to the property that the squares of the homology classes are even. Here, we will consider the simplest realization which has non-zero squares, i.e., we get the form

$$\tilde{H} = \begin{pmatrix} 2 & -1 \\ -1 & 2 \end{pmatrix}. \tag{23}$$

This form cannot be an intersection form because $\tilde{H}$ has determinant 3. Therefore, only the $\tilde{H}$ mod 2 reduction has the meaning to be an intersection. However, for the moment, we will consider $\tilde{H}$ and apply the same construction as for the E_8, i.e., we see $\tilde{H}$ as the Cartan matrix for a Lie algebra. In this case, we get the Lie algebra of $SU(3)$ or the color group. The whole discussion uses some hand-waving arguments, but it is a sign that the 3 $(S^2 \times S^2)$ part of the K3 surfaces is connected with the generations. Every part $S^2 \times S^2$ has one color group and realizes the electric charge spectrum $0, \pm\frac{1}{3}, \pm\frac{2}{3}, \pm 1$. Thus, every $S^2 \times S^2$ is the 4-dimensional expression for one generation. This result agrees with the discussion in [31] where we generate fermions from a Casson handle. Let us assume that the number of generations is given by the number of $S^2 \times S^2$ summands. How many generations are possible? Here, we have the surprising result: *if the underlying spacetime is a smooth manifold, then the minimal number of generations must be three!* A spacetime with only one or two generations is not a smooth manifold. Then, the K3 surface is the minimal model.

We will close this paper with another speculation, a global symmetry induced from the K3 surface. Starting point is the intersection form again. From the point of number theory, this form is an even unimodular positive-definite lattices of rank 24, the so-called Niemeier lattice. In 2010, Eguchi–Ooguri–Tachikawa observed that the elliptic genus of the K3 surface decomposes into irreducible characters of the N = 4 superconformal algebra. The corresponding q-series is a mock modular form related to the sporadic group M_{24}, the Mathieu group, a simple group of order 244823040. The whole theory is known as Mathieu moonshine or umbral moonshine [75]. The interesting point here is the maximal subgroup of M_{24}, the split extension of $PGL(3,4)$ by S_3. The group is the projective group of 3×3 matrices with values in the field of four elements. It seems that this maximal subgroup acts in some sense on the K3 surface, and we conjecture that this group acts on the $S^2 \times S^2$ part. If our idea of a relation between the three generations and the $3\,(S^2 \times S^2)$ part of the K3 surfaces is true, then we hope to get the mixing matrix for quarks and neutrinos from this action.

9. Conclusions and Outlook

In this paper, we presented a top-down approach to fermions and bosons, in particular the standard model. What was done in the paper?

- We constructed a spacetime, the K3 surface and derive some numbers like the cosmological constant or some energy scales and neutrino masses agreeing with experimental data.
- We derived from a representation of K3 surfaces by branched covering a simple picture: fermions are hyperbolic knot complements, whereas bosons are link complements (torus bundles).
- We obtained the gauge group from this picture (at least in principle).
- We derived the correct charge spectrum and obtained one generation.
- We conjectured about the number of generations and global symmetry (the $PGL(3,4)$) to get the mixing between the generations.

What are the consequences for physics? The model only has a few direct consequences. We introduced fermions and bosons in a geometric way. Except for the right-handed neutrino (needed for the see-saw mechanism to generate the masses), we only got the fermions and gauge bosons of the standard model. No extension is needed. The usage of torus bundles for the gauge bosons should generate additional relations for the corresponding scattering amplitudes. The appearance of the global symmetry $PGL(3,4)$ should be related to the mixing of quarks and neutrinos. In [40], we also discussed the appearance of an asymmetry between particles and anti-particles induced by the topology of the spacetime. This idea is also valid in this model, but we cannot match it to the observations.

Is there an outline on some new experiments derived by this model? Currently, this model makes some predictions about the neutrino masses, charge spectrum and the existence of a right-handed neutrino. However, these predictions can be checked by a better measurement in known experiments. Now, there are no new ideas about special experiments connected with this model.

Among these results, there are, of course, many open points of the kind: what is the color and weak charge? How can we implement the Higgs mechanism? What is mass? For the Higgs mechanism, we had found a possible scheme in our previous work [40,76], but it is only a beginning. Many aspects of this paper are related to the ideas of Furey and Gresnigt. It is a future project to extend it and bridge our approach with these ideas.

Funding: This research received no external funding.

Acknowledgments: I first want to thank the anonymous referees for all helpful remarks and comments to make this work more readable. Furthermore, I want to thank Carl Brans for an uncountable number of delightful discussions over so many years. I also want to give a special thanks to Chris Duston for discussions about branched coverings, in particular to draw my attention towards branched coverings. In addition, I want to acknowledge all of my discussions with Jerzy Krol over the years. Finally, I want to give a huge thank you to my family, in particular to my daughter Lucia, for painting the figures.

Conflicts of Interest: The author declares no conflict of interest.

References

1. Collaborations, F.G. Testing Einstein's Special Relativity with Fermi's Short Hard Gamma-Ray Burst GRB090510. *Nature* **2009**, *462*, 331–334.
2. Furey, C. Towards a Unified Theory of Ideals. *Phys. Rev. D* **2012**, *86*, 025024. [CrossRef]
3. Furey, C. Generations: Three Prints, in Colour. *JHEP* **2014**, *10*, 046. [CrossRef]
4. Furey, C. Charge Quantization from a Number Operator. *Phys. Lett. B* **2015**, *742*, 195–199. [CrossRef]
5. Furey, C. Standard Model Physics from an Algebra? Ph.D. Thesis, University of Waterloo, Waterloo, ON, Canada, 2015. Available online: https://arxiv.org/abs/1611.09182 (accessed on 1 October 2019).
6. Gresnigt, N. Braids, Normed Division Algebras, and Standard Model Symmetries. *Phys. Lett. B* **2018**, *783*, 212–221. [CrossRef]
7. Gresnigt, N. Braided Fermions from Hurwitz Algebras. *J. Phys. Conf. Ser.* **2019**, *1194*, 012040. [CrossRef]
8. Gillard, A.; Gresnigt, N. Three Fermion Generations with Two Unbroken Gauge Symmetries from the Complex Sedenions. *Eur. Phys. J. C* **2019**, *79*, 446. doi:10.1140/epjc/s10052-019-6967-1. [CrossRef]
9. Stoica, O. Leptons, Quarks, and Gauge from the Complex Clifford Algebra $\mathbb{C}\ell_6$. *Adv. Appl. Cliff. Alg.* **2018**, *28*, 52. doi:10.1007/s00006-018-0869-4. [CrossRef]
10. Thomson, W. On Vortex Motion. *Trans. R. Soc. Ed.* **1869**, *25*, 217–260. [CrossRef]
11. Misner, C.; Thorne, K.; Wheeler, J. *Gravitation*; Freeman: San Francisco, CA, USA, 1973.
12. Mielke, E. Knot Wormholes in Geometrodynamics? *Gen. Relat. Grav.* **1977**, *8*, 175–196. [CrossRef]
13. Giulini, D. Matter from Space. Based on a talk delivered at the conference "Beyond Einstein: Historical Perspectives on Geometry, Gravitation, and Cosmology in the Twentieth Century", September 2008 at the University of Mainz in Germany. To appear in the Einstein-Studies Series, Birkhaeuser, Boston. *arXiv* **2008**, arXiv:0910.2574.
14. Jehle, H. Topological characterization of leptons, quarks and hadrons. *Phys. Lett. B* **1981**, *104*, 207–211. [CrossRef]
15. Gavrilik, A.M. Quantum algebras in phenomenological description of particle properties. *Nucl. Phys. B (Proc. Suppl.)* **2001**, *102/103*, 298–305. doi:10.1016/S0920-5632(01)01570-5. [CrossRef]
16. Finkelstein, R. Knots and Preons. *Int. J. Mod. Phys. A* **2009**, *24*, 2307–2316. [CrossRef]
17. Bilson-Thompson, S. A Topological Model of Composite Preons. *arXiv* **2005**, arXiv:0503213v2.
18. Bilson-Thompson, S.; Markopoulou, F.; Smolin, L. Quantum Gravity and the Standard Model. *Class. Quant. Grav.* **2007**, *24*, 3975–3994. [CrossRef]
19. Duston, C. Exotic Smoothness in 4 Dimensions and Semiclassical Euclidean Quantum Gravity. *Int. J. Geom. Meth. Mod. Phys.* **2010**, *8*, 459–484. [CrossRef]
20. Duston, C.L. Topspin Networks in Loop Quantum Gravity. *Class. Quant. Grav.* **2012**, *29*, 205015. [CrossRef]
21. Duston, C.L. The Fundamental Group of a Spatial Section Represented by a Topspin Network. Based on Work Presented at the LOOPS 13 Conference at the Perimeter Institute. *arXiv* **2013**, arXiv:1308.2934.
22. Denicola, D.; Marcolli, M.; al Yasry, A. Spin Foams and Noncommutative Geometry. *Class. Quant. Grav.* **2010**, *27*, 205025. [CrossRef]
23. Asselmeyer-Maluga, T. Smooth Quantum Gravity: Exotic Smoothness and Quantum Gravity. In *At the Frontiers of Spacetime: Scalar-Tensor Theory, Bell's Inequality, Mach's Principle, Exotic Smoothness*; Asselmeyer-Maluga, T., Ed.; Springer: Basel, Switzerland, 2016.
24. Rolfson, D. *Knots and Links*; Publish or Prish: Berkeley, CA, USA, 1976.
25. Hilden, H. Every Closed Orientable 3-Manifold is a 3-Fold Branched Covering Space of S^3. *Bull. Am. Math. Soc.* **1974**, *80*, 1243–1244. [CrossRef]
26. Hirsch, U. Über Offene Abbildungen Auf Die 3-Sphäre. *Math. Z.* **1974**, *140*, 203–230. [CrossRef]
27. Montesinos, J. A Representation of Closed, Orientable 3-Manifolds as 3-Fold Branched Coverings of S^3. *Bull. Am. Math. Soc.* **1974**, *80*, 845–846. [CrossRef]
28. Piergallini, R. Four-Manifolds as 4-Fold Branched Covers of S^4. *Topology* **1995**, *34*, 497–508. [CrossRef]

29. Iori, M.; Piergallini, R. 4-Manifolds as Covers of S^4 Branched over Non-Singular Surfaces. *Geom. Topol.* **2002**, *6*, 393–401. [CrossRef]
30. Piergallini, R.; Zuddas, D. On Branched Covering Representation of 4-Manifolds. *J. Lond. Math. Soc.* **2018**, *99*, in press. [CrossRef]
31. Asselmeyer-Maluga, T.; Brans, C. How to Include Fermions Into General Relativity by Exotic Smoothness. *Gen. Relat. Grav.* **2015**, *47*, 30. [CrossRef]
32. Asselmeyer-Maluga, T.; Rosé, H. On the Geometrization of Matter by Exotic Smoothness. *Gen. Relat. Grav.* **2012**, *44*, 2825–2856. [CrossRef]
33. Brans, C. Localized exotic smoothness. *Class. Quant. Grav.* **1994**, *11*, 1785–1792. [CrossRef]
34. Brans, C. Exotic smoothness and physics. *J. Math. Phys.* **1994**, *35*, 5494–5506. [CrossRef]
35. Yau, S.T. On the Ricci Curvature of a Compact Kähler Manifold and the Complex Monge-Ampère Equation. *Commun. Pure Appl. Math.* **1978**, *31*, 339–411. [CrossRef]
36. LeBrun, C. Four-Manifolds Without Einstein Metrics. *Math. Res. Lett.* **1996**, *3*, 133–147. [CrossRef]
37. Asselmeyer-Maluga, T.; Król, J. On Topological Restrictions of the Spacetime in Cosmology. *Mod. Phys. Lett. A* **2012**, *27*, 1250135. [CrossRef]
38. Donaldson, S. An Application of Gauge Theory to the Topology of 4-Manifolds. *J. Diff. Geom.* **1983**, *18*, 269–316. [CrossRef]
39. Asselmeyer-Maluga, T.; Krol, J. How to Obtain a Cosmological Constant from Small Exotic $\mathbb{R}^4$. *Phys. Dark Universe* **2018**, *19*, 66–77. [CrossRef]
40. Asselmeyer-Maluga, T.; Krol, J. A Topological Approach to Neutrino Masses by Using Exotic Smoothness. *Mod. Phys. Lett. A* **2019**, *34*, 1950097. [CrossRef]
41. Asselmeyer-Maluga, T.; Krol, J. A Topological Model for Inflation. *arXiv* **2018**, arXiv:1812.08158.
42. Whitt, B. Fourth Order Gravity as General Relativity Plus Matter. *Phys. Lett.* **1984**, *145B*, 176–178. [CrossRef]
43. Starobinski, A. A New Type of Isotropic Cosmological Models Without Singularity. *Phys. Lett. B* **1980**, *91*, 99–102. [CrossRef]
44. Gompf, R.; Stipsicz, A. *4-Manifolds and Kirby Calculus*; American Mathematical Society: Providence, RI, USA, 1999.
45. Ade, P.A.; Aghanim, N.; Armitage-Caplan, C.; Arnaud, M.; Ashdown, M.; Atrio-Barandela, F.; Aumont, J.; Baccigalupi, C.; Banday, A.J.; Barreiro, R.B.; et al. Planck 2013 Results. XVI. Cosmological Parameters. *Astron. Astrophys.* **2013**, *571*, A16.
46. Ade, P.C.P.; Aghanim, N.; Arnaud, M.; Ashdown, M.; Aumont, J.; Baccigalupi, C.; Banday, A.J.; Barreiro, R.B.; Bartlett, J.G.; Bartolo, N. Planck 2015 Results. XIII Cosmological Parameters. *Astron. Astrophys.* **2016**, *594*, A13.
47. Bonvin, V.; Courbin, F.; Suyu, S.H.E.A. H0LiCOW-V. New COSMOGRAIL Time Delays of HE 0435-1223: H0 to 3.8 Per Cent Precision from Strong Lensing in a Flat ΛCDM Model. *Mon. Not. R. Astron. Soc.* **2016**, *465*, 4914–4930. [CrossRef]
48. Asselmeyer-Maluga, T.; Król, J. Inflation and Topological Phase Transition Driven by Exotic Smoothness. *Adv. HEP* **2014**, 14. [CrossRef]
49. Freedman, M. The topology of four-dimensional manifolds. *J. Diff. Geom.* **1982**, *17*, 357–454. [CrossRef]
50. Coxeter, H. Integral Caleay Numbers. *Duke Math. J.* **1946**, *13*, 561–578. [CrossRef]
51. Friedrich, T. On the Spinor Representation of Surfaces in Euclidean 3-Space. *J. Geom. Phys.* **1998**, *28*, 143–157. [CrossRef]
52. Ashtekar, A.; Engle, J.; Sloan, D. Asymptotics and Hamiltonians in a First, Order Formalism. *Class. Quant. Grav.* **2008**, *25*, 095020. [CrossRef]
53. Ashtekar, A.; Sloan, D. Action and Hamiltonians in Higher Dimensional General Relativity: First, Order Framework. *Class. Quant. Grav.* **2008**, *25*, 225025. [CrossRef]
54. Floer, A. An instanton Invariant for 3-manifolds. *Commun. Math. Phys.* **1988**, *118*, 215–240. [CrossRef]
55. Calegari, D. *Foliations and the Geometry of 3-Manifolds*; Oxford University Press: Oxford, UK, 2007.
56. Thurston, W. *Three-Dimensional Geometry and Topology*, 1st ed.; Princeton University Press: Princeton, NJ, USA, 1997.
57. Giulini, D. Properties of 3-Manifolds for Relativists. *Int. J. Theor. Phys.* **1994**, *33*, 913–930. [CrossRef]
58. Kalliongis, J.; McCullough, D. Isotopies of 3-Manifolds. *Top. Appl.* **1996**, *71*, 227–263. [CrossRef]

59. Hatcher, A.; McCullough, D. Finiteness of Classifying Spaces of Relative Diffeomorphism Groups of 3-Manifolds. *Geom. Top.* **1997**, *1*, 91–109. Available online: http://www.math.cornell.edu/~hatcher/Papers/bdiffrel.pdf (accessed on 1 October 2019). [CrossRef]
60. Giveon, A.; Kutasov, D. Brane Dynamics and Gauge Theory. *Rev. Mod. Phys.* **1999**, *71*, 983–1084. [CrossRef]
61. Hong, S.; McCullough, D. Mapping Class Groups of 3-Manifolds, Then and Now. In *Geometry and Topology Down Under*; Hodgson, C.D., Jaco, W.H., Scharlemann, M.G., Tillmann, S., Eds.; AMS: Providence, RI, USA, 2013; Volume 597.
62. Asselmeyer-Maluga, T.; Król, J. Abelian Gerbes, Generalized Geometries and Foliations of Small Exotic R^4. *arXiv* **2014**, arXiv:0904.1276v5.
63. Atiyah, M. On Framings of 3-Manifolds. *Topology* **1990**, *29*, 1–7. [CrossRef]
64. Hirzebruch, F. *Topological Methods in Algebraic Geometry*; Springer: Berlin/Heidelberg, Germany; New York, NY, USA, 1973.
65. Freed, D.; Gompf, R. Computer calculation of Witten's 3-manifold invariant. *Commun. Math. Phys.* **1991**, *141*, 79–117. [CrossRef]
66. Asselmeyer-Maluga, T.; Król, J. Quantum Geometry and Wild Embeddings as Quantum States. *arXiv* **2013**, arXiv:1211.3012.
67. Asselmeyer-Maluga, T. Hyperbolic Groups, 4-Manifolds and Quantum Gravity. *J. Phys. Conf. Ser.* **2019**, *1194*, 012009. [CrossRef]
68. Piergallini, R.; Zuddas, D. A Universal Ribbon Surface in B^4. *Proc. Lond. Math. Soc.* **2005**, *90*, 763–782. [CrossRef]
69. Bullock, D.; Przytycki, J. Multiplicative Structure of Kauffman Bracket Skein Module Quantization. *Proc. AMS* **1999**, *128*, 923–931. [CrossRef]
70. Bullock, D. A Finite Set of Generators for the Kauffman Bracket Skein Algebra. *Math. Z.* **1999**, *231*, 91–101. [CrossRef]
71. Prasolov, V.; Sossinisky, A. *Knots, Links, Braids and 3-Manifolds*; AMS: Providence, RI, USA, 1997.
72. Asselmeyer-Maluga, T.; Mader, R. Exotic R^4 and Quantum Field Theory. In Proceedings of the 7th International Conference on Quantum Theory and Symmetries (QTS7), Prague, Czech Republic, 7–13 August 2011; Al, C.B.E., Ed.; IOP Publishing: Bristol, UK, 2012; p. 012011. [CrossRef]
73. Maclachlan, C.; A.W., R. *The Arithmetic of Hyperbolic 3-Manifolds*; Springer Publisher: New York, NY, USA, 2003; Volume 219.
74. Casson, A. *Three Lectures on New Infinite Constructions in 4-Dimensional Manifolds*; Progress in Mathematics ed.; Notes by Lucian Guillou; Birkhauser Boston: Boston, MA, USA, 1986; Volume 62, First Published 1973.
75. Cheng, M.; Duncan, J.; Harvey, J. Umbral Moonshine and the Niemeier Lattices. *Res. Math. Sci.* **2014**, *1*, 3. [CrossRef]
76. Asselmeyer-Maluga, T.; Król, J. Higgs Potential and Confinement in Yang-Mills Theory on Exotic $\mathbb{R}^4$. *arXiv* **2013**, arXiv:1303.1632.

© 2019 by the author. Licensee MDPI, Basel, Switzerland. This article is an open access article distributed under the terms and conditions of the Creative Commons Attribution (CC BY) license (http://creativecommons.org/licenses/by/4.0/).

Article

An Investigation on the Prime and Twin Prime Number Functions by Periodical Binary Sequences and Symmetrical Runs in a Modified Sieve Procedure

Bruno Aiazzi *, Stefano Baronti, Leonardo Santurri and Massimo Selva

Institute of Applied Physics "Nello Carrara", IFAC-CNR, Research Area of Florence, 50019 Sesto Fiorentino, Italy; s.baronti@ifac.cnr.it (S.B.); l.santurri@ifac.cnr.it (L.S.); m.selva@ifac.cnr.it (M.S.)
* Correspondence: b.aiazzi@ifac.cnr.it; Tel.: +39-055-5226451

Received: 18 April 2019; Accepted: 4 June 2019; Published: 10 June 2019

Abstract: In this work, the Sieve of Eratosthenes procedure (in the following named Sieve procedure) is approached by a novel point of view, which is able to give a justification of the Prime Number Theorem (P.N.T.). Moreover, an extension of this procedure to the case of twin primes is formulated. The proposed investigation, which is named Limited INtervals into PEriodical Sequences (LINPES) relies on a set of binary periodical sequences that are evaluated in limited intervals of the prime characteristic function. These sequences are built by considering the ensemble of deleted (that is, 0) and undeleted (that is, 1) integers in a modified version of the Sieve procedure, in such a way a symmetric succession of runs of zeroes is found in correspondence of the gaps between the undeleted integers in each period. Such a formulation is able to estimate the prime number function in an equivalent way to the logarithmic integral function Li(x). The present analysis is then extended to the twin primes, by taking into account only the runs whose size is two. In this case, the proposed procedure gives an estimation of the twin prime function that is equivalent to the one of the logarithmic integral function $\text{Li}_2(x)$. As a consequence, a possibility is investigated in order to count the twin primes in the same intervals found for the primes. Being that the bounds of these intervals are given by squares of primes, if such an inference were actually proved, then the twin primes could be estimated up to infinity, by strengthening the conjecture of their never-ending.

Keywords: prime numbers; Prime Number Theorem (P.N.T.); modified Sieve procedure; binary periodical sequences; prime number function; prime characteristic function; limited intervals; logarithmic integral estimations; twin prime numbers

---

1. Introduction

The Sieve procedure is able to achieve heuristic justifications of the Prime Number Theorem (P.N.T.) [1]. Such a theorem gives the asymptotic trend of the prime number function $\pi(x)$, where $\pi(x)$ denotes the quantity of prime numbers p less or equal to $x \in \mathbb{R}$, that is,

$$\pi(x) = \text{ number of primes } p, \ p \leq x. \tag{1}$$

Let log(x) be the natural logarithm of x. If the real functions $A(x)$ and $B(x)$ are asympthotically equal, that is, $\lim_{x \to \infty} A(x)/B(x) = 1$, then we say that $A(x)$ and $B(x)$ are equivalent as $x \to \infty$, and we write $A(x) \sim B(x)$. Consequently, the P.N.T. can be written as

$$\pi(x) \sim x/\log(x). \tag{2}$$

After the infinitude of primes was recognized since ancient times, the estimation (2) was conjectured by Gauss [2] and Legendre [3] at the end of the 18*th* century. Gauss himself improved Equation (2), by considering the logarithmic integral function Li(x), which is defined as

$$\text{Li}(x) = \int_2^x \frac{dt}{\log t}. \tag{3}$$

Again, the function (3) is such that

$$\pi(x) \sim \text{Li}(x) \tag{4}$$

but the approximation (4) is much more precise than (2). In fact, it can be demonstrated that the piece $x/\log(x)$ is only the first term of the series expansion of (3). The aim of this work is to introduce a novel heuristic procedure (LINPES, Limited INtervals into PEriodical Sequences) that is equivalent to the Li(x) approximation, in the sense of Equation (4), apart from a simple multiplicative constant, by exploiting some binary periodic sequences, and related symmetrical runs. Pieces of these sequences compose limited intervals of the prime characteristic function $\xi_p(n)$, which is defined as

$$\xi_p(n) = \begin{cases} 1 & \text{if } n \text{ is prime} \\ 0 & \text{otherwise.} \end{cases} \tag{5}$$

As a matter of fact, a topic that is very much discussed nowadays in the literature just concerns the possible discovering of some regularities and periodicities in the distribution of the primes in certain intervals of the integer sequence [4]. In this work, the implications of the LINPES procedure are also investigated, in particular with an extension to the twin primes, whose distribution is given by a function known as twin prime function $\pi_2(x)$, which is similar to (1), that is,

$$\pi_2(x) = \text{number of pairs of twin primes } (p, p+2), \ p \leq x. \tag{6}$$

Unlike the case of primes, the infinitude of twin primes is still unproved. However, analogously to the P.N.T., the density of the twin primes has been conjectured [5], by considering that the probability to be a prime of an integer n is equal to $1/\log(n)$. Consequently, the probability that n and $n+2$ are both prime can be computed, in such a way the strong twin prime conjecture[6] gives an equivalence between the twin prime function $\pi_2(x)$ and the logarithmic integral function $\text{Li}_2(x)$, that is,

$$\pi_2(x) \sim C \, \text{Li}_2(x) \tag{7}$$

where $\text{Li}_2(x)$ is defined as

$$\text{Li}_2(x) = \int_2^x \frac{dt}{(\log t)^2} \tag{8}$$

and $C = 2\Pi_2 \simeq 1.3203$ is a multiplicative constant that takes into account the statistical dependence of the primes n and $n+2$ [5]. The related constant $\Pi_2 \simeq 0.6602$ is named twin prime constant, that is,

$$\Pi_2 = \prod_{p>2,\, p\,\text{prime}} \left(1 - \frac{1}{(p-1)^2}\right). \tag{9}$$

As it will be shown later, the proposed LINPES procedure is able to estimate the twin prime function in an equivalent way as the $\text{Li}_2(x)$ function, apart from a multiplicative constant. However, this is made by admitting that a basic relation, which is true for the primes, is also valid for the twin primes. In this case, the contribution of the present work will be a more probable assertion of the infinitude of twin primes.

Before starting our discussion, we itemize the variables utilized in this paper

- $\pi(x)$: prime number function (1)
- $\text{Li}(x)$: logarithmic integral function (3), which leads to an estimation of $\pi(x)$
- $\pi_2(x)$: twin prime number function (6)
- $\text{Li}_2(x)$: logarithmic integral function (8), which leads to an estimation of $\pi_2(x)$
- $\pi(N)$: prime number function computed in the fixed integer N
- p: generic prime number
- $p(n)$: arithmetic function that gives the succession of primes
- $\xi_p(n)$: arithmetic function that gives the characteristic function of primes (5)
- $R_s(n)$: number of residual integers in the $n-th$ step of the Sieve procedure
- $\pi_R(N)$: estimation of $\pi(N)$ given by the heuristic method of Section 2
- $\xi(k,n)$: approximation of $\xi_p(n)$ after the $k-th$ step of the Sieve procedure
- $\psi(k,n)$: periodic binary sequence obtained in the $k-th$ step of the modified Sieve procedure
- $T(k)$: period of the periodic binary sequence $\psi(k,n)$
- $J(k,n)$: sliding interval whose size is the same of $I(k)$ and whose initial point is given by n
- $S(k)$: size of the interval $I(k)$
- $R(k)$: number of residual runs of zeroes in each period $T(k)$
- $L(m,k)$: size of the $m-th$ run of zeroes in each period $T(k)$
- $I(k) = [p(k)^2, p(k+1)^2)$: interval of $\xi_p(n)$ where a piece of $\psi(k,n)$ is stored
- $D(k,n)$: local density of the residual runs of zeroes by moving a sliding interval $J(k,n)$ in $T(k)$
- $\overline{D}(k)$: average density of the residual runs of zeroes in the period $T(k)$
- $P(k)$: estimated number of primes in the interval $I(k)$ by using the proposed procedure
- $L(k)$: estimated number of primes in $I(k)$ by using the logarithmic integral function $\text{Li}(x)$
- $\pi(k)$: real number of primes in the interval $I(k)$
- $\pi_P(N)$: estimation of $\pi(N)$ by using the proposed procedure
- $\text{Li}(N)$: estimation of $\pi(N)$ by using the logarithmic integral function $\text{Li}(x)$
- $\tilde{\pi}_P(N)$: corrected version of the estimation $\pi_P(N)$
- $R_2(k)$: number of runs sized 2 in each period $T(k)$
- $\overline{D}_2(k)$: average density of the residual runs 2 in the period $T(k)$
- $P_2(k)$: estimated number of twin primes in the interval $I(k)$ by using the proposed procedure
- $\pi_{2P}(N)$: estimation of $\pi_2(N)$ by using the proposed procedure
- $\text{Li}_2(N)$: estimation of $\pi_2(N)$ by using the logarithmic integral function $\text{Li}_2(x)$
- $L_2(k)$: estimated number of twin primes in $I(k)$ by using the logarithmic integral function $\text{Li}_2(x)$
- $\tilde{\pi}_{2P}(N)$: corrected version of the estimation $\pi_{2P}(N)$
- $\pi_2(k)$: real number of twin primes in the interval $I(k)$.

This paper is organized as follows: Section 2 reports a well-known heuristic method, which is able to estimate the prime number function $\pi(x)$ in the sense of (2), apart from a multiplicative constant. Section 3 shows instead how the LINPES procedure is able to obtain an estimation of $\pi(x)$ that is equivalent to the logarithmic-integral function $\text{Li}(x)$. Section 4 extends the proposed procedure to the case of twin primes. Finally, future research and conclusive remarks are provided in Section 5.

2. A Heuristic Estimation of $\pi(x)$ Equivalent to the $x/\log(x)$ Function

In this section, a well-known heuristic method to justify the P.N.T. in a probabilistic way is briefly resumed, by starting from the Sieve procedure, which splits the primes from the composites in a list of integers up to a given number N. The Sieve procedure is the most common way to obtain the primes, and it is also presently a research topic in order to improve its efficiency [7]. Let $p(n)$ be the arithmetic function whose n-th element is the n-th prime, with $n \in \mathbb{N}$ [8,9]. The Sieve procedure can be summarized by the following steps:

- Step 1: List the integers in the interval $I_N = (1, N]$, with $N \in \mathbb{N}$, then put $n = 1$ and start from the lowest prime $p(n) = p(1) = 2$.
- Step 2: Cancel all the multiples of $p(n)$ not yet struck out, by starting from $p(n)^2$ up to N.
- Step 3: Go to the next remaining integer $q > p(n)$ in the list. If $q^2 > N$, the procedure ends, otherwise increase n to $n + 1$.
- Step 4: Put $p(n) = q$ and return to Step 2.

In order to directly compute the characteristic function of primes $\xi_p(n)$, we can memorize the status of each integer in a binary vector ranging from 1 to I_N. In practice, we associate the value 0 to an integer that has been struck out by the procedure, and the value 1 otherwise. Such a vector is initialized by all 1 values, because no integer is deleted when the procedure starts. Then, in each iteration of the Sieve procedure, a 0 value is assigned to the cells that identify the deleted integers (that is, the composite integers). At the end of the procedure, only the cells related to the prime numbers will retain the initial 1 value.

The Sieve procedure is able to obtain heuristic justifications of the relation (2) by considering purely probabilistic considerations [10]. To show this, let be N an integer whose order of magnitude is large enough to allow sufficiently robust statistics. In the first step ($n = 1$), the multiples of $p(1) = 2$ are struck out, starting from $p(1)^2 = 4$, and the number of deleted integers is approximately given by

$$\left\lfloor \frac{N}{2} \right\rfloor - 1 \simeq \frac{N}{2}. \tag{10}$$

Therefore, the quantity of residual integers is about $R_s(1) \simeq N/2$. In the following step ($n = 2$), the multiples of $p(2) = 3$ are struck out. Given the independence of the congruences modulo p, where p is a prime, about $1/3$ of the residual integers will be deleted (for the Chinese Remainder Theorem [9]). The updated number of the residual integers $R_s(2)$ will be given by

$$R_s(2) \simeq \left(1 - \frac{1}{2}\right) \times \left(1 - \frac{1}{3}\right) \times N. \tag{11}$$

In general, about $1 - 1/p(k)$ of the residual integers will be struck out in the $k-th$ step of the Sieve procedure. The procedure ends when the greatest prime number not exceeding $N^{1/2}$ is reached, that is, $p(K)$, where K is such that $p(K)^2$ is the greatest prime square lower than N. At this point, we obtain an estimation $\pi_R(N)$ of the number of residual integers $R_\varepsilon(K)$, and consequently of the quantity of primes $\pi(N)$, that is,

$$\pi_R(N) = \left(1 - \frac{1}{2}\right) \times \left(1 - \frac{1}{3}\right) \times \left(1 - \frac{1}{p(K)}\right) \times N = N \times \prod_{k=1}^{K} \left(1 - \frac{1}{p(k)}\right) = N \times \prod_{k=1}^{K} \frac{p(k) - 1}{p(k)}. \tag{12}$$

Let us apply the Merten's Third Theorem [11] to the reciprocal of the product structure (12), by taking the limit as $N \to \infty$, that is, as $K \to \infty$. We obtain

$$\lim_{K \to \infty} \prod_{k=1}^{K} \frac{p(k)}{p(k) - 1} \times \frac{1}{\log(p(K)^2)} = \frac{1}{2} \times e^{\gamma} \simeq \frac{1}{2} \times 1.7811 \simeq 0.8905 \tag{13}$$

where γ is the *Eulero-Mascheroni constant*. Consequently, we can get the limit of $\pi_R(N)$ as $N \to \infty$, that is, an approximation of the limit of $\pi(N)$, by considering

$$\lim_{N \to \infty} \pi_R(N) = \lim_{N \to \infty} N \times \prod_{k=1}^{K} \frac{p(k) - 1}{p(k)} = \lim_{N \to \infty} N \times \frac{c}{\log N} = \lim_{N \to \infty} \frac{cN}{\log N} \tag{14}$$

that is, $\pi_R(N) \sim \frac{cN}{\log N}$, with $c = 2e^{-\gamma} \simeq 1/0.8905 \simeq 1.1229$, and being $\lim_{N\to\infty} N = \lim_{K\to\infty} p(K)^2$. Noticeably, from the relations (2) and (14), the real quantity of prime numbers in the interval $I_N = [1, N]$, is overestimated, as $N \to \infty$, by a factor c, due to the previous approximations.

As a conclusion, this heuristic procedure gives a justification of the P.N.T. that is equivalent to the relation (2), except for the c constant [10,12]. In Section 3, the proposed LINPES procedure will be described, which gives a justification of the P.N.T. that is instead equivalent to the more precise estimation (4), by means of a procedure that is not purely probabilistic, but that is also featured by analytic considerations, which can be shared with other scientific sectors.

3. The LINPES Estimation of $\pi(x)$ Equivalent to the Li(x) Function

In this section, the novel heuristic LINPES procedure is described, by showing that it can give an estimation of the prime number function $\pi(x)$. To this end, an ensemble of periodic binary sequences will be considered in limited intervals of the prime characteristic function $\xi_p(n)$. Such a topic is of a great interest because the distribution of primes in short intervals has been deeply investigated in literature, up to the present [13,14]. The proposed procedure is also able to provide useful insights into the estimation of the trend of the twin prime number function $\pi_2(x)$. In this analysis, we denote in the following $p(0) = 1$ for convenience, even if the integer 1 is not considered to be a prime.

3.1. Periodic Binary Sequences Inside the Prime Characteristic Function $\xi_p(n)$

The occurrence of pieces of periodic binary sequences inside the prime characteristic function $\xi_p(n)$ is discussed here. To this end, both the Sieve procedure and a modified version of it are investigated step-by-step, where each step is labelled with the progressive index k, with $k = 0$ denoting the beginning of the two procedures. The difference between the modified and the true Sieve procedure is simply that in the Sieve procedure, in each step $k \geq 1$, only the multiplies of the prime $p(k)$ are struck out, but not the prime itself, whereas in the modified Sieve procedure the prime itself is also deleted. As previously stated, the status of each integer (0→deleted, 1→undeleted) is stored in a N-size vector, which is initialized with all 1 values. The outputs of the Sieve procedure and its modified version are denoted as $\xi(k,n)$ and $\psi(k,n)$, respectively, for each step $k > 0$. Consequently, the deletion of an integer from the true or the modified Sieve procedure simply means that a 0 value replaces a 1 value in the two previous sequences. In the case of the Sieve procedure, the sequence $\xi(k,n)$ is an approximation at the step k of the prime characteristic function $\xi_p(n)$.

At the beginning of the procedures ($k = 0$), we have two equal periodic sequences of all 1 values, that is, $\xi(0,n)$ and $\psi(0,n)$, whose period is $T(0) = 1$. In the first step of the modified Sieve procedure ($k = 1$), the multiples of $p(1) = 2$ are struck out, including $p(1)$ itself. Consequently, we obtain a sequence $\psi(1,n)$, which is still periodic, with alternating 1 and 0 symbols. The period of $\psi(1,n)$ is given by the prime value $p(1)$ itself, that is, $T(1) = 2$. In the following, $T(k)$ will denote the period of the sequence $\psi(k,n)$. Conversely, in the Sieve procedure, the prime $p(1)$ is not deleted. In this case, the output sequence $\xi(1,n)$ is not periodic, but includes a piece of the periodic sequence $\psi(1,n)$, by starting from the square $p(1)^2 = 4$. Before such a value, the previous sequence $\xi(0,n)$ is preserved, which coincides with $\psi(0,n)$. It follows that $\xi(1,n)$ is a mixed sequence, being composed by pieces of both $\psi(0,n)$ and $\psi(1,n)$, that is,

$$\xi(1,n) = \begin{cases} \psi(0,n) & \text{if } p(0)^2 \leq n < p(1)^2 \\ \psi(1,n) & \text{if } n \geq p(1)^2. \end{cases} \quad (15)$$

Similarly, in the second step of the modified Sieve procedure ($k = 2$), every multiple of $p(2) = 3$, which is not yet struck out, is deleted, including the prime itself, to give the new sequence $\psi(2,n)$. Therefore, this sequence comes from the deletion of all the multiplies of the primes $p(1)$ and $p(2)$, including the primes themselves. It follows that the sequence $\psi(2,n)$ is periodic, with a period equal to the product of $p(1)$ and $p(2)$, as it will be demonstrated in Theorem 1. If we consider the second step of the Sieve procedure, where the primes $p(1)$ and $p(2)$ have not been deleted, we obtain the sequence $\zeta(2,n)$. This is again a mixed sequence, where a piece of the periodic sequence $\psi(2,n)$ is introduced, by starting from the square $p(2)^2 = 9$, whereas the previous binary values are saved before this square. Consequently, we have

$$\zeta(2,n) = \begin{cases} \psi(0,n) & \text{if } p(0)^2 \leq n < p(1)^2 \\ \psi(1,n) & \text{if } p(1)^2 \leq n < p(2)^2 \\ \psi(2,n) & \text{if } n \geq p(2)^2. \end{cases} \qquad (16)$$

In general, the multiples of the prime $p(k)$, which are not yet struck out in the previous steps, are deleted in the *k-th* step of the modified Sieve procedure, including the prime $p(k)$ itself. Consequently, after performing all the first k steps, we obtain the periodic sequence $\psi(k,n)$, as shown in Theorem 1. In the case of the original Sieve procedure, after the *k-th* step, we obtain the sequence $\zeta(k,n)$, which is an approximation of the prime characteristic function until the prime $p(k)$. Such an approximation differs from the previous one $\zeta(k-1,n)$, only by starting from the square $p(k)^2$. In fact, after this point, a piece of the periodic sequence $\psi(k,n)$ is recognizable. It follows that $\zeta(k,n)$ can be eventually written as a mixed sequence, which is a generalization of Equations (15) and (16), that is,

$$\zeta(k,n) = \begin{cases} \psi(0,n) & \text{if } p(0)^2 \leq n < p(1)^2 \\ \psi(1,n) & \text{if } p(1)^2 \leq n < p(2)^2 \\ \ldots & \\ \psi(k-1,n) & \text{if } p(k-1)^2 \leq n < p(k)^2 \\ \psi(k,n) & \text{if } n \geq p(k)^2. \end{cases} \qquad (17)$$

By evaluating the expression (17), we can recognize that subsets of the periodic binary sequences $\psi(k,n)$ are present, for each k, in the related intervals $I(k) = [p(k)^2, p(k+1)^2)$ of the prime characteristic function. This happens until the end of the Sieve procedure, because each $k-th$ interval is not influenced by the deletions done in the following steps. We now show that the sequences $\psi(k,n)$ are periodic and that their periods are given by the product of all the primes up to $p(k)$.

Theorem 1. *Let be given the binary sequences $\psi(k,n)$, which are generated by the deletion of the multiplies of all the primes up to $p(k)$, including the primes themselves. Then, the sequences $\psi(k,n)$ are periodic, and their periods $T(k)$ are given by the product of all the primes up to $p(k)$, that is,*

$$T(k) = \prod_{i=1}^{k} p(i) \qquad (18)$$

Proof. The deletion of the multiplies of all the primes up to $p(k)$ gives all the sets, as a function of k, of reduced residue systems modulo $T(k)$, where $T(k)$ is given by Equation (18). Each set is composed by all the positive integers relatively prime to $T(k)$, that is, by all the numbers such that $\gcd(n, T(k)) = 1$. The quantity of integers in each set is given by the Euler phi function $\phi(T(k))$, which computes the number of positive integers less than $T(k)$ and relatively prime to $T(k)$. However, the sets of reduced residue systems are abelian groups, so that each of them is associated to a principal Dirichlet character function. This is an arithmetical function $\chi_1(k,n)$, which is nothing but $\psi(k,n)$, being defined as

$$\chi_1(k,n) = \begin{cases} 1 & \text{if } \gcd(n,T(k)) = 1 \\ 0 & \text{if } \gcd(n,T(k)) > 1. \end{cases} \qquad (19)$$

In [8], it is proven that $\chi_1(k,n)$ is a periodic sequence, and in particular that

$$\chi_1(k, n + T(k)) = \chi_1(k, n) \quad \forall n \tag{20}$$

This completes the proof. □

Table 1 reports the periods $T(k)$ of the sequences $\psi(k,n)$, $k = 0, \ldots, 7$, in comparison with the sizes $S(k) = p(k+1)^2 - p(k)^2$ of the intervals $I(k)$, where subsets of each $\psi(k,n)$ are recognizable. The pseudo-prime $p(0) = 1$ is put in brackets.

Table 1. Periods $T(k)$ of the sequences $\psi(k,n)$, for primes $p(k) \leq p(7)$, in comparison with the sizes $S(k)$ of the intervals $I(k)$. The ratios $S(k)/T(k)$ are rapidly decreasing as the prime $p(k)$ grows.

k	p(k)	p(k+1)	I(k)	S(k)	T(k)	S(k)/T(k)
(0)	(1)	2	[1, 4)	3	1	3.000000
1	2	3	[4, 9)	5	2	2.500000
2	3	5	[9, 25)	16	6	2.666667
3	5	7	[25, 49)	24	30	0.800000
4	7	11	[49, 121)	72	210	0.342857
5	11	13	[121, 169)	48	2 310	0.020779
6	13	17	[169, 289)	120	30 030	0.003996
7	17	19	[289, 361)	72	510 510	0.000141

By considering the ratios $S(k)/T(k)$, it is evident that the periods $T(k)$ increase much faster than the width of the intervals $S(k)$. This makes sense because the periodicity of the sequences $\psi(k,n)$ is hardly recognizable by simply investigating the subsets of each $\psi(k,n)$ in the intervals $I(k)$.

3.2. The Symmetric Sequences of the Runs of Zeroes in the Periods $T(k)$

In Section 3.1, the prime distribution has been represented as the intersection of an endless number of periodic binary sequences $\psi(k,n)$, whose periods $T(k)$ rapidly grow, and such that subsets of these sequences can be found in limited intervals $I(k)$ of the prime characteristic function $\xi_p(n)$. In particular, each of these intervals ranges between the squares of a prime $p(k)$ and of the successive $p(k+1)$. Consequently, the real primes in each interval $I(k)$ are given by the 1 values of the correspondent sequence $\psi(k,n)$. In order to complete this analysis, we now consider the gaps between these primes, by following an established trend in literature. In particular, we are interested to investigate the distributions of the runs of zeros $R(k)$ in each period $T(k)$, being the binary sequences $\psi(k,n)$ composed by isolated ones followed by strings, more or less large, of zeroes. It follows that the quantity $R(k)$ also gives the number of undeleted integers (i.e., isolated ones) in each period $T(k)$, because the quantity $T(k)$, for $k \geq 1$, is an even number, so that the last digit of each period is a zero.

Let us consider the Sieve procedure described step-by-step in Section 3.1 and the number of runs of zeroes $R(k)$ in each period $T(k)$ of the binary sequences $\psi(k,n)$. For $k = 0, 1$, we have only one run ($R(0) = R(1) = 1$), whose sizes are $L(1,0) = 1$ and $L(1,1) = 2$, respectively. For $k = 2$, the deletion of both the multiples of $p(1)$ and $p(2)$ give two runs ($R(2) = 2$) in the period $T(2) = 6$, whose sizes are $L(1,2) = 4$ and $L(2,2) = 2$, respectively, and so on. Table 2 reports the number of runs $R(k)$ and their sizes $L(m,k)$, for $k \leq 4$, where the index m identifies the specific run and k gives the step of the Sieve procedure. Noticeably, the runs of each period $T(k)$ are symmetrical around a symmetry center given by a run sized 4, except for a final run that is sized 2. Such a trend is expected to be a rule also for the successive steps.

Table 2. Runs of zeroes in the periods $T(k)$ of the sequences $\psi(k,n)$, for primes $p(k) \leq p(4)$. For each k, the number of runs $R(k)$ and their sizes $L(m,k)$ are reported, with $m = 1, \ldots, R(k)$. Let us notice the symmetry of the runs in each period $T(k)$. By starting from $k = 2$, the symmetry center is given by a run of length 4, whereas the final run of length 2 is out of symmetry.

k	$p(k)$	$T(k)$	$R(k)$	$L(m,k)$
(0)	(1)	1	1	1
1	2	2	1	2
2	3	6	2	4 2
3	5	30	8	6 4 2 4 2 4 6 2
4	7	210	48	10 2 4 2 4 6 2 6 4 2 4 6 6 2 6 4 2 6 4 6 8 4 2 4 2 4 8 6 4 6 2 4 6 2 6 6 4 2 4 6 2 6 4 2 4 2 10 2

3.3. The Relation Between the Primes in an Interval $I(k)$ and the Runs in a Period $T(k)$

For evidencing the relation between each period $T(k)$ and the correspondent number of runs of zeroes $R(k)$, we report in Table 3 the scores of $R(k)$ for $k \leq 7$.

Table 3. Periods $T(k)$ and related runs of zeroes $R(k)$ for the primes $p(k) \leq p(7)$. The special prime $p[0] = 1$ is put in round brackets.

k	$p(k)$	$T(k)$	$R(k)$
(0)	(1)	1	1
1	2	2	1
2	3	6	2
3	5	30	8
4	7	210	48
5	11	2310	480
6	13	30,030	5760
7	17	510,510	92,160

Such scores also give the number of the integers that have not been struck out by the modified Sieve procedure in the period $T(k)$, which in turn can be related to the number of undeleted integers (and consequently of the primes) in the correspondent interval $I(k)$. We will show in Theorem 2 that a correlation exists between $T(k)$ and $R(k)$, in such a way the number of primes in each interval $I(k)$ can be inferred. According on the theory of congruences, Theorem 2 gives the quantity of the integers that have not been struck out (i.e., $R(k)$) in each period $T(k)$, that is,

Theorem 2. *Let be given the periodic binary sequences $\psi(k,n)$ defined in Theorem 1, and whose periods are $T(k) = \prod_{i=1}^{k} p(i)$. Then, the number of undeleted integers, that is, the number of runs of zeroes $R(k)$, in a period $T(k)$, for $k \geq 1$, is given by*

$$R(k) = \prod_{i=1}^{k} (p(i) - 1), \qquad k \geq 1 \tag{21}$$

Proof. The number of undeleted integers in each period $T(k)$ is given by the number of integers in the reduced residue systems modulo $T(k)$, that is, the number of positive integers less than $T(k)$ and relatively prime to $T(k)$. Such a value is given by the Euler phi function $\phi(T(k))$, once computed in $T(k)$, that is [8]

$$\phi(T(k)) = T(k) \cdot \prod_{p|T(k)} \left(1 - \frac{1}{p}\right) = T(k) \cdot \prod_{p|T(k)} \left(\frac{p-1}{p}\right) = T(k) \cdot \frac{\prod_{i=1}^{k}(p(i)-1)}{\prod_{i=1}^{k} p(i)} = \prod_{i=1}^{k}(p(i)-1) \quad (22)$$

where $p(i)$, $i = 1, \ldots, k$, are the primes dividing $T(k)$. □

By starting from $p(4) = 7$, Table 1 shows that the interval $I(k)$ is included in the first period of the sequence $\psi(k, n)$. Consequently, a subset of the undeleted integers $R(k)$ in each period $T(k)$ lies in the correspondent interval $I(k)$, where they are just primes. Therefore, we can infer the quantity of primes $P(k)$ in each $I(k)$, by starting from the quantity $R(k)$ in the correspondent period $T(k)$. As a first approximation, a simple proportional relationship is investigated. Let us consider the local density $D(k, n)$ of the undeleted integers in the period $T(k)$, where $D(k, n)$ is computed in sliding intervals $J(k, n)$ whose size is the same of $I(k) = [p(k)^2, p(k+1)^2)$, that is, $p(k+1)^2 - p(k)^2$. In this context, the index n represents the starting point of each $J(k, n)$. If such intervals span the whole period $T(k)$, we assume that the density $D(k, n)$ is not a function of n. In this case, it is equal to the average density $\overline{D}(k)$ over $T(k)$, and we have

$$D(k,n) = \overline{D}(k) = \frac{R(k)}{T(k)} = \frac{\prod_{i=1}^{k}(p(i)-1)}{\prod_{i=1}^{k} p(i)} = \prod_{i=1}^{k} \frac{p(i)-1}{p(i)}, \quad k \geq 1 \quad (23)$$

It is noteworthy that the product structure in Equation (23) is the same as in Equation (12). Let us suppose that the previous assumption holds. Then, an estimation of the local density $D(k, n)$ in each interval $I(k)$ (that is, for $n = p(k)^2$), will be just the average density $\overline{D}(k)$ over the period $T(k)$. Consequently, we can write

$$D\left(k, p(k)^2\right) \simeq \overline{D}(k), \quad k \geq 1. \quad (24)$$

Therefore, by starting from Equation (23), we can estimate the quantity of primes $P(k)$ in each interval $I(k)$, for $k \geq 1$. To this end, the average density $\overline{D}(k)$ is multiplied by the size $S(k) = p(k+1)^2 - p(k)^2$, that is,

$$P(k) = \overline{D}(k) \cdot S(k) = (p(k+1)^2 - p(k)^2) \cdot \prod_{i=1}^{k} \frac{p(i)-1}{p(i)}, \quad k \geq 1. \quad (25)$$

Evidently, Equation (25) is analogous to Equation (12), apart from the size N of the global interval I_N, where $N \in I_K = [p(K)^2, p(K+1)^2)$, that is changed into the size $p(k+1)^2 - p(k)^2$ of the local interval $I(k)$.

3.4. The Novel LINPES Estimation of the Prime Number Function $\pi(x)$

Equation (25) gives a succession of estimations $P(k)$ of the real number of primes $\pi(k)$ in each interval $I(k) = [p(k)^2, p(k+1)^2)$. Therefore, the next step will be to blend all these scores to compute a global estimation $\pi_P(N)$ of the quantity of the primes up to N, where $N \in I(K)$, analogously to Equation (12). In theory, $\pi_P(N)$ is simply computable by adding all the contributions $P(k)$ of Equation (25), for $k = 1, \ldots, K$, where $p(K)$ is the greatest prime number not exceeding $N^{1/2}$. However, such a procedure includes the term $p(K+1)$, which is unknown. In order to overcome this issue, the computation of $\pi_P(N)$ has to involve only the terms up to $P(K-1)$, plus a final term $P(K, N)$, where the interval I_K is only partially considered. Consequently, we obtain

$$\pi_P(N) = \sum_{k=0}^{K-1} P(k) + P(K,N) = P(0) + \sum_{k=1}^{K-1}\left[\left(p(k+1)^2 - p(k)^2\right) \cdot \prod_{i=1}^{k} \frac{p(i)-1}{p(i)}\right] + P(K,N) \quad (26)$$

where $P(0) = p(1)^2 - p(0)^2$, and $P(K,N) = (N - p(K)^2) \cdot \prod_{i=1}^{K} \frac{p(i)-1}{p(i)}$. Let us notice that Equation (26) includes as many contributions as the primes are, where each term is given by a relation similar to Equation (12), with the global size N that is replaced by the size of the interval $I(k)$. Each contribution includes an average number of primes that is given by $\prod_{i=1}^{k} \frac{p(i)-1}{p(i)}$, so that the average distance $p(k+1) - p(k)$ between two consecutive primes is $\prod_{i=1}^{k} \frac{p(i)}{p(i)-1}$, which is of the order of magnitude of $\log(p(k))$. For the Cramér conjecture [15], this average distance is $p(k+1) - p(k) = \mathcal{O}(\log^2(p(k)))$. Another conjecture by Cramér, by starting from the Riemann's hypothesis, was $p(k+1) - p(k) = \mathcal{O}(\sqrt{p(k)}\log(p(k)))$ [12,16]. Consequently, the error given by neglecting the partial term $P(K,N)$ is smaller than the loading term of the Cramér conjectures, so that the partial term $P(K,N)$ could be omitted.

3.5. The Corrected LINPES Estimation by Using the Equivalence with the Li(x) Function

We want now to show that Equations (3) and (26) are related. To this end, we write the logarithmic integral function $\mathrm{Li}(N)$ as a summation of integrals, each of them is computed in the interval $I(k) = [p(k)^2, p(k+1)^2]$, that is,

$$\mathrm{Li}(N) = \int_{2}^{p(1)^2} \frac{dt}{\log t} + \sum_{k=1}^{K-1} \int_{p(k)^2}^{p(k+1)^2} \frac{dt}{\log t} + \int_{p(K)^2}^{N} \frac{dt}{\log t}, \qquad (27)$$

where the first term starts from 2 to cope with a possible improper integral, and $p(K)^2$ is the greatest square of a prime less than N. Consequently, the $\mathrm{Li}(N)$ function is expressed by Equation (27) as a succession of estimations $L(k)$, in a similar way to Equation (26), that is,

$$\mathrm{Li}(N) = L(0) + \sum_{k=1}^{K-1} L(k) + L(K,N), \qquad (28)$$

where $L(0) = \int_{2}^{p(1)^2} \frac{dt}{\log t}$, $L(K,N) = \int_{p(K)^2}^{N} \frac{dt}{\log t}$, and

$$L(k) = \int_{p(k)^2}^{p(k+1)^2} \frac{dt}{\log t}. \qquad (29)$$

We now apply the Mean Value Theorem to each interval $I(k)$ in Equation (27), that is,

$$\mathrm{Li}(N) = \frac{p(1)^2 - 2}{\log(\varsigma_0)} + \sum_{k=1}^{K-1} \frac{p(k+1)^2 - p(k)^2}{\log(\varsigma(k))} + \frac{N - p(K)^2}{\log(\varsigma_K)}, \qquad (30)$$

where $\varsigma_0 \in I(0)$, $I(0) = [p(0)^2, p(1)^2]$, $\varsigma(k) \in I(k)$, $k = 1, \ldots, K-1$, and $\varsigma_K \in I(K,N)$, $I(K,N) = [p(K)^2, N]$. In order to show the equivalence between the Equations (26) and (30), we also consider the lower bound $p(k)^2$ of the interval $I(k)$. By taking, in the two summations, the ratio between the two terms multiplying the interval size $S(k) = p(k+1)^2 - p(k)^2$, we can write

$$\frac{\prod_{i=1}^{k} \frac{p(i)-1}{p(i)}}{\frac{1}{\log(\varsigma(k))}} = \left(\frac{\prod_{i=1}^{k} \frac{p(i)}{p(i)-1}}{\log(\varsigma(k))} \right)^{-1} \qquad (31)$$

From Equation (13), we have

$$\lim_{k \to \infty} \frac{\prod_{i=1}^{k} \frac{p(i)}{p(i)-1}}{\log(\varsigma(k))} = \lim_{k \to \infty} \left[\frac{\prod_{i=1}^{k} \frac{p(i)}{p(i)-1}}{\log(p(k)^2)} \times \frac{\log(p(k)^2)}{\log(\varsigma(k))} \right] = \frac{1}{2} \times e^{\gamma} \times \lim_{k \to \infty} \frac{\log(p(k)^2)}{\log(\varsigma(k))} \qquad (32)$$

where $\varsigma(k) \in I(k) = [p(k)^2, p(k+1)^2)$, so that its maximum distance from $p(k)^2$ is $p(k+1)^2 - p(k)^2$. However, we know that the $k-th$ prime $p(k)$ is given asymptotically by $p(k) \sim k \log(k)$ [9]. Therefore, $p(k)^2 \sim k^2 \cdot \log(k)^2$ and $p(k+1)^2 \sim (k+1)^2 \cdot \log(k+1)^2 \sim k^2 \log(k)^2$, so that for each point $\varsigma(k) \in [p(k)^2, p(k+1)^2)$ we have $\varsigma(k) \sim k^2 \log(k)^2$. It follows that

$$\lim_{k \to \infty} \frac{\prod_{i=1}^{k} \frac{p(i)}{p(i)-1}}{\log(\varsigma(k))} = \frac{1}{2} \times e^{\gamma} \times \lim_{k \to \infty} \frac{\log(p(k)^2)}{\log(\varsigma(k))} = \frac{1}{2} \times e^{\gamma} = \frac{1}{c} \simeq 0.8905 \qquad (33)$$

and consequently Equation (31) gives, for each fixed k,

$$\frac{\prod_{i=1}^{k} \frac{p(i)-1}{p(i)}}{\frac{1}{\log(\varsigma(k))}} = c_I(k) \quad \text{where } \lim_{k \to \infty} c_I(k) = c = 2 \times e^{-\gamma} \simeq 1.1229. \qquad (34)$$

It follows that the trends of the two estimations (26) and (30) are the same as $k \to \infty$, apart from the constant coefficient c. Due to this multiplicative factor, the proposed estimation (26) overestimates the prime number function $\pi(N)$ with respect to Equation (30), and in this sense it is similar to the heuristic procedure described in Section 2. However, it has to be noticed that this last one is completely probabilistic, whereas the proposed method is also based on an analytical procedure, that is, the recognition of an infinite number of binary periodical sequences and related intervals of the prime characteristic function. In order to correct this discrepancy, we relax the conjecture of Section 3.3, in such a way the trend of the local density $D(k, n)$ becomes a function of n. Experimentally, the values of the local density $D(k, p_k^2)$ in the interval $I(k)$ are lower than those of the average density $\overline{D}(k)$. The following conjecture is then proposed, which links $D(k, p_k^2)$ and $\overline{D}(k)$ by means of the constant c of the Third Mertens' Theorem [11].

Conjecture 1. *The local density $D(k, n)$ of the undeleted integers in the period $T(k)$, if computed in sliding intervals whose size is the same of $I(k) = [p(k)^2, p(k+1)^2)$, is a function of the starting point n of the sliding interval. In particular, the average density $\overline{D}(k)$ is greater than the local density $D(k, p(k)^2)$ in the interval $I(k)$, in such a way the succession $c_I(k)$ of their ratios exceeds the unity. Moreover, the limit value as $k \to \infty$ of $c_I(k)$ is equal to the constant $c = 2 \cdot e^{-\gamma} \simeq 1.1229$ of the Third Mertens' Theorem, that is,*

$$\lim_{k \to \infty} \frac{\overline{D(k)}}{D(k, p_k^2)} = c. \qquad (35)$$

The typical trend of $D(k, n) = D(16, n) = D(n)$, for $k = 16$ and varying n, is plotted in Figure 1, together with the average density $\overline{D}(k) = \overline{D}(16) = \overline{D}$ in the period $T(k) = T(16)$. Let us notice that, as it will be discussed in the following, such a trend is less appreciable for small values of the primes.

Figure 1 can be explained as follows. Let us consider the sequences $\psi(k, n)$ defined in Section 3.1, where the multiples of the primes up to $p(k)$ have been struck out, included the primes themselves. In each of these sequences, all the undeleted integers are just primes in the range $[p(k+1), p(k+1)^2]$, whereas the undeleted integers greater than $p(k+1)^2$ can be indifferently primes or composites, because the multiples of the primes greater than $p(k)$ have not yet been struck out.

At the beginning of the modified Sieve procedure ($k = 0$), the local density $D(k, n)$ of the undeleted integers is not a function of n, because no integer has been still struck out. In the first step ($k = 1$), only the even integers (i.e., the multiplies of $p(1) = 2$) have been struck out, so that $D(k, n)$ is still a constant value up to infinity. Noticeably, the multipliers (i.e. the integers multiplying $p(1)$ to give the deleted multiplies) are equal to the undeleted integers when the procedure starts (i.e., all the integers). This rule also holds for the following steps, that is, the multipliers of the prime $p(k)$ in the $k-th$ step of the modified Sieve procedure are equal to the undeleted integers in the previous $(k-1)-th$ step. It follows that the multipliers of $p(2) = 3$ are all the odd integers, whose distribution is again uniform. Some of these multipliers (that is, 3, 5, 7) are just primes in the interval

$[p(2), p(2)^2)$, but they can also be composites beyond $p(2)^2$. In this case, the distribution of the composite multipliers exactly compensate the decreasing trend of the distribution of the multipliers that are also prime numbers. If the primes $p(k)$ are sufficiently small, such a compensation happens quickly, because it starts from $p(k)^2$. In these cases, the distribution of the local density $D(k,n)$ is still approximately uniform. However, as $p(k)$ grows, a transient state is noticeable, because, for such values of k and small values of n, the local density $D(k,n)$ is greater than the average density $\overline{D}(k)$. In fact, for such n values, only a portion of the multiplies of the primes $p(i), i = 1, \ldots, k$, have been struck out, because the deletion of the multiplies of the prime $p(i)$, $i < k$, starts only from $p(i)^2$, apart from the prime $p(i)$ itself. This means that the deletion of the multiplies of $p(i), i = 1, \ldots, k$, is completed only at the lower bound of the interval $I(k)$, that is, $p(k)^2$. Consequently, after this point, the transient state ends and the stationary state begins, where the local density $D(k,n)$ fluctuates around the average density $\overline{D}(k)$.

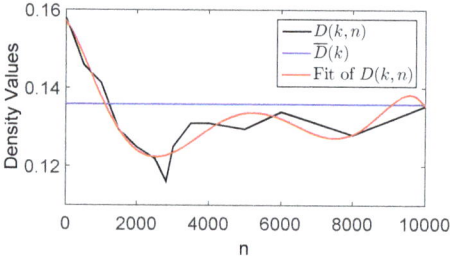

Figure 1. Typical trend (in black), with $k = 16$, $p(16) = 53$ and $p(17) = 59$, of the local density of the non-deleted integers $D(n)$ by varying n in sliding intervals whose size is $S(16) = 3481 - 2809 = 672$. Notice that it is shown only the initial part of the period $T(16)$, whose order of magnitude is 10^{19}, in such a way the symmetrical trend of the period falls outside the figure. The red line reports a polynomial fitting of the density $D(k,n)$, whereas the blue line concerns the average density $\overline{D}(k)$ in the period $T(k)$. The minimum value of the local density is just reached at the lower bound of the interval $I(k)$, that is, $p(16)^2 = 2809$.

Figure 1 shows the trend of the local density $D(k,n)$ in the case of $p(k) = 16$. Starting approximately from this value of k, we can notice a minimum value $D(k, p(k)^2)$ for the distribution of $D(k,n)$, which is located immediately after the transient state, that is, at the lower bound of the interval $I(k)$. Such a minimum value is about a 10 percent lower than the average density $\overline{D}(k)$. In fact, as previously explained, the multipliers of the prime $p(k)$ are just primes up to $p(k)^2$, whereupon they can be even composites. It follows that the distribution of the composite multipliers compensate the decreasing distribution of the multipliers that are prime numbers only starting from the multiple $p(k)^3 = p(k)^2 \cdot p(k)$. Therefore, as $k \to \infty$, such a compensation is delaying, in such a way the ratio between $\overline{D}(k)$ and $D(k,n)$ more and more grows up to the c value of Equation (35). As a matter of fact, if all the multipliers were primes, their distribution would decrease by following a logarithmic trend, so that $D(k,n)$ would augment with the same trend, by starting from the minimum value in the interval $I(k)$. In the real case, however, the compensation given by the composite multipliers has the effect that the local density does not grow indefinitely, but tends to the limit value $c \cdot D(k, p(k)^2)$. Let us notice that, if we stop the procedure to a finite value of k, the ratio between $\overline{D}(k)$ and $D(k,n)$ is $c_I(k) \cdot D(k, p(k)^2)$, where the succession $c_I(k)$ is increasing and tends to the limit value c as $k \to \infty$.

In order to evaluate the effect of the compensation delay for the small primes $p(k), k = 1, \ldots, 7$, in comparison with the case of $p(16) = 53$, Table 4 reports: a) the multipliers f_I such that the multiples $f_I \cdot p(k)$ lie in the interval $I(k) = [p(k)^2, p(k+1)^2)$, and b) the first multiplier that is a composite number, that is, $f_c = p(k)^2$, whose correspondent multiple is $p(k)^2 \cdot p(k) = p(k)^3$. Evidently, as k

grows, the difference between the upper bound $p(k+1)^2$ of $I(k)$ and $p(k)^3$ becomes so large that the compensation effect of the composite multipliers is no longer noticeable in the interval itself.

Table 4. Prime numbers $p(k)$, $k = 1, \ldots, 7$, and $k = 16$, and the related intervals $I(k)$, together with: a) the multipliers f_I such that the multiples $f_I \cdot p(k)$ lie inside the intervals $I(k)$; b) the first multiplier f_c that is a composite number. Let us notice that the difference between f_c and the multipliers f_I rapidly grows, so that the distance between the multiple $f_c \cdot p(k)$ and the upper bound of the interval $I(k)$ becomes larger and larger.

k	$p(k)$	$I(k)$	$f_I \mid (f_I \cdot p(k)) \in I(k)$	f_c	$f_c \cdot p(k)$
1	2	[4, 9)	2; 3	4	8
2	3	[9, 25)	3; 5; 7	9	27
3	5	[25, 49)	5; 7	25	125
4	7	[49, 121)	7; 11; 13; 17	49	343
5	11	[121, 169)	11; 13	121	1331
6	13	[169, 289)	13; 17; 19	169	2197
7	17	[289, 361)	17; 19	289	4913
16	53	[2809, 3481)	53; 59; 61	2809	148,877

Figure 2 shows the trend of the succession $c_I(k)$, as k approaches infinity. Evidently, such a succession tends to the constant value c. The x-axis is in a logarithmic scale, in such a way the values of $p(k)^2$ can be visualized up to 10^{15}.

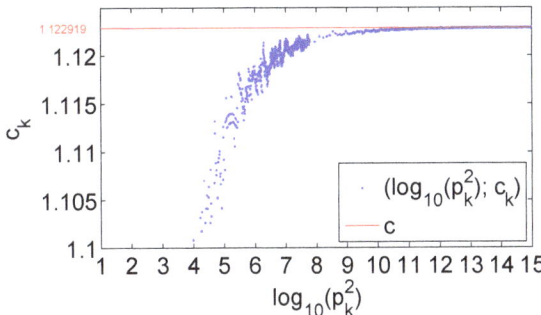

Figure 2. Trend of the succession $c_I(k)$ whose elements are the ratios between the average densities $\overline{D}(k)$ in the period $T(k)$ and the local densities $D(k, n)$ in the correspondent interval $I(k)$. For $k \to \infty$, such a succession asymptotically approximates the constant c. In the x-axis, a base-10 logarithmic scale has been chosen for a better visualization.

Finally, Table 5 highlights the equivalence between the proposed estimation (26) and the logarithmic-integral one (3). To this end, a number of linear regressions have been computed between the occurrences $P(k)$ (25) in each interval $I(k)$ of the proposed estimation versus the correspondent ones $L(k)$ (29) of the integral-logarithmic function. Each row of Table 5 is referred to the prime squares $p(k)^2$ ranging from a power-of-ten to the following one, except the first raw, which includes all the squares lower than 10^6, in order to elaborate a sufficient number of points. For each of these ranges, we report the coefficients m_1 and q_1 of the linear regressions $y_i = m_1 x_i + q_1$, together with the coefficient of determination R_1^2, which is a measure of the fitting between the two estimations. Evidently, the coefficient of determination tends very fast to its optimal value, that is 1, despite that the number of observations has increased. Let us notice that the intercept q_1 is practically negligible with respect to the full-scale level, whereas the slope m_1 is approaching the constant value $1/c$.

For comparison, Table 5 also reports the parameters and the coefficient of determination in the case of the linear regressions $y_i = m_2 x_i + q_2$ concerning the occurrences $P(k)$ versus the targets $\pi(k)$. These scores are defined as the number of primes in each interval $I(k)$. Even in this case, the fitting between $P(k)$ and $\pi(k)$ is impressive, as shown by the coefficient of determination R_2^2. Noticeably, the slope m_2 still approaches the value $1/c$, because the P.N.T. guarantees that the logarithmic-integral function and the prime number function goes to infinity in the same way.

Table 5. Parameters and coefficients of determination of the linear regressions $y_i = m_1 x_i + q_1$ of the proposed estimations $P(k)$ versus the logarithmic-integral ones $L(k)$, together with the parameters and coefficients of determination of the linear regressions $y_i = m_2 x_i + q_2$ of $P(k)$ versus the true number of primes $\pi(k)$. Each point is computed in an interval $I(k)$.

k	$p(k)^2$	m_1	q_1	R_1^2	m_2	q_2	R_2^2
[1, 168]	$(1, 10^6)$	0.894209	1.2846	0.9999932989	0.894649	0.2597	0.9996747582
[169, 446]	$(10^6, 10^7)$	0.892762	0.7754	0.9999985384	0.894052	−2.5697	0.9998452835
[447, 1229]	$(10^7, 10^8)$	0.891565	1.0381	0.9999997462	0.891906	−2.2200	0.9999418064
[1230, 3401]	$(10^8, 10^9)$	0.891025	2.1044	0.9999999196	0.891016	2.0534	0.9999821107
[3402, 9592]	$(10^9, 10^{10})$	0.890801	2.2963	0.9999999842	0.890751	5.6943	0.9999941659
[9593, 27,293]	$(10^{10}, 10^{11})$	0.890657	4.9719	0.9999999945	0.890664	2.8478	0.9999981622
[27,294, 78,498]	$(10^{11}, 10^{12})$	0.890606	5.7853	0.9999999989	0.890606	5.6440	0.9999993974
[78,499, 227,647]	$(10^{12}, 10^{13})$	0.890570	10.3672	0.9999999997	0.890569	13.0142	0.999998112
[227,648, 664,579]	$(10^{13}, 10^{14})$	0.890555	14.8795	0.9999999999	0.890555	13.7581	0.9999999398
[664,580, 1,951,957]	$(10^{14}, 10^{15})$	0.890546	20.1618	1.0000000000	0.890546	27.3660	0.9999999808

From the previous analysis, it follows that, for a given N, the proposed approximation $\pi_P(N)$ overestimates the prime number function $\pi(N)$ by a factor c_N, which can be computed by considering that we have an overestimation for each interval $I(k)$ that can be computed by considering a factor in the finite set $c_I(k)$, $k = 1, \ldots, K$, where K is such that $N \simeq p(K)^2$ (see Equation (34)). If $N \to \infty$, the overestimation factor c_N tends to the constant c. Being c_N unknown, an adjusted version (36) of (26) can be defined by means of the correction factor $1/c$, that is,

$$\tilde{\pi}_P(N) = \frac{1}{c} \cdot \left(P_0 + \sum_{k=1}^{K-1} P(k) + P_{K,N} \right) =$$
$$= \frac{1}{c} \cdot \left(p(1)^2 - p(0)^2 \right) + \frac{1}{c} \cdot \sum_{k=1}^{K-1} \left[\left(p(k+1)^2 - p(k)^2 \right) \cdot \prod_{i=1}^{k} \frac{p(i)-1}{p(i)} \right] + \frac{1}{c} \cdot \left(N - p(K)^2 \right) \cdot \prod_{i=1}^{K} \frac{p(i)-1}{p(i)}. \tag{36}$$

Clearly, the corrected version $\tilde{\pi}_P(N) = \frac{1}{c} \cdot \pi_P(N)$ is able to give better estimations than $\pi_P(N)$ as N approaches infinity. In order to give a quantitative assessment, Table 6 reports the scores of $\pi_P(N)$ (26) and of its adjusted version $\tilde{\pi}_P(N)$ (36), in comparison with the logarithmic integral estimation $\text{Li}(N)$ (27), and with the prime number function $\pi(N)$. The range of each row of Table 6 starts from a power-of-ten and ends to the following one up to 10^{15}.

It can be noticed that the scores of $\tilde{\pi}_P(N)$ slightly underestimate both the true number of primes $\pi(N)$ and the logarithmic integral function $\text{Li}(N)$, which, in turn, is such that the sign of its difference with $\tilde{\pi}_P(N)$ changes infinitely many times [17,18], by showing some irregularities in the distribution of the primes [19], which have been investigated by considering differences in some subsets of the primes themselves [20]. Concerning the previous underestimation, this is due to the fact that the limit value c is an upper bound for the succession $c_I(k)$. Evidently, $\tilde{\pi}_P(N)$ would be perfectly accurate if the terms $c_I(k)$ were available for the computation of (36), by considering the real number of primes in each interval $I(k)$.

Table 6. The proposed estimation $\pi_P(N)$ and its adjusted version $\tilde{\pi}_P(N)$ in comparison with the logarithmic integral estimation $\text{Li}(N)$, and the prime number function $\pi(N)$. The scores of $\text{Li}(N)$ have been computed by using the MATLAB® toolbox. The scores of $\pi_P(N)$ and $\tilde{\pi}_P(N)$ have been rounded to the nearest integer.

$N = 10^i$	$\pi(N)$	$\text{Li}(N)$	$\pi_P(N)$	$\tilde{\pi}_P(N)$
10^1	4	6	4	4
10^2	25	30	27	24
10^3	168	178	181	161
10^4	1229	1246	1348	1 200
10^5	9592	9630	10,639	9 474
10^6	78,498	78,628	87,688	78 090
10^7	664,579	664,918	744,175	662,715
10^8	5,761,455	5,762,209	6,460,497	5,753,306
10^9	50,847,534	50,849,235	57,056,721	50,811,064
10^{10}	455,052,511	455,055,615	510,796,987	454,883,106
10^{11}	4,118,054,813	4,118,066,401	4,623,402,885	4,117,306,712
10^{12}	37,607,912,018	37,607,950,281	42,226,535,908	37,604,250,381
10^{13}	346,065,536,839	346,065,645,810	388,584,655,120	346,048,624,432
10^{14}	3,204,941,750,802	3,204,942,065,692	3,598,796,310,868	3,204,857,671,495
10^{15}	29,844,570,422,669	29,844,571,475,288	33,512,578,849,645	29,844,157,918,447

4. An Extension of the Procedure to the Twin Prime Numbers

4.1. Preliminary Concepts

Two prime numbers p and q are twin primes if $|p - q| = 2$, which is the lowest possible distance between primes, apart from $p = 2$ and $q = 3$, where $|p - q| = 1$. Let us note that two consecutive pairs of twin primes do not ever occur, apart from the case $\{3,5\}$ and $\{5,7\}$. In fact, one number in the sequence $\{n, n+2, n+4\}$ is certainly a multiple of 3. The gaps between consecutive primes have been extensively investigated in literature [13,15,21]. However, differently from the primes, *it is presently unknown whether there are infinitely many pairs of twin primes*. In any case, a preliminary counting shows that the twin primes are relatively abundant into the sequence of primes, and, consequently, it is reasonable to infer the so-called twin prime conjecture, which states that *there are infinitely many pairs of twin primes*. This conjecture is strengthened by the fact that the distribution of the primes does not change abruptly. Recently, significant progress has been made by showing that $\liminf\limits_{k \to \infty} [p(k+1) - p(k)] = \ell < \infty$, that is, a finite upper bound exists for the limit inferior of the difference between consecutive primes. In particular, Zhang found that $\ell \leq 7 \cdot 10^7$ [22], and this bound has been successively improved by Maynard to $\ell \leq 600$ [23]. Finally, the Polymath's project, whose aim is to collect all the various efforts that try to put the bound lower as much as possible, has reached the value of $\ell \leq 246$ [24]. Evidently, in order to demonstrate the twin prime conjecture, a bound of $\ell = 2$ should be obtained. In this work, we try to give a contribution to the discussion of this conjecture, by following a different strategy, that is, by exploiting the concepts previously introduced for the primes. Consequently, as for the primes, the approach is not merely probabilistic, but also analytic, so constituting a possible significant step for further advancements, as in the case of approaches based on periodic functions [25]. The distribution of the twin primes is commonly characterized by using the twin prime function $\pi_2(x)$ (6). Such a distribution decays more rapidly than the distribution of the primes. In fact, Brun demonstrated in 1919 [26] that, if S_T is the set of twin primes given by $S_T = \{p : p \text{ prime and } p+2 \text{ prime}\}$, the related series of the reciprocals converges to the finite limit $B \simeq 1.9022$ [1], that is,

$$\sum_{p \in S_T} \left(\frac{1}{p} + \frac{1}{p+2} \right) = B \tag{37}$$

regardless of the fact of whether the number of summation terms is infinite or not, whereas the same summation instead diverges for the primes.

Analogously to the P.N.T., a possible function for approximating the twin prime function $\pi_2(x)$ has been proposed [5] as the logarithmic integral function $\text{Li}_2(x)$ (8). As for the primes, we want to obtain an equivalent procedure and investigate possible consequences.

4.2. A Possible Relation Between the Twin Primes in the Intervals and the Undeleted Integers in the Periods

In Section 3.2, the distribution of the runs into each period $T(k)$ has been investigated. In the present analysis, the same investigation can be made for the particular case in which the size of the runs is 2. Evidently, such an investigation can potentially give an estimation of the quantity of twin primes, similarly to the one given by the Equation (26) for the primes. In fact, we will suggest that the number of the runs sized 2 in the interval $I(k)$ is equal to the quantity of twin primes in the same interval. Such a number is equal to the number of $\{101\}$ sequences, if the sequence $\{10\}$ is completely included in the interval. However, such a sequence cannot occur across two intervals, because each interval, apart from the first one, ends with an even number (that is, a 0), because it is followed by a square of an odd prime (that is, another 0), which is an odd number. For the sake of clarity, in the following we denote the runs sized 2 as runs 2. Let us notice that this procedure can be extended to run-lengths of whatever size, by following the *Hardy-Littlewood conjecture B* [6]. Such a topic will be the object of future explorations.

Table 7 reports the number $R_2(k)$ of the runs 2 in each period $T(k)$ for $p(k)$, $k = 0, \ldots, 7$. As for the total number of runs $R(k)$ (21) in the same period, a correlation can be found between $R_2(k)$ and the prime number $p(k)$. In particular, the scores of Table 7 suggest the following conjecture for $R_2(k)$

$$R_2(k) = \prod_{i=2}^{k}(p(i) - 2), \quad k \geq 2. \tag{38}$$

Table 7. Number of runs 2, denoted as $R_2(k)$, that are included in the periods $T(k)$, for $p(k)$, $k = 0, \ldots, 7$. These scores are compared with the total number of runs $R(k)$. The special prime $p(0) = 1$ is put in round brackets.

k	p(k)	T(k)	R(k)	$R_2(k)$
(0)	(1)	1	1	0
1	2	2	1	1
2	3	6	2	1
3	5	30	8	3
4	7	210	48	15
5	11	2310	480	135
6	13	30,030	5760	1485
7	17	510,510	92,160	22,275

Equation (38) can be investigated by taking the modified Sieve procedure. At the start of the procedure ($k = 0$), we have no run 2. In the first step ($k = 1$), the multiples of $p(1) = 2$ are struck out, in such a way the sequence $\psi(1, n)$ is made by runs 2 only. In particular, a single run 2 is included in the period $T(1) = 2$, so that $R_2(1) = 1$. For $k = 2$, we delete the multiples of $p(2) = 3$, so that the period $T(2) = 6$ becomes three times greater. This implies that the number of runs 2 could increase from 1 to 3, but the deletion in the point $n = 3$ vanishes two of these runs. Let us notice that the cancellation of one multiple vanishes two runs 2 only in this step, being all the runs 2 consecutive, but this does not happen in the following steps, where only one run 2, or even none, is deleted at the time. It follows that $R_2(2) = 1$, as in the previous step. On the whole, we obtain that the deleted runs 2 in the period $T(2)$ are a fraction $2/3 = 2/p(2)$ of the total number of runs 2 in the same period if no cancellations were made.

Similarly, for $k = 3$, the multiples of $p(3) = 5$ are struck out, so that the period $T(3)$ becomes five times greater. It follows that the number of runs 2 would grow from 1 to 5, but two cancellations (for $n = 5, 25$) vanish two of the five runs 2. Consequently, we obtain $R_2(3) = 3$ and the fraction of the deleted runs 2 is $2/5 = 2/p(3)$ of the total runs in this period if no cancellation were made. In this step, all the cancellations imply the deletion of one run 2, but this will not also be a rule for the following steps. In fact, for $k = 4$, we have eight cancellations in the period $T(4)$, but only six of them stroke out a run 2. However, the fraction of the deleted runs 2 in the period is still given by $6/21 = 2/7 = 2/p(4)$ of the pre-existing ones before the cancellations, being $R_2(4) = 3 \cdot 7 - 6 = 15$.

In the case of primes, it follows from the relation (21) that we struck out, in each step, a fraction $1/p(k)$ of the total number of runs in the period $T(k)$ if no cancellations were made, which is given by the product of the prime $p(k)$ by the actual number of runs in the previous period $T(k-1)$. By considering the scores of Table 7, a similar relation can be conjectured for the runs 2 in the case of twin primes, in order to link the number of cancelled runs 2 and the total number of runs 2 in the period $T(k)$ if no cancellations were made. Unfortunately, in general, the actual number of the deleted runs 2 is not easily computable, by starting from the total number of cancellations in $T(k)$. However, in the same way of the primes, our conjecture is that the deletion of the multiples of $p(k)$ has the effect to exactly cancel a fraction $2/p(k)$ of the runs 2 in the period $T(k)$.

If this conjecture holds, Equation (38) follows by induction. In fact, it is true for $p(2) = 3$. Let us suppose that Equation (38) holds for $p(k-1)$ and show that it is also true for $p(k)$. By the induction hypothesis, the number of runs 2 in the period $T(k-1)$ is given by $R_2(k-1) = \prod_{i=2}^{k-1}(p(i) - 2)$. We must show that the number of runs 2 in the period $T(k)$ is $R_2(k) = \prod_{i=2}^{k}(p(i) - 2)$. Given $R_2(k-1)$, the number of runs 2 in the new period $T(k)$ becomes $p(k) \cdot R_2(k-1)$, because $T(k)$ is $p(k)$ times greater than $T(k-1)$. By taking the previous conjecture, a fraction $2/p(k)$ of the runs 2 is struck out, in such a way we have a fraction of residual runs 2 given by $(p(k) - 2)/p(k) \cdot R_2(k-1) = (p(k) - 2)/p(k) \cdot \prod_{i=2}^{k-1}(p(i) - 2) = \prod_{i=2}^{k}(p(i) - 2) = R_2(k)$.

4.3. A Heuristic Estimation of $\pi_2(x)$ Equivalent to the $Li_2(x)$ Approximation

From Equation (38), we can give an estimation $\pi_{2P}(N)$ of the twin prime function $\pi_2(x)$, which is equivalent to the approximation given by the $Li_2(x)$ function (8). Such an estimation can be viewed as a generalization of Equation (26) to the case of the twin primes. To this end, analogously to Equation (23) for the primes, we compute the average density $\overline{D_2}(k)$ of the number of runs 2 in a period $T(k)$. By starting from the total number of runs 2 $R_2(k)$ in the period $T(k)$, the average density $\overline{D_2}(k)$ is given by the relation

$$\overline{D_2}(k) = \frac{R_2(k)}{T(k)} = \frac{\prod_{i=2}^{k}(p(i) - 2)}{\prod_{i=1}^{k} p(i)} = \frac{1}{2} \times \prod_{i=2}^{k} \frac{p(i) - 2}{p(i)}, \quad k \geq 2. \tag{39}$$

As for the primes, we can initially approximate the local density $D_2(k, n)$ in the interval $I(k)$ as the average density $\overline{D_2}(k)$, that is, $D_2(k, p(k)^2) \simeq \overline{D_2}(k)$. In this case, the estimated number of twin primes $P_2(k)$ in $I(k)$, for $k \geq 2$, is given by

$$P_2(k) = \overline{D_2}(k) \times S(k) = (p(k+1)^2 - p(k)^2) \times \frac{1}{2} \times \prod_{i=2}^{k} \frac{p(i) - 2}{p(i)}, \quad k \geq 2 \tag{40}$$

The total estimation $\pi_{2P}(N)$ is then obtained by adding all the contributions $P_2(k)$, that is,

$$\pi_{2P}(N) = \sum_{k=0}^{K-1} P_2(k) + P_2(K, N) = P_2(0) + P_2(1) + \sum_{k=2}^{K-1} \left[(p(k+1)^2 - p(k)^2) \cdot \frac{1}{2} \cdot \prod_{i=2}^{k} \frac{p(i)-2}{p(i)} \right] + P_2(K, N) \tag{41}$$

where $P_2(0) = p(1)^2 - p(0)^2$, $P_2(1) = \frac{1}{2} \cdot (p(2)^2 - p(1)^2)$, $P_2(K, N) = (N - p(K)^2) \cdot \frac{1}{2} \cdot \prod_{i=2}^{K} \frac{p(i)-2}{p(i)}$, and K is the greatest prime number not exceeding $N^{1/2}$. As for the primes, Equation (41) overestimates the true $\pi_2(N)$ scores, because the local density $D_2(k,n)$ is not actually constant in the period $T(k)$, but it is a function of n. However, the offset of the local density in the interval $I(k)$ with respect to the average density is greater than for the primes. Experimentally, each $P_2(k)$ value (40) overtakes the true quantity of twin primes computed in $I(k)$ of about 20%, that is, more or less a double of the percentage previously found for the primes, and reported in Figure 1, even if the trends of the local densities are similar. Quantitatively, the ratio between the average density $\overline{D_2}(n)$ and the local density $D_2(k, p(k)^2)$ seems to approximate the constant c^2 as $k \to \infty$, that is, the square of c.

To evidence this statement, let us consider the estimation given by the $\text{Li}_2(x)$ function, that is, $C\,\text{Li}_2(x)$, for $x = N$, from Equation (8), that is, $C\,\text{Li}_2(N)$, as a summation of integrals, each of them is computed in the interval $I(k) = [p(k)^2, p(k+1)^2)$

$$C\,\text{Li}_2(N) = C \int_2^{p(1)^2} \frac{dt}{\log^2 t} + C \sum_{k=1}^{K-1} \int_{p(k)^2}^{p(k+1)^2} \frac{dt}{\log^2 t} + C \int_{p(K)^2}^{N} \frac{dt}{\log^2 t} \tag{42}$$

being $p(K)^2$ the greatest square of a prime less than N. Similarly to Equation (28), we can write Equation (42) as a succession of estimations $L_2(k)$ in each interval $I(k)$, that is,

$$C\,\text{Li}_2(N) = C\,L_2(0) + C \sum_{k=1}^{K-1} L_2(k) + C\,L_2(K,N), \tag{43}$$

where $L_2(0) = \int_2^{p(1)^2} \frac{dt}{\log^2 t}$, $L_2(K,N) = \int_{p(K)^2}^{N} \frac{dt}{\log^2 t}$ and

$$L_2(k) = \int_{p(k)^2}^{p(k+1)^2} \frac{dt}{\log^2 t}. \tag{44}$$

Then, we apply the Mean Value Theorem for Integrals to Equation (42) in each interval $I(k)$

$$C\,\text{Li}_2(N) = C \frac{p(1)^2 - 2}{\log^2(\varsigma_0)} + C \sum_{k=1}^{K-1} \frac{p(k+1)^2 - p(k)^2}{\log^2(\varsigma(k))} + C \frac{N - p(K)^2}{\log^2(\varsigma_K)}, \tag{45}$$

where the point ς_0 belongs to the interval $I(0) = [p(0)^2, p(1)^2)$, $\varsigma(k)$ belongs to the interval $I(k)$, $k = 1, \ldots, K-1$, and ς_K belongs to the interval $I(K,N) = [p(K)^2, N)$. As for the primes, we have to consider the lower bound $p(k)^2$ of the interval $I(k)$. Let us take the ratio between the two terms multiplying the size $S(k) = p(k+1)^2 - p(k)^2$, in the summations of the Equations (41) and (45), so that we obtain

$$\frac{\frac{1}{2} \cdot \prod_{i=2}^{k} \frac{p(i)-2}{p(i)}}{\frac{C}{\log^2(\varsigma(k))}} = \left(\frac{2C \cdot \prod_{i=2}^{k} \frac{p(i)}{p(i)-2}}{\log^2(\varsigma(k))} \right)^{-1} \tag{46}$$

If we consider the lower bound $p(k)^2$ of the interval $I(k)$, we have

$$\frac{2C \cdot \prod_{i=2}^{k} \frac{p(i)}{p(i)-2}}{\log^2(\varsigma(k))} = \frac{2C \cdot \prod_{i=2}^{k} \frac{p(i)}{p(i)-2}}{\log^2(p(k)^2)} \cdot \frac{\log^2(p(k)^2)}{\log^2(\varsigma(k))} \tag{47}$$

Let us notice that the ratio $\frac{p(i)-2}{p(i)}$ can be split as

$$\frac{p(i)-2}{p(i)} = \frac{p(i)-2}{(p(i)-1)^2} \times \frac{(p(i)-1)^2}{p(i)} = \frac{p(i)^2 - 2p(i)}{(p(i)-1)^2} \times \frac{(p(i)-1)^2}{p(i)^2} = \frac{(p(i)-1)^2 - 1}{(p(i)-1)^2} \times \frac{(p(i)-1)^2}{p(i)^2}$$
$$\Longrightarrow \frac{p(i)-2}{p(i)} = \frac{p(i)-1}{p(i)} \times \frac{p(i)-1}{p(i)} \times \left(1 - \frac{1}{(p(i)-1)^2}\right) \tag{48}$$

Consequently, we obtain

$$\prod_{i=2}^{k} \frac{p(i)}{p(i)-2} = \prod_{i=2}^{k} \left[\frac{p(i)}{p(i)-1} \times \frac{p(i)}{p(i)-1} \times \frac{1}{1-\frac{1}{(p(i)-1)^2}} \right] \quad (49)$$

Then, we define

$$\begin{cases} C(k) = 2 \times \prod_{i=2}^{k} \left(1 - \frac{1}{(p(i)-1)^2}\right), & k \geq 2 \\ C(1) = C(0) = 1. \end{cases} \quad (50)$$

From Equation (49) and considering that $\lim_{k \to \infty} \frac{\log^2(p(k)^2)}{\log^2(\varsigma(k))} = 1$ (see Section 3.5), the limit, as $k \to \infty$, of the ratio (47) is given by

$$\lim_{k \to \infty} \frac{2C \times \prod_{i=2}^{k} \frac{p(i)}{p(i)-2}}{\log^2(p(k)^2)} \times \frac{\log^2(p(k)^2)}{\log^2(\varsigma(k))} = \lim_{k \to \infty} \frac{2C \times \prod_{i=2}^{k} \frac{p(i)}{p(i)-2}}{\log^2(p(k)^2)} = \lim_{k \to \infty} \frac{2C \times \frac{2}{C(k)} \times \prod_{i=2}^{k} \left[\frac{p(i)}{p(i)-1} \times \frac{p(i)}{p(i)-1}\right]}{\log^2(p(k)^2)}. \quad (51)$$

We noticed in the Equation (33) that

$$\lim_{k \to \infty} \frac{\prod_{i=1}^{k} \frac{p(i)}{p(i)-1}}{\log(p(k)^2)} = \frac{1}{2} \times e^{\gamma} \simeq \frac{1}{c} \simeq 0.8905. \quad (52)$$

Evidently, we have

$$\lim_{k \to \infty} \prod_{i=2}^{k} \frac{p(i)}{p(i)-1} = \frac{1}{2} \times \lim_{k \to \infty} \prod_{i=1}^{k} \frac{p(i)}{p(i)-1} \quad (53)$$

and, consequently, from Equation (9),

$$\lim_{k \to \infty} C(k) = C. \quad (54)$$

Finally, from Equation (51), we obtain the limit of the ratio (47)

$$\lim_{k \to \infty} \frac{4 \times \prod_{i=2}^{k} \left[\frac{p(i)}{p(i)-1} \times \frac{p(i)}{p(i)-1}\right]}{\log^2(p(k)^2)} = 4 \times \left(\frac{1}{2c}\right)^2 = \frac{1}{c^2} \simeq 0.8905^2 = 0.7931 \quad (55)$$

and Equation (46) gives

$$\frac{\frac{1}{2} \times \prod_{i=2}^{k} \frac{p(i)-2}{p(i)}}{\frac{C}{\log^2(\varsigma(k))}} = c_{2I}(k) \qquad \text{where } \lim_{k \to \infty} c_{2I}(k) = c^2 \simeq 1.2609. \quad (56)$$

For a given N, the proposed approximation $\pi_{2P}(N)$ overestimates the twin prime number function $\pi_2(N)$ by a factor c_{2N}, which can be computed by considering that we have an overestimation for each interval $I(k)$ that can be computed by considering a factor in the finite set $c_{2I}(k)$, $k = 1, \ldots, K$, where K is such that $N \simeq p(K)^2$. Equations (55) and (56) show that the succession $c_{2I}(k)$ tends to the constant c^2 as $N \to \infty$. Consequently, we can define a corrected version $\tilde{\pi}_{2P}(N)$ (57) of the proposed estimation $\pi_{2P}(N)$, by multiplying Equation (41) by the factor $1/c^2 \simeq 0.7931$, that is,

$$\tilde{\pi}_{2P}(N) = \frac{1}{c^2} \times \left(P_2(0) + P_2(1) + \sum_{k=2}^{K-1} P_2(k) + P_2(K, N) \right) = \frac{1}{c^2} \times \left(p(1)^2 - p(0)^2 \right) + \frac{1}{2c^2} \times \left(p(2)^2 - p(1)^2 \right) + \\ + \frac{1}{2c^2} \times \sum_{k=2}^{K-1} \left[\left(p(k+1)^2 - p(k)^2 \right) \times \prod_{i=2}^{k} \frac{p(i)-2}{p(i)} \right] + \frac{1}{2c^2} \times \left(N - p(K)^2 \right) \times \prod_{i=2}^{K} \frac{p(i)-2}{p(i)}. \quad (57)$$

As for the primes, Equation (57) is expected to improve the estimation of $\pi_2(N)$ as N approaches infinity. This is evidenced in the scores of Table 8, where a comparison is made between the proposed estimation $\pi_{2P}(N)$ and its adjusted version $\tilde{\pi}_{2P}(N)$ with the estimation $C\operatorname{Li}_2(N)$ given by the logarithmic integral function (8) and the twin prime number function $\pi_2(N)$. The ranges of N are the same as Table 6.

Table 8. The proposed estimation $\pi_{2P}(N)$ and its adjusted version $\tilde{\pi}_{2P}(N)$ in comparison with the logarithmic integral estimation $C\operatorname{Li}_2(N)$ and the prime number function $\pi_2(N)$. The scores of the logarithmic integer function have been computed by using the MATLAB® toolbox. The scores of $\pi_{2P}(N)$ and $\tilde{\pi}_{2P}(N)$ have been rounded to the nearest integer.

$N = 10^i$	$\pi_2(N)$	$C\operatorname{Li}_2(N)$	$\pi_{2P}(N)$	$\tilde{\pi}_{2P}(N)$
10^1	2	2	4	3
10^2	8	11	12	10
10^3	35	43	48	38
10^4	205	212	250	198
10^5	1224	1246	1522	1207
10^6	8169	8246	10,252	8131
10^7	58,980	58,751	73,579	58,353
10^8	440,312	440,365	553,514	438,977
10^9	3,424,506	3,425,306	4,312,478	3,420,314
10^{10}	27,412,679	27,411,414	34,537,569	27,390,848
10^{11}	224,376,048	224,368,862	282,810,653	224,289,776
10^{12}	1,870,585,220	1,870,559,864	2,358,205,655	1,870,231,592
10^{13}	15,834,664,872	15,834,598,303	19,964,600,235	15,833,405,367
10^{14}	135,780,321,665	135,780,264,892	171,202,650,560	135,776,370,890
10^{15}	1,177,209,242,304	1,177,208,491,858	1,484,356,543,022	1,177,204,581,001

The connection between the $\pi_{2P}(N)$ estimation (41) and the $C\operatorname{Li}_2(N)$ estimation (42) is investigated in Table 9, by considering the parameters and the coefficient of determination of the linear regressions $y_i = m_1 x_i + q_1$ between the occurrences of $P_2(k)$ (40) versus those of $CL_2(k)$, where $L_2(k)$ is given by (44), in each interval $I(k)$. As for the primes, an excellent fitting is given by the linear relationship between $P_2(k)$ and $CL_2(k)$. This is confirmed by the coefficient of determination R_1^2, which rapidly tends to 1 as k grows. On the other hand, the intercept q_1 is negligible, whilst the slope m_1 approaches the limit value $1/c^2$.

The fitting of the linear regressions $y_i = m_2 x_i + q_2$ between the occurrences of $P_2(k)$ (40) versus those of the twin prime number function $\pi_2(k)$, if computed in the same interval $I(k)$, is also reported in Table 9. Even if less impressive than in the case of Table 5 for the primes, the goodness of the fitting is clearly shown by the coefficient of determination R_2^2, which is practically at its best value. As for m_1, the slope m_2 seems to approximate the limit value $1/c^2$.

Table 9. Parameters and coefficients of determination of the linear regressions $y_i = m_1 x_i + q_1$ of the proposed estimations for the twin primes $P_2(k)$ versus the logarithmic-integral ones $CL_2(k)$, together with the parameters and coefficients of determination of the linear regressions $y_i = m_2 x_i + q_2$ of $P_2(k)$ versus the true number of twin primes $\pi_2(k)$. Each point is computed in an interval $I(k)$.

$p(k)^2$	m_1	q_1	R_1^2	m_2	q_2	R_2^2
$(1, 10^6)$	0.799120	0.3205	0.9999525356	0.784981	0.6900	0.9819305687
$(10^6, 10^7)$	0.797052	0.1191	0.9999935927	0.807148	-0.8767	0.9947532680
$(10^7, 10^8)$	0.794901	0.1435	0.9999989157	0.792818	0.7199	0.9983424066
$(10^8, 10^9)$	0.793935	0.2629	0.9999996567	0.794232	-0.5788	0.9992892724
$(10^9, 10^{10})$	0.793529	0.2602	0.9999999326	0.793309	1.2422	0.9998094846
$(10^{10}, 10^{11})$	0.793273	0.5082	0.9999999770	0.793336	-0.0376	0.9999152638
$(10^{11}, 10^{12})$	0.793180	0.5523	0.9999999955	0.793186	0.6827	0.9999711368
$(10^{12}, 10^{13})$	0.793115	0.9036	0.9999999988	0.793125	0.0544	0.9999902660
$(10^{13}, 10^{14})$	0.793088	1.2179	0.9999999997	0.793089	0.9592	0.9999967269
$(10^{14}, 10^{15})$	0.793072	1.5449	0.9999999999	0.793072	2.4024	0.9999988979

In summary, the proposed approach estimates the true number of twin primes by considering the number of runs 2 in each interval $I(k) = [p(k)^2, p(k+1)^2)$, in such a way each estimation $P_2(k)$ fits the correspondent one given by $CL_2(k)$. Consequently, in the case the conjecture (38) holds, we can infer that the distribution of the twin primes follows the same trend in all the intervals $I(k)$. Because these intervals are a function of the squares of both the prime $p(k)$ and its successive one, it follows that, *being the primes are a never-ending succession, the unproved hypothesis of the infinitude of the twin primes would be further strengthened.*

5. Conclusions and Future Developments

In this work, an original heuristic procedure in order to obtain the distribution of the prime number function $\pi(x)$ is proposed and investigated, which gives estimations of the scores of $\pi(x)$ equivalently to the logarithmic integral function $\mathrm{Li}(x)$. However, this approach is not fully probabilistic, but it is also based on analytical concepts, that is, a set of infinitely many binary periodic sequences is found by means of a modified Sieve procedure, whose periods have a subset that is included in limited and disjoint intervals $I(k)$ of the prime characteristic function. In each period $T(k)$, these binary sequences define a succession of 1 values, which are separated by runs of consecutive zeroes. Starting from the number of runs of zeroes in a period $T(k)$, an estimation of the total number of primes can be found, which is linked to the logarithmic integral estimation by the constant c of the Third Mertens' Theorem. Noticeably, the succession of the runs of zeroes, whose elements are the gaps between two consecutive primes, is symmetric in each period $T(k)$. As a result, the proposed LINPES procedure estimates the prime number function in each interval $I(k)$, whose bounds are the squares of a prime number and of the successive one. As a particular case, this procedure is also specialized to the case of the twin primes, in such a way only the runs sized 2 are considered in each period. Consequently, a heuristic relation for the number of these runs in a period $T(k)$ is formulated, whose trend is linked to the relation previously found for the total number of runs in the case of primes. Therefore, such a relation gives an estimation of the twin prime number function $\pi_2(x)$ in each interval $I(k)$, which is equivalent to the estimation of the logarithmic integral function $\mathrm{Li}_2(x)$, by means of the square of the constant c. Being the bounds of these intervals given by squares of primes, their number is infinite. As a consequence, the proposed procedure could give a contribution to the presumed infinity of the succession of the twin primes. Future developments will further investigate the relation of the number of runs 2 in a period $T(k)$, together with the symmetry of the succession of the runs of zeroes.

Author Contributions: Conceptualization, B.A.; Methodology, B.A., S.B., L.S. and M.S.; Formal Analysis, L.S.; Investigation, B.A., S.B., L.S. and M.S.; Data Curation, B.A., L.S. and M.S.; Writing—Original Draft Preparation, B.A.; Writing—Review & Editing, B.A. and M.S.; Visualization, M.S.; Supervision, S.B.

Funding: This research received no external funding.

Acknowledgments: The author would thank the site https://primes.utm.edu/lists/small/millions/ for providing the prime numbers that have been used for the computations in this work.

Conflicts of Interest: The authors declare no conflict of interest.

References

1. Crandall, R.; Pomerance, C. *Prime Numbers: A Computational Perspective*; Springer: New York, NY, USA, 2001.
2. Gauss, C.F. Letter to Encke, dated 24 December (1849). *Werke Kng. Ges. Wiss. Gottingen* **1863**, *2*, 444–447.
3. Legendre, A.M. *Essai sur la thèorie des Nombres*; Duprat: Paris, France, 1798.
4. Torquato, S.; Zhang, G.; de Courcy-Ireland, M. Uncovering multiscale order in the prime numbers via scattering. *J. Stat. Mech.* **2018**. [CrossRef]
5. Goldston, D.A. Are There Infinitely Many Twin Primes? Available online: http://arxiv.org/pdf/0710.2123.pdf (accessed on 17 April 2019).
6. Hardy, G.H.; Littlewood, J.E. Some problems of "Partito Numerorum", III: On the expression of a number as a sum of primes. *Acta Math.* **1923**, *44*, 1–70. [CrossRef]

7. Helfgott, H.A. An improved Sieve of Eratosthenes. Available online: https://arxiv.org/abs/1712.09130 (accessed on 17 April 2019).
8. Apostol, T. M. *Introduction to Analytic Number Theory*; Springer: New York, NY, USA, 1976.
9. Fine, B.; Rosenberger, G. *Number Theory: An Introduction via the Distribution of Primes*; Birkhäuser: Boston, MA, USA, 2007.
10. Montgomery, H. A heuristic for the Prime Number Theorem. *Math. Intell.* **2006**, *28*, 6–9. [CrossRef]
11. Mertens, F. Ein Beitrag zur analytischen Zahlentheorie. *J. Reine Angew. Math.* **1874**, *78*, 46–62.
12. Granville, A. Harald Cramér and the Distribution of Prime Numbers. Available online: https://www.dartmouth.edu/~chance/chance_news/for_chance_news/Riemann/cramer.pdf (accessed on 17 April 2019).
13. Selberg, A. On the normal density of primes in small intervals, and the difference between consecutive primes. *Arch. Math. Naturvid.* **1943**, *47*, 87–105.
14. Languasco, A.; Zaccagnini, A. Short intervals asymptotic formulae for binary problems with prime powers. *J. Théorie Nombres Bordeaux* **2018**, *30*, 609–635. [CrossRef]
15. Cramer, H. On the order of magnitude of the difference between consecutive prime numbers. *Acta Arith.* **1936**, *2*, 23–46. [CrossRef]
16. Cramer, H. On the distribution of primes. *Proc. Camb. Phil. Soc.* **1920**, *20*, 272–280.
17. Bays, C.; Hudson, R. H. A new bound for the smallest x with $\pi(x) > \mathrm{Li}(x)$. *Math. Comput.* **2000**, *69*, 43–56. [CrossRef]
18. Saouter, Y.; Demichel, P. A sharp region where $\pi(x) - \mathrm{Li}(x)$ is positive. *Math. Comput.* **2010**, *79*, 2395–2405. [CrossRef]
19. Bays, C.; Hudson, R.H. Zeroes of the Dirichlet L-functions and irregularities in the distribution of primes. *Math. Comput.* **1999**, *69*, 861–866. [CrossRef]
20. Granville, A.; Martin, G. Prime number races. *Am. Math. Mon.* **2006**, *113*, 1–33. [CrossRef]
21. Pintz, J. Very large gaps between consecutive primes. *J. Num. Theor.* **1997**, *63*, 286–301. [CrossRef]
22. Zhang, Y. Bounded gaps between primes. *Ann. Math.* **2014**, *179*, 1121–1174. [CrossRef]
23. Maynard, J. Small gaps between primes. *Ann. Math.* **2015**, *181*, 383–413. [CrossRef]
24. Polymath, D.H.J. The "Bounded Gaps between Primes" Polymath Project: A Retrospective. Available online: https://arxiv.org/abs/1409.8361 (accessed on 17 April 2019).
25. Bagchi, B. A promising approach to the twin prime problem. *Reson* **2003**, *8*, 26–31. [CrossRef]
26. Brun, V. La série $1/5 + 1/7 + 1/11 + 1/13 + 1/17 + 1/19 + 1/29 + 1/31 + 1/41 + 1/43 + 1/59 + 1/61 + \ldots$ où les dénominateurs sont "nombres premiers jumeaux" est convergent ou finie. *Bull. Sci. Math.* **1919**, *43*, 100–104.

© 2019 by the authors. Licensee MDPI, Basel, Switzerland. This article is an open access article distributed under the terms and conditions of the Creative Commons Attribution (CC BY) license (http://creativecommons.org/licenses/by/4.0/).

Article

On the Number of Witnesses in the Miller–Rabin Primality Test

Shamil Talgatovich Ishmukhametov *, Bulat Gazinurovich Mubarakov and Ramilya Gakilevna Rubtsova

Institute of Computational Mathematics and Information Technology, Kazan Federal University, Kremlevskya St. 35, Kazan 420008, Russia; mubbulat@mail.ru (B.G.M.); Ramilya.Rubtsova@kpfu.ru (R.G.R.)
* Correspondence: Shamil.Ishmukhametov@kpfu.ru

Received: 24 February 2020; Accepted: 28 March 2020; Published: 1 June 2020

Abstract: In this paper, we investigate the popular Miller–Rabin primality test and study its effectiveness. The ability of the test to determine prime integers is based on the difference of the number of primality witnesses for composite and prime integers. Let $W(n)$ denote the set of all primality witnesses for odd n. By Rabin's theorem, if n is prime, then each positive integer $a < n$ is a primality witness for n. For composite n, the power of $W(n)$ is less than or equal to $\varphi(n)/4$ where $\varphi(n)$ is Euler's Totient function. We derive new exact formulas for the power of $W(n)$ depending on the number of factors of tested integers. In addition, we study the average probability of errors in the Miller–Rabin test and show that it decreases when the length of tested integers increases. This allows us to reduce estimations for the probability of the Miller–Rabin test errors and increase its efficiency.

Keywords: prime numbers; primality test; Miller–Rabin primality test; strong pseudoprimes; primality witnesses

1. Introduction

The MillerRabin primality test is an algorithm that checks whether a given number is prime or composite. Its original version, due to Gary L. Miller, was deterministic and relied on the unproved extended Riemann Hypothesis [1]. Michael O. Rabin modified it to obtain a probabilistic algorithm [2].

Definition 1. *Let m be a positive integer represented as* $m = 2^s \cdot u$ *where u is odd. We introduce two auxiliary functions* $bin(m) = s$ *and* $odd(m) = u$.

Definition 2. *Let n be an odd natural,* $n > 9$. *An integer* $a, 1 \leq a < n$, *is called a primality witness for n if it is co-prime to n and one of the following conditions holds:*

$$
\begin{aligned}
&1.\ a^{odd(n-1)} \equiv 1 \bmod n, \\
&2.\ a^{odd(n-1)2^i} \equiv -1 \bmod n \text{ for some } i,\ 0 \leq i < bin(n-1),
\end{aligned}
\quad (1)
$$

(We replaced original Rabin's definition of the compositeness witnesses by the opposite relation). For generality, we count 1 and $n-1$ as primality witnesses and call them trivial witnesses since they satisfy (1) for any n.

Let $W(n)$ denote the set of all primality witnesses for n. The Rabin theorem [2] asserts that if number n is prime then each non-zero integer, $a < n$ is a primality witness for n, and therefore, the number of all witnesses $|W(n)| = n-1$. For composite n, it satisfies inequality $|W(n)| \leq \varphi(n)/4$ where $\varphi(n)$ is Euler's totient function. Since Rabin did not consider 1 as a witness, then he stated the strict inequality $|W(n)| < \varphi(n)/4$.

Later, Gary Miller [1] developed a primality test that takes any integer a, $1 < a < n$, checks if a is not a factor of n (otherwise, n is trivially composite), and whether a is a primality witness for n, that is, lies in the set $W(n)$. If the answer is positive, then n is probable prime with probability exceeding 3/4. If we need in a more exact result, we should repeat this procedure several times taking different numbers $a < n$.

The researchers refer to this algorithm as to the Miller and Rabin primality test. We abbreviate it to *MR test*.

Definition 3. *Parameters a which are used in Miller's algorithm are called* bases. *They are chosen randomly from interval* $[1; n-1]$. *If, for a given odd integer, n relation (1) holds at a base a, we say, n passes the MR test at base a. Otherwise, we call a a compositeness witness for n and deduce that n is certainly composite.*

The probability of error after k successful iterations becomes less than $1/4^k$. The only type of error in the Rabin' procedure is defining a composite integer as prime.

More details on the Miller–Rabin test can be found in Chapter 3 of text-book [3] by Crandall and Pomerance. We abbreviate Miller–Rabin test as MR test.

Definition 4. *Composite integers qualifying by MR test as probable prime at a base a are called strong pseudoprimes relative to base a. Composite integers being probably prime relative to all a from a set A of bases are called strong probable prime relative to set of bases A.*

Investigation of pseudoprime integers has a long history in the Computational Number Theory. We outline main advantages in this direction in the next section.

2. Some History Remarks

Fist attempts to find fast primality algorithms were based on Fermat's Little Theorem asserting that for prime n and for any positive integer a, the following relation holds

$$a^n \equiv a \bmod n \qquad (2)$$

Indeed, many composite integers do not satisfy (2) and can be discarded after the first check. Composite n that satisfy (2) are called Fermat pseudoprimes relative to base a.

It is important to note that all strong pseudoprimes relative to a base a are also Fermat pseudoprimes relative to a.

We can decrease the number of false decisions by Fermat's test by checking the relation (2) with several different a. However, this does not allow us to completely avoid false conclusions since so-called Carmichael numbers exist.

Integer n is called a Carmichael number if it satisfies (2) for all a. Carmichael numbers appear relatively rarely and the least Carmichael number is $561 = 3 \cdot 7 \cdot 11$. It is known that Carmichael numbers are exactly those integers which satisfy Korselt's criterion:

Korselt Criterion (1899). A positive compositeinteger n is a Carmichael number if and only if n is square-free, and for all prime divisors p of n, it is true that $p - 1 | n - 1$.

One of the interesting problems is to find for a given odd integer n the least witness. In 1994 Alford, Granville and Pomerance proved [4] that such witnesses exceed $(\log n)^{1/(3 \log \log \log n)}$ for infinitely many n. We also show that there are finite sets of odd composites which do not have a reliable witness, namely a common witness for all of the numbers in the set.

MR test discards a Carmichael number n, if the base was chosen from $[1; n-1] \backslash W(n)$.

Let us fix a base a and let n_a be a least composite integer that the MR Test accepts at the base a. Then, any odd $n < n_a$ for which a is a primality witness, is definitely prime. This means that when we know n_a, we can definitely check any $n < n_a$ for primality using only one round of the MR procedure. The corresponding integer n_a is small. But if we take a set A of several different bases a and find a least composite n_A for which all $a \in A$ are primality witness, this n_A can be very large. Candidates for bases a can be any positive integers that are not squares. However, historically, candidates for special bases are chosen from the set of primes.

Let P_k denote the set of the first k primes $P_k = \{2, 3, 5, 7, \ldots, p_k\}$, and let ψ_k be a least strong pseudoprime relative to P_k for a $k \geq 1$. Function ψ_k is well defined and is exponentially computable. Its computation began already 40 years ago.

First four values of ψ_k have been found by C. Pomerance, J. Selfridge, and S.Waggstaff [5] in 1980.

A systematic calculation of ψ_k for larger k has been initiated by J. Jaeschke [6] who elaborated basic algorithms helpful for searching for strong pseudoprimes of different forms. In 1993 Jaeschke calculated ψ_k for $5 \leq k \leq 8$ and proposed upper bounds for ψ_k at $9 \leq k \leq 11$.

F. Arnault in papers [7,8] described another algorithm to search for Carmichael numbers and strong pseudoprimes integers.

Jaeschke' hypothesis have been improved in 2001 by Z. Zang [9] who constructed a lesser 19-digits decimal integer $Q_{11} = 3825123056546413051$ bounding above ψ_{11}. Z.Zang conjectures that values ψ_k for $9 \leq k \leq 11$ are equal to each other and coincide with Q_{11}.

In 2012 J. Jiang and Y. Deng [10] confirmed Zang's Hypothesis by showing that $Q_{11} = \psi_9 = \psi_{10} = \psi_{11}$.

The last record is reached by J. Sorenson and J. Webster [11] in 2016 . They found ψ_{12} and ψ_{13}, where $\psi_{13} = 3317044064679887385961981 \approx 3.3 \cdot 10^{24}$. So at the moment we can successfully determine prime integers less than $3.3 \cdot 10^{24}$ by only 13 rounds of the MR test. But this bound is much less than integers used in Cryptography. For example, DSS algorithm uses prime integers of length 256 bits ($\approx$80 decimal digits).

Another branch of investigations in connected with the problem of distribution of Fermat pseudoprimes and strong pseudoprimes. Let $F(n)$ denote set

$$F(n) = \{a \bmod n : a^{n-1} \equiv 1 \bmod n\}.$$

Clearly, $F(n) \supseteq W(n)$.

In 1985 P. Erdos and C. Pomerance [12] studied an asymptotic behavior of average function

$$A(x) = \frac{1}{x} \sum_{n \leq x} {}'|F(n)|$$

where sum is counted over odd integers. They showed using complex number-theoretical calculations that $A(x)$ is a growing function bounded below by $x^{15/23}$.

Our average function $Avg(x)$ looks close to $A(x)$ but we show that for almost all composite n $W(n)$ consists of only two elements 1 and $n-1$ and function $Avg(x)$ tends to zero with x tending to infinity.

Average number of errors in the MR test was also studied in 1993 by I. Damgard, P. Landrock and C Pomerance. In paper [13] they studied an average probability of the false decision by the MR test in the following procedure:

Fix $k > 0$ and $t > 0$ and choose randomly k-bit odd integer n. Check it with t rounds of MR test with randomly chosen bases from $[1; n-1]$. If n was discarded during the procedure (that is, found $a \notin W(n)$), take another n. Continue until n was found passed t rounds. Let $p_{k,t}$ be the probability that the procedure returns a composite integer.

The authors found explicit upper bounds for various k and t. In particular they proved that $p_{k,1} \leq k^2 4^{2-\sqrt{k}}$ for $k \geq 2$. Their results show that the probability of false decisions of the MR test depends on the length of tested numbers and it decreases if the length of the numbers increases.

3. Counting Number of Witnesses

In this section we deduce exact formulas for the number of primality witnesses for different types of composite integers.

We begin our investigation with a little proposition improving Rabin's estimate.

Theorem 1. *If $a \in W(n)$, then $n - a \in W(n)$.*

Proof. Let $k = ord_n(a)$. If k is odd, then $a^{odd(n-1)} \mod n = 1$, and $(n-a)^{odd(n-1)} \equiv -1 \mod n$, therefore, $n - a$ is also a witness.

If k is even, then $a^{k/2} \equiv -1 \mod n$. If $k/2$ is even, then $(n-a)^{k/2} \equiv a^{k/2} \equiv -1 \mod n$, and $(n-a)$ is a witness.

Finally, if $k/2$ is odd, then $(n-a)^{k/2} \equiv -a^{k/2} \equiv 1 \mod n$. Since $k/2 \mid odd(n-1)$, then $a^{odd(n-1)} \equiv 1 \mod n$, and $(n-a)$ again is a witness.

This completes the proof. □

Corollary 1. *(The Improved Rabin Theorem). Let n be a natural, and A be an arbitrary set of bases less than n, co-prime to n, such that for any $a \in A$, $n - a$ is not in A. If all bases $a \in A$ are primality witnesses of n, then n is probable prime with probability of error less than or equal to $1/16^k$.*

Indeed, when we found a primality witness a for integer n, we get two primality witnesses for n, namely, a and $n - a$. So, this reduces the probability of error by a factor of $4^2 = 16$.

Let $N_w(n) = |W(n)|$ be the power of number of primality witnesses $W(n)$. As mentioned earlier, for prime n $N_w(n) = n - 1$, and for composite n $N_w \leq \varphi(n)/4$.

Below we estimate function $N_w(n)$ more exactly. First we formulate a theorem restricting possible witnesses for a composite n.

Theorem 2. *Let $n = u \cdot v$ for co-prime factors u and v (possibly, composite), and $a \in W(n)$. Then,*

1. $ord_u(a) \mid GCD(\varphi(u), (u - \varphi(u))v - 1)$,
2. $ord_v(a) \mid GCD(\varphi(v), (v - \varphi(v))u - 1)$, (3)
3. $bin(ord_u(a)) = bin(ord_v(b))$.

Proof. 1. Since a is a primality witness for n then $a^{n-1} \equiv 1 \mod n$ and $a^{n-1} \equiv 1 \mod u$. Besides, $n - 1 = uv - 1 = \varphi(u)v + (u - \varphi(u))v - 1$, so

$$1 \equiv a^{n-1} \equiv a^{\varphi(u)v + (u-\varphi(u))v - 1} \equiv a^{(u-\varphi(u))v - 1} \mod u,$$

since $a^{\varphi(u)} \equiv 1 \mod u$ by Euler's Theorem.

2. By symmetry.

3. If $ord_u(a)$ is odd, then $a^{odd(n-1)} \equiv 1 \mod n$ (otherwise, a satisfies the second clause of the MRT, and $ord_u(a)$ should be even). Then $a^{odd(n-1)} \equiv 1 \mod v$ and $ord_v(a)$ is odd.

If $bin(ord_u(a)) = i$ for $0 < i < bin(n-1)$, then a is a witness by second clause of the MRT, so $a^{odd(n-1)2^{i-1}} \equiv -1 \mod n$, $a^{odd(n-1)2^{i-1}} \equiv -1 \mod v$, and $a^{odd(n-1)2^i} \equiv 1 \mod v$, so $ord_v(a) = odd(n-1)2^i$ and $bin(ord_v(a))$ is equal to i.

The theorem is proved. □

Example 1. *Let $n = 15 \cdot 19 = 285$, and $a \in W(n)$. By Theorem 2:*

1. $ord_u(a) \mid GCD(\varphi(u), (u - \varphi(u))v - 1) = GCD(8, 132) = 4$,
2. $ord_v(a) \mid GCD(\varphi(v), (v - \varphi(v))u - 1) = GCD(18, 14) = 2$,
3. $bin(ord_u(a)) = bin(ord_v(b))$.

So, possible a satisfies $(ord_u(a), ord_v(a)) = (1,1)$, or, $(ord_u(a), ord_v(a)) = (2,2)$, so $n = 285$ has only trivial witnesses 1 and $n - 1$.

Theorem 3. *Let $n = p^k$ be a degree of prime p, then $N_w(n) = p - 1$.*

Proof. Let a be a witness for $n = p^k$, then $ord_a(n) \mid GCD(\varphi(n), n-1) = GCD(p^{k-1}(p-1), p^k - 1) = p - 1$.

Besides, any a satisfying $a^{p-1} \bmod n = 1$ is a witness of n. Indeed, let $a^{p-1} \bmod n = 1$. Then, $m = ord_n(a)$ is a factor of $n - 1 = p^k - 1$. Let $n - 1 = 2^s \cdot t$ for odd t, therefore, $m = 2^{s_1} \cdot t_1$, where $s_1 \leq s$ and t_1 is a factor of t.

If $s_1 = 0$, then $a^{t_1} \bmod n = 1$, $a^t \bmod n = 1$ and a is a witness by the first clause of the MRT. Otherwise, let $0 \leq r \leq s_1$ be such that $a^{t_1 2^r} \equiv -1 \bmod n$. Then $a^{t 2^r} \equiv -1 \bmod n$ and a is a witness by the second clause of the MRT. This completes the proof. □

We call integer n *semiprime* if it is a product of two distinct primes $n = pq$, $p < q$. Semiprimes are close to primes, and we prove below that they have a maximal number of primality witnesses among composite numbers.

Theorem 4. *Number of witnesses of semiprime $n = pq$ is equal to*

$$N_w(pq) = (odd(d))^2 \cdot (4^{bin(d)} + 2)/3, \qquad (4)$$

where $d = GCD(p - 1, q - 1)$.

We begin with example of application of this formula.

Example 2. *Let $n = 11 \cdot 31 = 341$. Then $d = GCD(p - 1, q - 1) = 10 = 5 \cdot 2^1$, $odd(d) = 5$, $s = bin(d) = 1$. By the theorem,*

$$N_w(31) = 5^2 \cdot (4 + 2)/3 = 50.$$

Proof. Let $d = GCD(p - 1, q - 1)$. Applying Theorem 2 to $n = pq$ we obtain

1. $ord_p(a) \mid d$, $ord_q(a) \mid d$,
2. $bin(ord_u(a)) = bin(ord_v(b))$.

We distribute all n-witnesses a into $s + 1$ classes W_i, $0 \leq i \leq s$, where class W_i consists of a with $bin(ord_p(a)) = bin(ord_q(a)) = i$.

Class W_0 contains such a that both $ord_p(a)$ and $ord_q(a)$ are odd. Let $a \in W_0$, and $(i, j) = (ord_p(a), ord_q(a))$. Numbers i and j are factors of $u = odd(d)$ by the choice of a. Conversely, each integer $a < n$ satisfying $ord_p(a) \mid u$, $ord_q(a) \mid u$, is a witness of n and lies in W_0.

Let fix a pair (i, j), $i \mid d, j \mid d$. By Euler's theorem, in Z_p there are exactly $\varphi(i)$ elements of multiplicative order i, and in Z_q there are $\varphi(j)$ elements of multiplicative order j, so there exist exactly $\varphi(i) \cdot \varphi(j)$ pairs (x, y), $0 < x < p$, $0 < y < q$, such that $(ord_p(x), ord_q(y)) = (i, j)$. But for each such pair (x, y) there exists a unique $a < n$ with $(a \bmod p, a \bmod q) = (x, y)$, so there is a injective correspondence between witnesses a of n with odd orders $ord_p(a), ord_q(a)$, and pairs (x, y) with $x \mid u$, $y \mid u$. Therefore, the power of W_0 is equal to

$$|W_0| = \sum_{x \mid u, \, y \mid u} \varphi(x) \cdot \varphi(y) = \left(\sum_{x \mid u} \varphi(x)\right)\left(\sum_{y \mid u} \varphi(y)\right) = u^2,$$

since by a known theorem of Euler for any natural m $\sum_{v \mid m} \varphi(v) = m$.

The next class W_1 has the same power u^2 since is consists of witnesses a with $bin(ord_p(a)) = bin(ord_q(a)) = 1$, and

$$|W_1| = \sum_{x|d,\ y|d} \varphi(2x) \cdot \varphi(2y) = u^2,$$

since $\varphi(2z) = \varphi(z)$ for odd z.

The power of class W_i is equal to

$$\sum_{x|d,\ y|d} \varphi(2^i x) \cdot \varphi(2^i y) = 4^{i-1} u^2.$$

Therefore, the number of all witnesses $N_w(n) = u^2(1 + 1 + 4 + \ldots + 4^{s-1}) = u^2 \cdot (4^s + 2)/3$. This completes the proof. □

Corollary 2. *(Rabin's theorem for semiprimes). The number of witnesses of $n = pq$, $p \leq q$, is less or equal to $\varphi(n)/4$.*

Proof. If $p = q$, then $N_w(n) = p - 1$ by Theorem 3, and $\varphi(n)/4 = p(p-1)/4$, so $N_w(n) < \varphi(n)/4$ at $p \geq 5$.

Let $p < q$. Ratio $N_w(n)/n$ reaches its maximum when $GCD(p-1; q-1) = p-1$, $q = 2p-1$, and $bin(p-1) = 1$. Indeed, $odd(n)$ is diminishing in two times when $bin(p-1)$ is added by 1, and the whole expression in (4) becomes less. Then, $max\ odd(d) = (p-1)/2$, so

$$max\ N_w(pq) = N_w(p(2p-1)) = \frac{(p-1)^2}{2} = \frac{\varphi(n)}{4}.$$

□

Example 3. *Let $n = 7 \cdot 13 = 91$. $N_w(91) = 3^2 \cdot 2 = 18 = \varphi(91)/4$.*

Now we study function $N_w(n)$ at products of k distinct primes. The general result for such products is formulated below:

Theorem 5. *Let $n = p_1 \cdot p_2 \cdot \ldots \cdot p_k$ be the product of k distinct primes. Then*

$$N_w(n) = u_1 \cdot u_2 \cdot \ldots \cdot u_k \cdot \left(1 + \frac{2^{ks} - 1}{2^k - 1}\right),\ \text{where}$$

$$s = min\{bin(d_1), bin(d_2), \ldots, bin(d_k)\},\ d_i = GCD\left(p_i - 1;\ \prod_{j \neq i} p_j - 1\right),$$

$u_i = odd(d_i)$.

Let us begin with an example $n = 7 \cdot 13 \cdot 31 = 2821$. The corresponding restrictions are listed below:

1. $ord_p(a)\ |\ d_1 = GCD(p-1; qr-1) = 6$, $u_1 = 3$,
2. $ord_q(a)\ |\ d_2 = GCD(q-1; pr-1) = 12$, $u_2 = 3$,
3. $ord_r(a)\ |\ d_3 = GCD(r-1; pq-1) = 30$, $u_3 = 15$,
4. $bin(ord_p(a)) = bin(ord_q(b)) = bin(ord_r(b))$.

Since $s = min\{bin(d_1), bin(d_2), bin(d_3)\} = min\{1, 2, 1\} = 1$, we obtain

$$N_w(2821) = 3 \cdot 3 \cdot 15 \left(1 + \frac{2^3 - 1}{2^3 - 1}\right) = 270$$

(compare with $\varphi(n)/4 = 6 \cdot 12 \cdot 30/4 = 540$).

Proof. Let $u_i = odd(d_i)$ and k-tuple $(x_1, x_2, \ldots, x_k)$ contains components $x_i \mid u_i$, $1 \leq i \leq k$. There are $\varphi(x_1) \cdot \ldots \cdot \varphi(x_k)$ witnesses of n with $ord_{p_i}(a) = x_i$ for $1 \leq i \leq k$. So,

$$|W_0| = \sum_{(x_1, x_2, \ldots, x_k), x_i \mid u_i} \varphi(x_1) \cdot \ldots \cdot \varphi(x_k) =$$

$$= \left(\sum_{x \mid u_1} \varphi(x)\right) \cdot \left(\sum_{x \mid u_2} \varphi(x)\right) \ldots \left(\sum_{x \mid u_k} \varphi(x)\right) = u_1 \cdot u_2 \cdot \ldots \cdot u_k.$$

As in the previous theorem, the power of class W_1 is equal to power of $W_0 = u_1 \cdot u_2 \cdot \ldots \cdot u_k$, while the power of the each further class W_{i+1} is equal to the power of the previous one multiplied by $\varphi(2^k) = 2^{k-1}$ since each additive $\varphi(2^i x_1) \cdot \ldots \cdot \varphi(2^i x_k)$ in the previous class corresponds to additive $\varphi(2^{i+1} x_1) \cdot \ldots \cdot \varphi(2^{i+1} x_k)$ and their ratio r_i is

$$r_i = \frac{\varphi(2^{i+1} x_1) \cdot \ldots \cdot \varphi(2^{i+1} x_k)}{\varphi(2^i x_1) \cdot \ldots \cdot \varphi(2^i x_k)} = 2^k.$$

The proof is complete. □

4. Frequency Function

In this part we introduce a notion of *frequency function* that characterizes the probability to find at one attempt a primality witness for a given integer n.

Let define frequency function $Fr(n)$ as follows

$$Fr(n) = \frac{N_w(n)}{\varphi(n)}.$$

According to Rabin's theorem, $Fr(n) = 1$ for prime n, and $Fr(n) \leq 1/4$ for composite n. We study distribution of values $Fr(n)$ for semiprime integers $n = pq$, $p < q$.

1. We begin our research with case $q - 1 = k(p - 1)$ for $k \geq 2$. Numbers of this type appear frequently among strong pseudoprimes. Let rewrite p and q in form $p = 2^s u + 1$, $q = 2^s ku + 1$, where u is odd, $s \geq 1$, and consider different s:

Case 1. $s = 1$, $u = odd(d) = (p-1)/2$, $N_w(pq) = 2u^2 = (p-1)^2/2$,

$$Fr(n) = \frac{(p-1)^2/2}{(p-1)(q-1)} = \frac{2u^2}{2u \cdot 2ku} = \frac{1}{2k}.$$

Function $Fr(n)$ reaches its maximum $1/4$ at $k = 2$: $(p, q) = (2u+1, 4u+1)$. Since, both p and q are prime then $u \equiv 0 \mod 3$, so $(p, q) = (6t+1, 12t+1)$, $t \geq 1$. Such pairs form a sequence

$$(7, 13), (19, 37), (31, 61), (37, 73), \ldots.$$

Case 2. $s = 2$, $u = odd(d) = (p-1)/4$, $N_w(pq) = 6u^2$, and

$$Fr(n) = \frac{6u^2}{(p-1)(q-1)} = \frac{6u^2}{4u \cdot 4ku} = \frac{3}{8k}.$$

Maximum of $Fr(n)$ is now $3/16 = 0.1875$ at $k = 2$.

Case 3. $s \geq 1$, At arbitrary s we have

$$Fr(n) = \frac{(1 + (4^s - 1)/3)u^2}{(p-1)(q-1)} = \frac{(1 + (4^s - 1)/3)u^2}{2^s u \cdot 2^s ku} = \frac{1}{3ku^2 \cdot 2^{2s-1}} + \frac{1}{3k}.$$

Thus, function $Fr(n)$ at semiprimes $n = pq$, $q - 1 = k(p - 1)$, is located in the interval

$$\frac{1}{3k} < Fr(n) \leq \frac{1}{2k}, \; k \geq 2. \tag{5}$$

2. Now, we turn to a common case $n = pq$:

$$p = 1 + k_1 u, \; q = 1 + k_2 u, \; GCD(k_1, k_2) = 1, \; u = t2^s, \; t \text{ odd}.$$

For such n

$$N_w(n) = t^2(4^s + 2)/3, \; \varphi(n) = k_1 k_2 t^2 4^s, \; Fr(n) = \frac{4^s + 2}{3k_1 k_2 \cdot 4^s}.$$

So,

$$\frac{1}{3k_1 k_2} < Fr(n) \leq \frac{1}{2k_1 k_2}$$

Conclusion. Function $Fr(n)$ at semiprimes $n = pq$ depends mostly on values k_1 and k_2 in representation $p = k_1 u + 1$, $q = k_2 u + 1$. $Fr(n)$ takes maximal values close to $1/4$ only at small k_1 and k_2. This completely corresponds to experimental data. Among values ψ_k the most expected are pseudoprimes of form $u = (u+1)(2u+1)$ with minimal values $k_1 = 1$ and $k_2 = 2$.

An important question connecting with efficiency of MRT is the average frequency of witnesses for composite numbers. As earlier, we study this problem for semiprime integers.

Let fix any prime p and a board B. We count average frequency of integers pq, $q > p$, $pq \leq B$. For convenience, we assume that $B = p(p + (p-1)k)$ for a positive $k \in \mathbf{Z}$.

For simplicity we explain all deductions at example $p = 11$. Every prime q has $d = GCD(p - 1, q - 1)$ equal either 2, or 10.

Let $d = 10$. Corresponding q lie in the set $\{21, 31, 41, 51, 61, 71, 81, 91, 101, \ldots, 10k + 11\}$, where $10k + 11 = B/p$. Each third integer in the sequence is a multiple of 3, some others are multiples of 7, 11 etc. Since q should be prime we need to remove them from the sequence. The rest consists of integers

$$Q_B = \{31, 41, 61, 71, 101, 113 \ldots\}. \tag{6}$$

We assume that primes $q \in Q_B$ are distributed uniformly in the interval $[1, B/p]$. Then the average frequency can be estimated as

$$Avg(Fr(n)) \approx \frac{1}{k}\left(\frac{1}{4} + \frac{1}{6} + \ldots + \frac{1}{2k}\right) = \frac{1}{2k}\left(1 + \frac{1}{2} + \frac{1}{3} + \ldots + \frac{1}{k}\right)$$

(we remind that $Fr(p(i(p-1) + p) = 1/2(i+1))$.

The expression in the last brackets is a partial sum of the Harmonic Series. Its value is

$$\sum_{i=1}^{k} \frac{1}{i} < \sum_{i=1}^{k+1} \frac{1}{i} = \ln k + \gamma + \varepsilon_n,$$

where $\gamma = 0.5772\ldots$ is the Euler–Mascheroni constant and $\lim_{k \to \infty} \varepsilon_n = 0$. Constant γ and additive ε_n can be ignored so

$$Avg(Fr(n)) < \frac{\ln k}{2k}$$

Since $(p-1)k+1 = B/p$, then $k > B/p^2 - 1$ and $\ln k < \ln B$, so

$$Avg(Fr(n)) < \frac{\ln B}{2(B-p^2)} \cdot p^2 \tag{7}$$

Let us move now to primes q of type $d = GCD(p-1, q-1) = 2$. They lie in the sequence

$$q \in \{13, 15, 17, 19, 23, , 25, 27, 29, \ldots, 2m+1\}$$

where $2m + 1 = B/p$, $q = 2i + 1$, $GCD(i, 5) = 1$. When we remove composite integers, the rest contains at least half members.

Integers $n = pq$ with $GCD(p-1, q-1) = 2$ have only trivial witnesses 1 and $n-1$ so their frequency function takes values

$$Fr(n) = \frac{2}{(p-1)(q-1)}.$$

Assuming that such n are distributed uniformly in the interval $[p^2; B]$ we estimate the average frequency by expression

$$Avg(Fr) \approx \left(2 \sum_{p \leq k \leq m} \frac{1}{(p-1)(2k+1)}\right) / \left(\frac{2m+1-p}{2}\right) <$$

$$\frac{4}{(2m+1-p)(p-1)} \cdot \frac{1}{2} \cdot \sum_{i=(p+1)/2}^{m} \frac{1}{i} < \frac{2}{(2m+1-p)(p-1)} \cdot \ln m$$

Substituting in the last expression $2m + 1 = B/p$ we get

$$Avg(Fr) < \frac{2p \ln B}{(B-p^2)(p-1)} \tag{8}$$

Expressions (7) and (8) give upper bounds for two types of integers $n = pq$. In the second case the estimation is lesser so average estimation for the united class of all $n = pq \leq B$, $p < q$, can be set by the upper bound of (7). This assertion does not depend on a special $p = 11$ so we can state the following theorem.

Theorem 6. *Let p be a prime and B satisfy $B > p^2$. Then the average frequency of witnesses in the class of semiprimes $n = pq \leq B$, $q > p$, has an upper bound*

$$Avg(Fr(n)) < \frac{p^2 \ln B}{2(B-p^2)}$$

Note than limit of the average function is 0 as $B \to \infty$. This explains the phenomenon that the number of false conclusions in the Miller–Rabin test decreases when length of tested integers increases.

5. Numbers with Maximal Frequency of Witnesses

In this section we study composite n with maximal frequency $Fr(n) = 1/4$. Let $n = p_1 p_2 \ldots p_k$ be the product of k different primes.

We begin with case $k = 2$. As we see from the previous section, integers $n = pq$ have maximal frequency only in case when $q = 2p - 1$. Such pairs appear comparatively often, and their quantity is diminishing together with their size.

Table 1 contains number of semiprimes with maximal frequency in intervals $[(i-1) \cdot 10^5; i \cdot 10^5;]$, $1 \leq i < 10$.

Table 1. Distribution of semiprimes with maximal frequency below 10^6.

1	2	3	4	5	6	7	8	9	10
670	494	448	412	424	386	393	358	370	343

Case $k = 3$ is more interesting. In order function $Fr(pqr)$ reached its maximum = 0.25, we need satisfaction of four requirements:

$$
\begin{aligned}
&1.\ GCD(p-1; qr-1) = p-1,\\
&2.\ GCD(q-1; pr-1) = q-1,\\
&3.\ GCD(r-1; pq-1) = r-1.\\
&4.\ bin(p-1) = bin(q-1) = bin(r-1) = 1.
\end{aligned}
\qquad (9)
$$

Such triples exist, and an example of it was already given in Rabin's paper [2] $n = 487 \cdot 1531 \cdot 2683 = 2000436751$. Rabin himself estimated $Fr(n)$ as 0.2493, but the difference is due to the fact that he did not include 1 in the list of witnesses.

Such triples appear much more seldom and have a form

$$n = (2k_1 + 1)u \cdot (2k_2 + 1)u \cdot (2k_3 + 1)u \text{ for } u \in N.$$

We arranged the search of such triples at a computer and found 160 such integers not exceeding $2 \cdot 10^{14}$. The least triple we found is

$$n = 19 \cdot 199 \cdot 271 = 1024651.$$

The largest found triple has a form $n = (u+1)(3u+1)(5u+1)$ at $u = 24102$:

$$n = 24103 \cdot 72307 \cdot 120511 = 21002\,84533\,02331.$$

Let us study the form $\langle u, 3u, 5u \rangle$ and find restrictions on u in order to $n = (u+1)(3u+1)(5u+1)$ satisfies first 3 conditions of (9). The first requirement is satisfied automatically. The second and third requirement are listed below:

$$(3u+1) - 1 \mid (u+1)(5u+1) - 1 \;\to\; u \equiv 0 \bmod 3.$$

$$(5u+1) - 1 \mid (u+1)(3u+1) - 1 \;\to\; 3u + 4 \equiv 0 \bmod 5,$$

so $u = 6 + 15t$ for $t \geq 1$. If we add requirements $p \equiv q \equiv r \equiv 3 \bmod 4$ we obtain

$$15t + 7 \equiv 3 \bmod 4 \;\to\; t \equiv 1 \bmod 4,\ u = 6 + 15(1 + 4t_1) = 21 + 60t_1.$$

Let now consider products of k primes where $k \geq 4$. The maximum of frequency of such products is $1/2^{k-1}$, since it is reached when for any $i \leq k$ $(p_i - 1)/2$ is odd, and $(p_i - 1) \mid \prod(p_{j \neq i} - 1)$. Then,

$$Fr(p) = 2 \cdot \prod_{i=1}^{k} \frac{p_i - 1}{2} = \frac{\varphi(n)}{2^{k-1}}.$$

A quick search of tuples $n = pqrt$ below 10^{12} gave 70 examples of them. The least 4-tuple was

$$n = 19 \cdot 31 \cdot 127 \cdot 547 = 40917241,$$

while the largest was

$$n = 19 \cdot 127 \cdot 14071 \cdot 29347 = 99\,64281\,70081.$$

Some computational results on distribution of strong semiprime integers can be found in [14].

6. Conclusions

In this section we will summarize the main results of the paper.

1. We found exact formulas for the number of witnesses for composite n with different number of factors.
2. We introduced the frequency function $Fr(n)$ characterizing the probability to find at one attempt a primality witness for a given n and found exact bounds for distribution of this function for semiprime integers n.
3. Like as Damgard, Landrock, and Pomerance in [13], we studied an average values of $Fr(n)$ at intervals $[1;x]$ for semiprime integers $n = pq$, $n \leq x$, with fixed p and showed that it bounded above by $p^2 \log x / 2(x - p^2)$.
   Since such integers have maximal values of $F(n)$ among all composites, this opens a way in future investigations to find exact upper bounds for average values of frequency function among all k-bit odd integers for any k.
4. Finally, we described possible forms of composites with maximal values of frequency function for products of k distinct primes at $k \geq 2$ and using computer calculations found their examples and their quantity at initial intervals of set of all naturals.

Author Contributions: S.T.I. gave impetus to the research and proved Theorems 1 and 2. B.G.M. proved other theorems and propositions, and R.G.R. developed software for testing results. All authors have read and agreed to the published version of the manuscript.

Funding: This research was funded by RFBR grant 18-47-16005. This investigation was supported by the grant of Scientific and Educational Mathematical Center of the Volga Federal District, agreement No. 075-02-2020-1478.

Conflicts of Interest: The authors declare no conflict of interest.

References

1. Miller Gary, L. Riemann's Hypothesis and Tests for Primality. *J. Comput. Syst. Sci.* **1976**, *13*, 300–317. [CrossRef]
2. Rabin, M. Probabilistic algorithm for testing primality. *J. Number Theory* **1980**, *12*, 128–138. [CrossRef]
3. Crandall, R.; Pomerance, C. *The Prime Numbers: A Computational Perspertive*, 2nd ed., Springer: Berlin, Germany, 2005; 604p.
4. Alford, W.R.; Granville, A.; Pomerance, C. On the difficulty of finding reliable witnesses. In *Algorithmic Number Theory*, First Internat. Symp., ANTS-I; Lecture Notes in Computer Science; Springer: Berlin/Heidelberg, Germany, 1994; p. 116.
5. Pomerance, C.; Selfridge, J.L.; Wagstaff, S.S., Jr. The pseudoprimes to $25 \cdot 10^9$. *Math. Comput.* **1980**, *35*, 1003–1026.
6. Jaeschke, G. On Strong Pseudoprimes to Several Bases. *Math. Comput.* **1993**, *61*, 915–926. [CrossRef]
7. Arnault, F. Rabin-Miller primality test: composite numbers which pass it. *J. Symb. Comput.* **1995**, *64*, 355–361. [CrossRef]
8. Arnault, F. Constructing Carmichael numbers which are strong pseudoprimes to several bases. *J. Symb. Comput.* **1995**, *20*, 151–161. [CrossRef]
9. Zhang, Z. Finding strong pseudoprimes to several bases. *Math. Comput.* **2001**, *70*, 863–872. [CrossRef]
10. Jiang, J.; Deng, Y. Strong pseudoprimes to the first 9 prime bases. *arXiv* **2012**, arXiv:1207.0063v1.
11. Sorenson, J.; Webster, J. Strong pseudoprimes to twelve prime bases. *arXiv* **2015**, arXiv:1509.00864v1.
12. Erdos, P.; Pomerance, C. On the number of false witnesses for a composite number. *Math. Comput.* **1986**, *46*, 259–279. [CrossRef]

13. Damgard, I.; Landrock, P.; Pomerance, C. Average case error estimates for the strong probable prime test. *Math. Comput.* **1993**, *61*, 177194. [CrossRef]
14. Ishmukhametov, S.; Mubarakov, B. On practical aspects of the Miller–Rabin primality test. *Lobachevskii J. Math.* **2013**, *34*, 304–312 [CrossRef]

© 2020 by the authors. Licensee MDPI, Basel, Switzerland. This article is an open access article distributed under the terms and conditions of the Creative Commons Attribution (CC BY) license (http://creativecommons.org/licenses/by/4.0/).

Article

On a Generalization of a Lucas' Result and an Application to the 4-Pascal's Triangle †

Atsushi Yamagami * and Kazuki Taniguchi

Department of Information Systems Science, Soka University, Tokyo 192-8577, Japan; e1658229@soka-u.jp
* Correspondence: yamagami@soka.ac.jp
† This research did not receive any specific grant from funding agencies in the public, commercial, or not-for-profit sectors.

Received: 29 January 2020; Accepted: 10 February 2020; Published: 16 February 2020

Abstract: The Pascal's triangle is generalized to "the k-Pascal's triangle" with any integer $k \geq 2$. Let p be any prime number. In this article, we prove that for any positive integers n and e, the n-th row in the p^e-Pascal's triangle consists of integers which are congruent to 1 modulo p if and only if n is of the form $\dfrac{p^{em}-1}{p^e-1}$ with some integer $m \geq 1$. This is a generalization of a Lucas' result asserting that the n-th row in the (2-)Pascal's triangle consists of odd integers if and only if n is a Mersenne number. As an application, we then see that there exists no row in the 4-Pascal's triangle consisting of integers which are congruent to 1 modulo 4 except the first row. In this application, we use the congruence $(x+1)^{p^e} \equiv (x^p+1)^{p^{e-1}} \pmod{p^e}$ of binomial expansions which we could prove for any prime number p and any positive integer e. We think that this article is fit for the Special Issue "Number Theory and Symmetry," since we prove a symmetric property on the 4-Pascal's triangle by means of a number-theoretical property of binomial expansions.

Keywords: the p^e-Pascal's triangle; Lucas' result on the Pascal's triangle; congruences of binomial expansions

MSC: 11A99.

---

1. Introduction

As it is known, Pascal's triangle is constructed in the following way: Write the first row "1 1". Then each member of each subsequent row is given by taking the sum of the just above two members, regarding any blank as 0.

Example 1. *Here is the Pascal's triangle from the first row to the 7-th row:*

```
              1  1
             1  2  1
            1  3  3  1
           1  4  6  4  1
          1  5  10 10 5  1
         1  6  15 20 15 6  1
        1  7  21 35 35 21 7  1
```

Remark 1. For any integers $n \geq 1$ and $r \geq 0$, we put

$$_nC_r := \frac{n!}{r!(n-r)!} = \frac{n(n-1)\cdots(n-r+1)}{r\cdots 1},$$

where we put $0! = 1$. Then it is well-known that the n-th row in the Pascal's triangle is equal to the sequence

$$_nC_0, {_nC_1}, \ldots, {_nC_{n-1}}, {_nC_n}$$

consisting of $n+1$ terms.

In ([1], Section 1.4), the construction above is generalized as follows:

Definition 1. Let $k \geq 2$ be any integer. The k-Pascal's triangle is constructed in the following way: Write the first row "$\overbrace{1\,1\,\cdots\,1}^{k}$". Then each member of each subsequent row is given by taking the sum of the just above k members regarding the blank as 0.

Example 2. In the case where $k = 4$, the 4-Pascal's triangle from the first row to the 5-th row is the following:

```
            1  1  1  1
         1  2  3  4  3  2  1
      1  3  6 10 12 12 10  6  3  1
   1  4 10 20 31 40 44 40 31 20 10  4  1
1  5 15 35 65 101 135 155 155 135 101 65 35 15 5 1
```

Remark 2. (1) In ([1], Section 1.4), for any integers $k \geq 2$ and $n \geq 1$, it is mentioned that the n-th row in the k-Pascal's triangle consists of $n(k-1)+1$ integers

$$_nC_0^{(k)}, {_nC_1^{(k)}}, \ldots, {_nC_{n(k-1)-1}^{(k)}}, {_nC_{n(k-1)}^{(k)}}$$

satisfying the equation

$$(x^{k-1} + x^{k-2} + \cdots + x + 1)^n$$
$$= {_nC_0^{(k)}} x^{n(k-1)} + {_nC_1^{(k)}} x^{n(k-1)-1} + \cdots + {_nC_{n(k-1)-1}^{(k)}} x + {_nC_{n(k-1)}^{(k)}}$$

of polynomials with indeterminate x and integral coefficients. A detailed proof of this fact is described in ([2], Lemma 1.1).

(2) In ([1], Section 9.10), the following formula for $_nC_i^{(k)}$ is described:

$$_nC_i^{(k)} = \sum_{j=0}^{\left[\frac{i}{k}\right]} (-1)^j {_{n+i-jk-1}C_{n-1}} \cdot {_nC_j},$$

where $\left[\frac{i}{k}\right]$ is the greatest integer that is less than or equal to $\frac{i}{k}$.

In Example 1, we can see that the n-th row consists of odd integers when n is equal to the Mersenne number 1, 3 or 7. Actually, Lucas showed the following

Theorem 1 ([3], Exemple I in Section 228). Let $n \geq 1$ be any integer. Then $_nC_r$ is odd for any $0 \leq r \leq n$ if and only if n is a Mersenne number, i.e., n is of the form $2^m - 1$ with some integer $m \geq 1$.

In Section 2 in this article, we generalize the Lucas' result above as the following

Theorem 2. *Let p be any prime number and e any positive integer. For any integer $n \geq 1$, the n-th row in the p^e-Pascal's triangle consists of integers which are congruent to 1 modulo p if and only if n is of the form $\dfrac{p^{em}-1}{p^e-1}$ with some integer $m \geq 1$.*

Remark 3. (1) *Theorem 2 is a generalization of ([2], Theorem 0.2) which is in the case where $e = 1$.*
(2) *We can see that Example 2 gives a partial example of Theorem 2 in the case where $p = 2$, $e = 2$ and $m = 1, 2$.*

As an application of Theorem 2, we can prove that ([2], Conjecture 0.3) holds for $k = 4$, i.e., there exists no row in the 4-Pascal's triangle consisting of integers which are congruent to 1 modulo 4 except the first row as follows:

By Theorem 2, in the case where $k = 4$, we see that for any integer $n \geq 1$, the n-th row in the 4-Pascal's triangle consists of odd integers if and only if n is of the form $\dfrac{4^m-1}{3}$ with some integer $m \geq 1$.

Moreover, we can see an essential property of the $\dfrac{4^m-1}{3}$-th row in the 4-Pascal's triangle for any integer $m \geq 2$ as in the following theorem proved in Section 3.2:

Theorem 3. *For any integer $m \geq 2$, the $\dfrac{4^m-1}{3}$-th row in the 4-Pascal's triangle is congruent to the sequence*

$$\overbrace{1133\cdots 1133}^{2^{2m-3}}\overbrace{3311\cdots 3311}^{2^{2m-3}}$$

modulo 4, which consists of the repeated 1133's and 3311's whose numbers are the same 2^{2m-3}.

Therefore we can obtain the following

Corollary 1. *([2], Conjecture 0.3) holds for $k = 4$, i.e., there exists no row in the 4-Pascal's triangle consisting of integers which are congruent to 1 modulo 4 except the first row.*

Remark 4. (1) *By Example 2, in the case where $m = 2$, we can see that the 5-th row in the 4-Pascal's triangle is congruent to the sequence*

$$1\ 1\ 3\ 3\ 1\ 1\ 3\ 3\ 3\ 3\ 1\ 1\ 3\ 3\ 1\ 1$$

modulo 4, which matches the assertion of Theorem 3.
(2) *It seems that one could obtain the forms of the sequenece to which the $\left(\dfrac{4^m-1}{3} \pm \ell\right)$-th row in the 4-Pascal's triangle is congruent modulo 4 for some positive integers ℓ by means of Theorem 3. We would like to do these calculations in the future.*

In the proof of Theorem 3 in Section 3.2, we shall use the following lemma proved in Section 3.1:

Lemma 1. *For any prime number p and any positive integer e, we have the following coefficient-wise congruence*

$$(x+1)^{p^e} \equiv (x^p+1)^{p^{e-1}} \pmod{p^e}$$

of binomial expansions with indetermiate x.

2. A Proof of Theorem 2

Although Theorem 2 can be proved by the same argument as the proof of ([2], Theorem 0.2), we shall describe its detailed proof here to make this article self-contained.

Let n and e be any positive integers and p be any prime number.

Firstly, we assume that n is of the form $n = \dfrac{p^{em}-1}{p^e-1}$ with some integer $m \geq 1$. In the algebra $\mathbb{F}_p[x]$ of polynomials of one variable x with coefficients in the finite field $\mathbb{F}_p = \mathbb{Z}/p\mathbb{Z}$ of p elements, we see that for any positive integer ℓ,

$$(x-1)^{p^\ell - 1} = \frac{(x-1)^{p^\ell}}{x-1} = \frac{x^{p^\ell}-1}{x-1}$$
$$= x^{p^\ell - 1} + x^{p^\ell - 2} + \cdots + x + 1.$$

Therefore we see that

$$(x^{p^e-1} + x^{p^e-2} + \cdots + x + 1)^n = (x^{p^e-1} + x^{p^e-2} + \cdots + x + 1)^{\frac{p^{em}-1}{p^e-1}}$$
$$= ((x-1)^{p^e-1})^{\frac{p^{em}-1}{p^e-1}}$$
$$= (x-1)^{p^{em}-1}$$
$$= x^{p^{em}-1} + x^{p^{em}-2} + \cdots + x + 1$$
$$= x^{n(p^e-1)} + x^{n(p^e-1)-1} + \cdots + x + 1$$

in $\mathbb{F}_p[x]$. By Remark 2 (1), this implies that the n-th row in the p^e-Pascal's triangle consists of integers which are congruent to 1 modulo p as desired.

Conversely, we now assume that n is of the form

$$n = 1 + p^e + \cdots + p^{e(m-1)} + k$$

with some integers $m \geq 1$ and $1 \leq k \leq p^{em} - 1$. Moreover, we assume that we have

$$(x^{p^e-1} + x^{p^e-2} + \cdots + x + 1)^n = x^{n(p^e-1)} + x^{n(p^e-1)-1} + \cdots + x + 1$$

in $\mathbb{F}_p[x]$ to obtain some contradiction. Since the left hand side is equal to $(x-1)^{n(p^e-1)}$ and the right hand side is equal to $\dfrac{x^{n(p^e-1)+1}-1}{x-1}$, we then have the equality

$$(x-1)^{n(p^e-1)+1} = x^{n(p^e-1)+1} - 1$$

in $\mathbb{F}_p[x]$. Since $n = \dfrac{p^{em}-1}{p^e-1} + k$, this implies that

$$(x-1)^{p^{em}+k(p^e-1)} = x^{p^{em}+k(p^e-1)} - 1.$$

Let $v_p(a)$ be the p-adic valuation of any non-zero integer a, i.e., $p^{v_p(a)} \mid a$ and $p^{v_p(a)+1} \nmid a$. Since $1 \leq k \leq p^{em} - 1$, we see that $v_p(k) < em$ and then

$$v_p(p^{em} + k(p^e - 1)) = v_p(k).$$

Therefore we can put

$$p^{em} + k(p^e - 1) = p^{v_p(k)}t$$

with some positive integer t which is prime to p. Then we have

$$(x-1)^{p^{v_p(k)}t} = x^{p^{v_p(k)}t} - 1 = (x^t - 1)^{p^{v_p(k)}}$$

which implies that

$$(x-1)^{p^{v_p(k)}(t-1)} = (x^{t-1} + x^{t-2} + \cdots + x + 1)^{p^{v_p(k)}},$$

since $\mathbb{F}_p[x]$ is an integral domain. Since $p^{v_p(k)} < p^{em}$, we see that $t \geq 2$. Therefore substituting $x = 1$ leads a contradiction $t = 0$ in $\mathbb{F}_p$ as desired, and Theorem 2 is proved.

3. An Application to the 4-Pascal's Triangle

By Theorem 2, in the case where $p = 2$ and $e = 2$, we see that for any integer $n \geq 1$, the n-th row in the 4-Pascal triangle consists of odd integers if and only if n is of the form $\dfrac{4^m - 1}{3}$ with some integer $m \geq 1$.

In this section, we shall prove Theorem 3 asserting that for any integer $m \geq 2$, the $\dfrac{4^m - 1}{3}$-th row in the 4-Pascal's triangle is congruent to the sequence

$$\underbrace{1133\cdots 1133}_{2^{2m-3}}\underbrace{3311\cdots 3311}_{2^{2m-3}}$$

modulo 4. Here we should note that 2^{2m-3} is the number of 1133's and 3311's, respectively.

Then Theorems 2 and 3 imply that ([2], Conjecture 0.3) holds in the case where $k = 4$, i.e., there exists no row in the 4-Pascal's triangle consisting of integers which are congruent to 1 modulo 4 except the first row as we have seen in Corollary 1.

3.1. On a Congruence of Binomial Expansions

Before proving Theorem 3, we shall prove Lemma 1 on a congruence of binomial expansions in this subsection.

Let p be any prime number and e any positive integer. In order to prove the congruence

$$(x+1)^{p^e} \equiv (x^p + 1)^{p^{e-1}} \pmod{p^e}$$

of binomial expansions with indeterminate x, it suffices to see the following two congruences hold:

(1) For any integer $1 \leq \ell \leq p^e - 1$ which is prime to p,

$$_{p^e}C_\ell \equiv 0 \pmod{p^e}.$$

(2) In the case where $e \geq 2$, for any integers $0 \leq f \leq e-2$ and i such that $1 \leq ip^f \leq p^{e-1} - 1$ and $(i, p) = 1$,

$$_{p^e}C_{ip^{f+1}} \equiv {}_{p^{e-1}}C_{ip^f} \pmod{p^e}.$$

Firstly, we shall prove the part (1). In the case where $\ell = 1$, we see that

$$_{p^e}C_1 = p^e \equiv 0 \pmod{p^e}.$$

Moreover, in the case where $2 \leq \ell \leq p^e - 1$, we see that

$$_{p^e}C_\ell = \frac{p^e}{\ell} \prod_{j=1}^{\ell-1} \frac{p^e - j}{j}.$$

Since $v_p(p^e - j) = v_p(j)$ for any $1 \leq j \leq \ell - 1 < p^e$ and ℓ is prime to p, we then see that

$$v(_{p^e}C_\ell) = e - v_p(\ell) + \sum_{j=1}^{\ell-1} v_p\left(\frac{p^e - j}{j}\right)$$

$$= e + \sum_{j=1}^{\ell-1} (v_p(p^e - j) - v_p(j))$$

$$= e.$$

Therefore $_{p^e}C_\ell \equiv 0 \pmod{p^e}$, and part (1) is proved.

Secondly, we shall prove part (2). We see that

$$_{p^e}C_{ip^{f+1}} - {}_{p^{e-1}}C_{ip^f}$$

$$= \frac{p^e}{ip^{f+1}} \cdot \frac{\prod_{j=0}^{i-1}\left(\prod_{1 \leq k \leq p^{f+1}-1,\, (k,p)=1} (k + jp^{f+1} + (p^e - ip^{f+1}))\right)}{\prod_{j=0}^{i-1}\left(\prod_{1 \leq k \leq p^{f+1}-1,\, (k,p)=1} (k + jp^{f+1})\right)} \cdot {}_{p^{e-1}}C_{ip^f - 1}$$

$$- \frac{p^{e-1}}{ip^f} \cdot {}_{p^{e-1}}C_{ip^f - 1}$$

$$= \frac{p^{e-f-1}}{i} \cdot {}_{p^{e-1}}C_{ip^f - 1} \left(\frac{\prod_{j=0}^{i-1}\left(\prod_{1 \leq k \leq p^{f+1}-1,\, (k,p)=1} (k + jp^{f+1} + (p^e - ip^{f+1}))\right)}{\prod_{j=0}^{i-1}\left(\prod_{1 \leq k \leq p^{f+1}-1,\, (k,p)=1} (k + jp^{f+1})\right)} - 1\right)$$

and that

$$\prod_{j=0}^{i-1}\left(\prod_{1 \leq k \leq p^{f+1}-1,\, (k,p)=1} (k + jp^{f+1} + (p^e - ip^{f+1}))\right)$$

$$\equiv \prod_{j=0}^{i-1}\left(\prod_{1 \leq k \leq p^{f+1}-1,\, (k,p)=1} (k + jp^{f+1})\right)$$

$$\equiv \left(\prod_{1 \leq k \leq p^{f+1}-1,\, (k,p)=1} k\right)^i \pmod{p^{f+1}}.$$

Since $(i, p) = 1$, we then see that $_{p^e}C_{ip^{f+1}} - {}_{p^{e-1}}C_{ip^f}$ is divisible by $p^{e-f-1} \cdot p^{f+1} = p^e$ as desired.

3.2. A Proof of Theorem 3

Now we shall prove Theorem 3 by means of Lemma 1 with $p = 2$ and $e = 2$, i.e., the congruence of binomial expansions

$$(x+1)^4 \equiv (x^2 + 1)^2 \pmod{4}. \quad \cdots (*)$$

By Remark 2 (1), proving Theorem 3 is equivalent to proving that for any integer $m \geq 2$, the coefficient-wise congruence

$$(x^3 + x^2 + x + 1)^{\frac{4^m-1}{3}}$$
$$\equiv x^{4^m-1} + x^{4^m-2} - x^{4^m-3} - x^{4^m-4} + \cdots + x^{\frac{4^m}{2}+3} + x^{\frac{4^m}{2}+2} - x^{\frac{4^m}{2}+1} - x^{\frac{4^m}{2}}$$
$$- x^{\frac{4^m}{2}-1} - x^{\frac{4^m}{2}-2} + x^{\frac{4^m}{2}-3} + x^{\frac{4^m}{2}-4} - \cdots - x^3 - x^2 + x + 1 \pmod{4} \quad \cdots \quad (**)$$

holds with indeterminate x by the induction on m.

Before doing this, we see the following

Lemma 2. *The polynomial in the right hand side of the congruence relation* $(**)$ *can be decomposed as*

$$(x+1)(x^2-1)(x^4+1) \cdots (x^{2^{2m-2}}+1)(x^{2^{2m-1}}-1).$$

Proof. By a direct calculation, we can see that there exists some positive integer ℓ such that the polynomial in the right hand side of the congruence relation $(**)$ can be decomposed as

$$(x+1)(x^{4^m-2} - x^{4^m-4} + x^{4^m-6} - x^{4^m-8} + \cdots + x^{\frac{4^m}{2}+6} - x^{\frac{4^m}{2}+4} + x^{\frac{4^m}{2}+2} - x^{\frac{4^m}{2}}$$
$$- x^{\frac{4^m}{2}-2} + x^{\frac{4^m}{2}-4} - x^{\frac{4^m}{2}-6} + x^{\frac{4^m}{2}-8} - \cdots - x^6 + x^4 - x^2 + 1)$$
$$= (x+1)(x^2-1)(x^{4^m-4} + x^{4^m-8} + \cdots + x^{\frac{4^m}{2}+4} + x^{\frac{4^m}{2}}$$
$$- x^{\frac{4^m}{2}-4} - x^{\frac{4^m}{2}-8} - \cdots - x^4 - 1)$$
$$= \cdots$$
$$= (x+1)(x^2-1)(x^4+1) \cdots (x^{2^\ell}+1)(x^{3 \cdot 2^{\ell+1}} + x^{2 \cdot 2^{\ell+1}} - x^{2^{\ell+1}} - 1)$$
$$= (x+1)(x^2-1)(x^4+1) \cdots (x^{2^\ell}+1)(x^{2^{\ell+1}}+1)(x^{2^{\ell+2}}-1).$$

Since the degree of the polynomial in the right hand side of the congruence relation $(**)$ is equal to $4^m - 1$, we then see that

$$4^m - 1 = 1 + 2 + 2^2 + \cdots + 2^\ell + 2^{\ell+1} + 2^{\ell+2}$$
$$= 2^{\ell+3} - 1,$$

which implies that $\ell = 2m - 3$ as desired. □

Let us start to prove Theorem 3 by the induction on $m \geq 2$. Firstly, in the case where $m = 2$, since

$$(x^2 + 1)^4 \equiv (x^4 + 1)^2 \pmod{4}$$

and

$$(x+1)^4 \equiv (x^2+1)^2 \equiv x^4 + 2x^2 + 1 \equiv x^4 - 2x^2 + 1$$
$$\equiv (x^2 - 1)^2 \pmod{4}$$

by the congruence relation $(*)$, we see that

$$(x^3 + x^2 + x + 1)^5 \equiv (x+1)(x^2+1)(x+1)^4(x^2+1)^4$$
$$\equiv (x+1)(x^2+1)(x^2-1)^2(x^4+1)^2$$
$$\equiv (x+1)(x^2-1)(x^4+1)(x^8-1) \pmod{4}.$$

Therefore the congruence relation $(**)$ holds for $m = 2$ by Lemma 2.

Secondly, we assume that the congruence relation $(**)$ holds for some $m \geq 2$. By the congruence relation $(*)$, we see that

$$\begin{aligned}(x+1)^{4^m} &\equiv (x+1)^{2^{2m}} \equiv ((x+1)^4)^{2^{2m-2}} \\ &\equiv (x^2+1)^{2^{2m-1}} \equiv ((x^2+1)^4)^{2^{2m-3}} \\ &\equiv (x^{2^2}+1)^{2^{2m-2}} \\ &\equiv \cdots \\ &\equiv (x^{2^{2m-1}}+1)^2 \\ &\equiv (x^{\frac{4^m}{2}}+1)^2 \pmod{4}.\end{aligned}$$

By Lemma 2, we then see that

$$\begin{aligned}&(x^3+x^2+x+1)^{\frac{4^{m+1}-1}{3}} \\ &\equiv (x^3+x^2+x+1)^{\frac{4^m-1}{3}+4^m} \\ &\equiv (x+1)(x^2-1)(x^4+1)\cdots(x^{2^{2m-2}}+1)(x^{2^{2m-1}}-1)(x^2+1)^{4^m}(x+1)^{4^m} \\ &\equiv (x+1)(x^2-1)(x^4+1)\cdots(x^{2^{2m-2}}+1)(x^{\frac{4^m}{2}}-1)(x^{4^m}+1)^2(x^{\frac{4^m}{2}}+1)^2 \\ &\equiv (x+1)(x^2-1)(x^4+1)\cdots(x^{2^{2m-2}}+1)(x^{2^{2m-1}}+1)(x^{2^{2m}}+1)(x^{2 \cdot 4^m}-1) \\ &\equiv (x+1)(x^2-1)(x^4+1)\cdots(x^{2^{2m}}+1)(x^{2^{2m+1}}-1) \pmod{4},\end{aligned}$$

i.e., the congruence relation $(**)$ also holds for $m+1$ as desired. This proves Theorem 3.

Author Contributions: Conceptualization, A.Y.; Investigation, A.Y. and K.T.; Writing—original draft, A.Y. All authors have read and agreed to the published version of the manuscript.

Acknowledgments: The first author is very grateful to the second author, who is one of his students at Soka University, for giving some interesting talks regarding his calculations of the k-Pascal's triangles with some specified composite numbers k in seminars held in 2019 at Soka University.

Conflicts of Interest: The authors declare no conflict of interest.

References

1. Matsuda, O.; Tsuyama Math Club in National Institute of Technology, Tsuyama College. 11 *kara Hajimaru sūgaku—k-Pasukaru Sankakkei, k-Fibonatchi Sūretsu, chōōgonsū;* [Mathematics that begins with 11—*k*-Pascal's Triangles, *k*-Fibonacci Sequences and the Super Golden Numbers]; Tokyo Tosho Co. Ltd.: Tokyo, Japan, 2008. (In Japanese)
2. Yamagami, A.; Harada, H. On a generalization of a Lucas' result on the Pascal triangle and Mersenne numbers. *JP J. Algebra Number Theory Appl.* **2019**, *42* 159–169. [CrossRef]
3. Lucas, É. *Théorie des Nombres*; Gauthier-Villars et Fils; Libraires, du Bureau des Longitudes, de l'École Polytechnique, Quai des Grands-Augustins: Paris, France, 1891; Volume 55.

© 2020 by the authors. Licensee MDPI, Basel, Switzerland. This article is an open access article distributed under the terms and conditions of the Creative Commons Attribution (CC BY) license (http://creativecommons.org/licenses/by/4.0/).

Article

Algebraic Numbers as Product of Powers of Transcendental Numbers

Pavel Trojovský

Department of Mathematics, Faculty of Science, University of Hradec Králové,
500 03 Hradec Králové, Czech Republic; pavel.trojovsky@uhk.cz; Tel.: +42-049-333-2801

Received: 17 June 2019; Accepted: 4 July 2019; Published: 8 July 2019

Abstract: The elementary symmetric functions play a crucial role in the study of zeros of non-zero polynomials in $\mathbb{C}[x]$, and the problem of finding zeros in $\mathbb{Q}[x]$ leads to the definition of algebraic and transcendental numbers. Recently, Marques studied the set of algebraic numbers in the form $P(T)^{Q(T)}$. In this paper, we generalize this result by showing the existence of algebraic numbers which can be written in the form $P_1(T)^{Q_1(T)} \cdots P_n(T)^{Q_n(T)}$ for some transcendental number T, where $P_1, \ldots, P_n, Q_1, \ldots, Q_n$ are prescribed, non-constant polynomials in $\mathbb{Q}[x]$ (under weak conditions). More generally, our result generalizes results on the arithmetic nature of z^w when z and w are transcendental.

Keywords: Baker's theorem; Gel'fond–Schneider theorem; algebraic number; transcendental number

1. Introduction

The name "transcendental", which comes from the Latin word "transcendĕre", was first used for a mathematical concept by Leibniz in 1682. Transcendental numbers in the modern sense were defined by Leonhard Euler (see [1]).

A complex number α is called algebraic if it is a zero of some non-zero polynomial $P \in \mathbb{Q}[x]$. Otherwise, α is transcendental. Algebraic numbers form a field, which is denoted by $\overline{\mathbb{Q}}$. The transcendence of e was proved by Charles Hermite [2] in 1872, and two years later Ferdinand von Lindeman [3] extended the method of Hermite's proof to derive that π is also transcendental. It should be noted that Lindemann proved the following, much more general statement: The number e^α, where α is any non-zero algebraic number, is always transcendental (see [4]). In 1900, Hilbert raised the question of the arithmetic nature of the power α^β of two algebraic numbers α and β (it was the seventh problem in his famous list of 23 problems, which he presented at the International Congress of Mathematicians in Paris). The complete solution to this problem was found independently by Gel'fond and Schneider (see [5], p. 9) in 1934. Their results can be formulated as the following theorem (the ideas of the Gel'fond–Schneider proof were used partially in, e.g., [6–8]).

Theorem 1. *The Gel'fond–Schneider Theorem: Let α and β be algebraic numbers, with $\alpha \neq 0$ and $\alpha \neq 1$, and let β be irrational. Then α^β is transcendental.*

The Gel'fond–Schneider Theorem classifies the arithmetic nature of x^y when both x, y are algebraic numbers (because x^y is an algebraic number when y is rational). Nevertheless, when at least one of these two numbers is transcendental, anything is possible (see Table 1 below).

Table 1. Possible results for the power x^y when x or y is transcendental.

Value of x	Class of Numbers	Value of y	Class of Numbers	Power x^y	Class of Numbers
2	algebraic	$\log 3/\log 2$	transcendental	3	algebraic
2	algebraic	$i \log 3/\log 2$	transcendental	3^i	transcendental
e^i	transcendental	π	transcendental	-1	algebraic
e	transcendental	π	transcendental	e^π	transcendental
$2^{\sqrt{2}}$	transcendental	$\sqrt{2}$	algebraic	4	algebraic
$2^{\sqrt{2}}$	transcendental	$i\sqrt{2}$	algebraic	4^i	transcendental

In all the previous examples we have $x \neq y$ (in fact, we used the fact that the logarithm function is the inverse of the exponential function many times). Also, in the cases in which x and y are both transcendental (in the previous table), these numbers are possibly (though it's not proved) algebraically independent. So, what happens if we consider numbers of the form x^x with x transcendental? Is it possible that some of these numbers are algebraic? We remark that the numbers e^e and π^π are expected (but not proved) to be transcendental. In fact, it is easy to use the Gel'fond–Schneider Theorem to prove that every prime number can be written in the form T^T for some transcendental number T (for a more general result, see [9]). In this direction, a natural question arises: Given arbitrary, non-constant polynomials $P, Q \in \mathbb{Q}[x]$, is there always a transcendental number T such that $P(T)^{Q(T)}$ is algebraic? Note that $P(T)$ and $Q(T)$ are algebraically dependent transcendental numbers (so they do not come from our table). Marques [10] showed that the answer for the previous question is yes. More generally, he proved that for any fixed, non-constant polynomials $P(x), Q(x) \in \mathbb{Q}[x]$, the set of algebraic numbers of the form $P(T)^{Q(T)}$, with T transcendental, is dense in some connected subset of either $\mathbb{R}$ or $\mathbb{C}$. A generalization of this result for rational functions with algebraic coefficients was proved by Jensen and Marques [11]. However, the previous results do not apply, e.g., to prove the existence of algebraic numbers which can be written in the form $(T^2+1)^T \cdot T^{T^2+T+1}$, with T transcendental.

In this paper, we will solve this kind of problem completely by proving a multi-polynomial version of the previous results. The following theorem states our result more precisely.

Theorem 2. *Let $P_1, \ldots, P_n, Q_1, \ldots, Q_n \in \mathbb{Q}[x]$ be non-constant polynomials, such that the leading coefficients of the Q_j's have the same sign. Then the set of algebraic numbers of the form $P_1(T)^{Q_1(T)} \cdots P_n(T)^{Q_n(T)}$, with T transcendental, is dense in some open subset of the complex plane. In fact, this dense set can be chosen to be $\{Q(1 + \sqrt[p]{2}) : Q \in K\}$, for some dense set $K \subseteq \mathbb{Q}(\sqrt{-1}) \setminus \{0\}$, $K \cap \mathbb{Q} = \emptyset$, and any prime number $p > 2 \cdot (\max_{1 \leq j \leq n} \{\deg Q_j\})!$.*

The proof of the above theorem combines famous classical theorems concerning transcendental numbers (like the Baker's Theorem on linear forms in logarithms and the Gel'fond–Schneider Theorem) and certain purely field-theoretic results. We point out that, in a similar way, we can prove Theorem 2 for rational functions with algebraic coefficients, but we choose to prove this simpler case in order to avoid too many technicalities, which can obscure the essence of the main idea.

2. Proof of Theorem 2

2.1. Auxiliary Results

Before we proceed to the proof of Theorem 2, we will need the following three lemmas. The first two lemmas come from the work of Baker on linear forms of logarithms of algebraic numbers (see [5], Chapter 2):

Lemma 1 (Cf. Theorem 2.4 in [5]). *If $\alpha_1, \alpha_2, \ldots, \alpha_n$ are algebraic numbers other than 0 or 1, $\beta_1, \beta_2, \ldots, \beta_n$ are algebraic with $1, \beta_1, \beta_2, \ldots, \beta_n$ linearly independent over $\mathbb{Q}$, then $\alpha_1^{\beta_1} \alpha_2^{\beta_2} \cdots \alpha_n^{\beta_n}$ is transcendental.*

Lemma 2 (Cf. Theorem 2.2 in [5]). *Any non-vanishing linear combination of logarithms of algebraic numbers with algebraic coefficients is transcendental.*

Let $\mathcal{F}$ be a family of polynomials. Hereafter, we will denote by $\mathcal{R}_\mathcal{F}$ the set of all the zeros of the polynomials in $\mathcal{F}$. The last of these lemmas is a purely field-theoretical result.

Lemma 3. *Let n be any positive integer and let $\mathcal{F}$ be a family of polynomials in $\mathbb{Q}[x]$ for which there exists a positive integer ℓ such that all polynomials in $\mathcal{F}$ have degree at most ℓ. Then for all prime numbers $p > \ell!$, the following holds:*

$$(1 + \sqrt[p]{2})^n \notin \mathbb{Q}(\mathcal{R}_\mathcal{F}). \tag{1}$$

Proof of Lemma 3. Set $\mathcal{F} = \{F_1, F_2, \ldots\}$, $K_n = \mathbb{Q}(\mathcal{R}_{F_1 \cdots F_n})$ and $t_n = [K_n : \mathbb{Q}]$. Since $K_n \subseteq K_{n+1}$, then $t_{n+1} = \ell_n t_n$, for some positive integer ℓ_n. Note that $\ell_n = [K_{n+1} : K_n] = [K_n(\mathcal{R}_{F_{n+1}}) : K_n] \leq (\deg F_{n+1})! \leq \ell!$. We claim that $(1 + \sqrt[p]{2})^n \notin \mathbb{Q}(\mathcal{R}_\mathcal{F})$ for all integers $n \geq 1$. For the contrary, there exist positive integers m and s such that $(1 + \sqrt[p]{2})^m \in K_s$. Then the degree of $(1 + \sqrt[p]{2})^m$ (which is p) divides t_s. However, $t_s = \ell_{s-1} \cdots \ell_1 t_1$ and $p > \ell! \geq \max_{j \in [1, s-1]} \{\ell_j, t_1\}$, which gives an absurdity. This completes the proof. □

With these lemmas in hand, we can proceed to the proof of our main outcome.

2.2. The Proof

In order to simplify our presentation, we use the familiar notation $[a, b] = \{a, a+1, \ldots, b\}$, for integers $a < b$.

Of course, it is enough to prove our theorem for the case that $P_1, \ldots, P_n$ are multiplicatively independent. For that, we take an open, simply connected subset Ω of $\mathbb{C}$, such that $P_j(x) \notin \{0, 1\}$ for all $x \in \Omega$ and $j \in [1, n]$. Choosing, for example, the principal branch of the multi-valued logarithm function, the function $f(x) := \prod_{j=1}^n P_j(x)^{Q_j(x)}$ is well defined and analytic in Ω. Moreover, $f(x)$ is a non-constant function. In fact, if f were constant then $f'(x) = 0$ in Ω and so

$$\sum_{j=1}^n Q_j'(x) \log P_j(x) + \sum_{j=1}^n \frac{Q_j(x) P_j'(x)}{P_j(x)} = 0, \tag{2}$$

for all $x \in \Omega$. We claim that $g(x) := \sum_{j=1}^n Q_j(x) P_j'(x)/P_j(x)$ is not the zero function in Ω. In fact, otherwise $G(x) := P_1(x) \cdots P_n(x) g(x)$ would be the zero polynomial, but the formal polynomial G has degree $\leq t := \max_{j \in [1,n]} \{m_1 + \cdots + m_n + t_j - 1\}$, where for all $j \in [1, n]$, m_j and t_j are the degree of P_j and Q_j, respectively. Now, if $t_{i_1} = \cdots = t_{i_s} = \max_{j \in [1,n]} \{t_j\}$, we get the relation $\sum_{j=1}^s m_{i_j} b_{i_j} = 0$ (the coefficient of x^t in G must be zero), where for all $j \in [1, n]$, b_j is the leading coefficient of Q_j. However $\sum_{j=1}^s m_{i_j} b_{i_j} \neq 0$, since $m_j > 0$ and b_j have the same sign. This gives a contradiction. Thus, there exists $\beta \in \Omega \cap \overline{\mathbb{Q}}$ such that $g(\beta) \neq 0$. Substituting then $x = \beta$ in (2), we have that $\sum_{j=1}^n Q_j'(\beta) \log P_j(\beta)$ is a nonzero algebraic number which contradicts Lemma 2. Hence f is a non-constant function.

Since f is a non-constant analytic function and Ω is an open connected set, $f(\Omega)$ is an open connected subset of $\mathbb{C}$. Let $\mathcal{F}$ be the family of polynomials $\{Q_i(x) - d : i \in [1, n], d \in \mathbb{Q}\} \cup \{x^2 + 1\}$. Clearly, each polynomial in $\mathcal{F}$ has degree $\leq 2\ell := 2\max\{\deg Q_1, \ldots, \deg Q_n\}$. Thus, the conditions to apply Lemma 3 are fulfilled. Hence, for $p > 2\ell!$, we have that the set $\mathcal{P} := \{r(1 + \sqrt[p]{2}) : r \in \mathbb{Q}(\sqrt{-1}) \setminus \{0\}\}$ forms a dense subset of $\mathbb{C}$ and no positive integer power of its elements lies in $\mathbb{Q}(\mathcal{R}_\mathcal{F})$. Since $f(\Omega)$ is open, $f(\Omega) \cap \mathcal{P}$ is dense in $f(\Omega)$. Now, it remains to prove that every number in this intersection can be written in the desired form. For that, let $\alpha := r(1 + \sqrt[p]{2}) \in f(\Omega) \cap \mathcal{P}$, then

$$\alpha = f(T) = \prod_{j=1}^n P_j(T)^{Q_j(T)}, \tag{3}$$

where $T \in \Omega$. Therefore, it is enough to prove that T is a transcendental number. To get a contradiction, suppose the contrary; i.e., that T is algebraic. Then $P_1(T), \ldots, P_n(T), Q_1(T), \ldots, Q_n(T)$ are also algebraic numbers. By the choice of Ω, Lemma 1 ensures the existence of a nontrivial $\mathbb{Q}$-relation among $1, Q_1(T), \ldots, Q_n(T)$ (this implies, in particular, that the degree of T is at most ℓ). Without loss of generality we can assume that $a_n Q_n(T) = a_0 + \sum_{j=1}^{n-1} a_j Q_j(T)$, where a_j is an integer, with $a_n > 0$. Therefore, identity (3) becomes

$$\alpha^{a_n} = P_n(T)^{a_0} \left(P_1(T)^{a_n} P_n(T)^{a_1}\right)^{Q_1(T)} \cdots \left(P_{n-1}(T)^{a_n} P_n(T)^{a_{n-1}}\right)^{Q_{n-1}(T)}.$$

Note that $\alpha^{a_n} P_n(T)^{-a_0}$ is an algebraic number and $P_j(T)^{a_n} P_n(T)^{a_j} \neq 0$ for $j \in [1, n-1]$. We claim that $P_j(T)^{a_n} P_n(T)^{a_j} \neq 1$ for some $j \in [1, n-1]$. In fact, otherwise we would have $\alpha^{a_n} = P_n(T)^{a_0} \in \mathbb{Q}(T)$ and so $(1 + \sqrt[p]{2})^{a_n} \in \mathbb{Q}(T, \sqrt{-1})$ has degree at most 2ℓ. However, this gives an absurdity since the degree of $(1 + \sqrt[p]{2})^{a_n}$ is $p > 2\ell!$. Thus, sometimes $P_j(T)^{a_n} P_n(T)^{a_j}$ is an algebraic number different from 0 and 1, so we can apply Lemma 1 again to get a $\mathbb{Z}$-relation $b_{n-1} Q_{n-1}(T) = b_0 + \sum_{j=1}^{n-2} b_j Q_j(T)$, where b_j is an integer, with $b_{n-1} > 0$. Analogously, one can iterate this process $n-1$ times to conclude that

$$\alpha^q = A(T) \left(P_1(T)^{c_1} \cdots P_n(T)^{c_n}\right)^{Q_1(T)}, \tag{4}$$

where $A(T) \in \mathbb{Q}(P_1(T), \ldots, P_n(T))$ and q, c_j's $\in \mathbb{Z}$, with $q > 0$. If $\prod_{j=1}^n P_j(T)^{c_j} = 1$, we would arrive at the same absurdity as before since $\mathbb{Q}(P_1(T), \ldots, P_n(T)) \subseteq \mathbb{Q}(T)$. Thus $\prod_{j=1}^n P_j(T)^{c_j} \in \overline{\mathbb{Q}} \backslash \{0, 1\}$, so by the Gel'fond–Schneider Theorem we deduce that $Q_1(T)$ is a rational number, say r/s, with some integers r and s, $s > 0$. Hence, T belongs to $\mathcal{R}_{Q_1(x) - r/s} \subseteq \mathcal{R}_{\mathcal{F}}$. But then $\alpha^{qs} = A(T)^s P_1(T)^{rc_1} \cdots P_n(T)^{rc_n}$ (see (4)) and thus $(1 + \sqrt[p]{2})^{qs} \in \mathbb{Q}(\mathcal{R}_{\mathcal{F}})$, contradicting the choice of p in Lemma 3. In conclusion, T must be transcendental, and this completes the proof. □

3. Conclusions

In this paper, we use analytic (complex analysis), algebraic (Galois' extensions and symmetry) and transcendental tools (Baker's theory) to prove, in particular, the existence of infinitely many algebraic numbers of the form $P_1(T)^{Q_1(T)} \cdots P_n(T)^{Q_n(T)}$, where T is a transcendental number and $P_1, \ldots, P_n, Q_1, \ldots, Q_n$ are previously fixed rational polynomials (under some weak technical conditions).

Funding: The author was supported by Project of Excelence PrF UHK, University of Hradec Králové, Czech Republic 01/2019.

Acknowledgments: The author thanks anonymous referees for their careful corrections and their valuable comments that helped to improve the paper's quality and readability.

Conflicts of Interest: The author declares no conflict of interest.

References

1. Erdös, P.; Dudley, U. Some remarks and problems in number theory related to the work of Euler. *Math. Mag.* **1983**, *56*, 292–298. [CrossRef]
2. Hermite, C. Sur la fonction exponentielle. *C. R. Acad. Sci. Paris I* **1873**, *77*, 18–24.
3. Lindemann, F. Über die Zahl π. *Math. Ann.* **1882**, *20*, 213–225. [CrossRef]
4. Marques, D. On the arithmetic nature of hypertranscendental functions at complex points. *Expo. Math.* **2011**, *29*, 361–370. [CrossRef]
5. Baker, A. *Transcendental Number Theory*; Cambridge Mathematical Library, Cambridge University Press: Cambridge, UK, 1990.
6. Bundschuh, P.; Waldschmidt, M. Irrationality results for theta-functions by Gel'fond–Schneider method. *Acta Aritmetica* **1989**, *53*, 289–307.
7. Srivastava, H.M.; Araci, S.; Khan, W.A.; Acikgoz, M. A Note on the Truncated-Exponential Based Apostol-Type Polynomials, *Symmetry* **2019**, *11*, 538. [CrossRef]

8. Waldschmidt, M. Auxiliary functions in transcendental number theory, *Ramanujan J.* **2009**, *20*, 341–373. [CrossRef]
9. Sondow, J.; Marques, D. Algebraic and transcendental solutions of some exponential equations. *Ann. Math. Inform.* **2010**, *37*, 151–164.
10. Marques, D. Algebraic numbers of the form $P(T)^{Q(T)}$, with T transcendental. *Elem. Math.* **2010**, *65*, 78–80. [CrossRef]
11. Jensen, C.U.; Marques, D. Some field theoretic properties and an application concerning transcendental numbers. *J. Algebra Appl.* **2010**, *9*, 493–500. [CrossRef]

© 2019 by the author. Licensee MDPI, Basel, Switzerland. This article is an open access article distributed under the terms and conditions of the Creative Commons Attribution (CC BY) license (http://creativecommons.org/licenses/by/4.0/).

Article
On Some Formulas for Kaprekar Constants

Atsushi Yamagami * and Yūki Matsui

Department of Information Systems Science, Soka University, Tokyo 192-8577, Japan
* Correspondence: yamagami@soka.ac.jp

Received: 3 June 2019 ; Accepted: 2 July 2019; Published: 5 July 2019

Abstract: Let $b \geq 2$ and $n \geq 2$ be integers. For a b-adic n-digit integer x, let A (resp. B) be the b-adic n-digit integer obtained by rearranging the numbers of all digits of x in descending (resp. ascending) order. Then, we define the *Kaprekar transformation* $T_{(b,n)}(x) := A - B$. If $T_{(b,n)}(x) = x$, then x is called a b-adic n-digit Kaprekar constant. Moreover, we say that a b-adic n-digit Kaprekar constant x is *regular* when the numbers of all digits of x are distinct. In this article, we obtain some formulas for regular and non-regular Kaprekar constants, respectively. As an application of these formulas, we then see that for any integer $b \geq 2$, the number of b-adic odd-digit regular Kaprekar constants is greater than or equal to the number of all non-trivial divisors of b. Kaprekar constants have the symmetric property that they are fixed points for recursive number theoretical functions $T_{(b,n)}$.

Keywords: Kaprekar constants; Kaprekar transformation; fixed points for recursive functions

MSC: 2010: 11A99; 11P99

---

1 Introduction	1
2 Proofs of Theorems and Corollaries in the Introduction	9
2.1 A Proof of Theorem 1	9
2.2 A Proof of Corollary 1	12
2.3 A Proof of Theorem 2	12
2.4 A Proof of Corollary 2	14
3 On n-Digit Regular Kaprekar Constants with Specified n	15
3.1 Some Formulas for All n-Digit Regular Kaprekar Constants with Specified n	15
3.2 Some Observations on $v_{\text{reg}}(b,n)$ with Specified n	27
References	31

1. Introduction

Let $\mathbb{Z}$ be the set of all rational integers. In this article, the symbol $[\alpha]$ with any rational number α stands for the greatest integer that is less than or equal to α.

For integers $b \geq 2$ and $n \geq 2$, we denote by $\mathbb{Z}(b,n)$ the set of all b-adic n-digit integers, i.e.,

$$\mathbb{Z}(b,n) = \{x \in \mathbb{Z} \mid 0 \leq x \leq b^n - 1\}$$
$$= \{a_{n-1}b^{n-1} + \cdots + a_1 b + a_0 \mid 0 \leq a_0, a_1, \ldots, a_{n-1} \leq b-1\}.$$

For any:

$$x = a_{n-1}b^{n-1} + \cdots + a_1 b + a_0 \in \mathbb{Z}(b,n)$$

with $0 \leq a_0, a_1, \ldots, a_{n-1} \leq b-1$, we denote the b-adic expression of x by:

$$x = (a_{n-1} \cdots a_1 a_0)_b.$$

In the case where $b = 10$, we omit the subscript as:

$$x = a_{n-1} \cdots a_1 a_0$$

as usual if any confusion occurs with the product of $a_0, a_1, \ldots, a_{n-1}$.

Definition 1. *Let $c_{n-1} \geq \cdots \geq c_1 \geq c_0$ be the rearrangement of the numbers $a_0, a_1, \ldots, a_{n-1}$ of all digits of $x \in \mathbb{Z}(b, n)$ in descending order. We define the Kaprekar transformation as:*

$$T_{(b,n)} : \mathbb{Z}(b,n) \to \mathbb{Z}(b,n); \quad x \mapsto (c_{n-1} \cdots c_1 c_0)_b - (c_0 c_1 \cdots c_{n-1})_b.$$

Definition 2. *(1) For any $x \in \mathbb{Z}(b, n)$, we say that x is a b-adic n-digit Kaprekar constant if $T_{(b,n)}(x) = x$.*

(2) We see immediately that zero is a b-adic n-digit Kaprekar constant for any $b \geq 2$ and $n \geq 2$, which we call the trivial Kaprekar constant. Then, we denote by $\nu(b, n)$ the number of all b-adic n-digit non-trivial Kaprekar constants. By Ref. [1] (Proposition 1.3), we see that:

$$\nu(b,n) \leq {}_{b-1+\left[\frac{n}{2}\right]}C_{\left[\frac{n}{2}\right]} - 1,$$

where we put:

$${}_rC_s := \frac{r!}{s!(r-s)!} = \frac{r(r-1)\cdots(r-s+1)}{s \cdots 1}$$

for any integers $r > s > 0$.

(3) We say that a b-adic n-digit non-trivial Kaprekar constant $x = (a_{n-1} \cdots a_1 a_0)_b$ is regular when $a_i \neq a_j$ for any $i \neq j$. We denote by $\nu_{\text{reg}}(b, n)$ (resp. $\nu_{\text{non-reg}}(b, n)$) the number of all b-adic n-digit regular (resp. non-regular) Kaprekar constants. By the definition, we see immediately that:

$$\nu(b,n) = \nu_{\text{reg}}(b,n) + \nu_{\text{non-reg}}(b,n)$$

and if $b < n$, then $\nu_{\text{reg}}(b, n) = 0$ and $\nu(b, n) = \nu_{\text{non-reg}}(b, n)$.

Example 1. *Kaprekar [2,3], who was the initiator of this research, discovered that $\nu(10, 4) = 1$, and the only non-trivial 10-adic four-digit Kaprekar constant is: 6174.*

Example 2. Here is the list of all b-adic n-digit non-trivial Kaprekar constants for $2 \leq b \leq 15$ and $2 \leq n \leq 7$. Note that, in the list below, we omit the subscript b. Further, the symbol $-$ means that $\nu(b,n) = 0$, and non-trivial Kaprekar constants with the symbol $*$ are regular.

n	2	3	4	5	6	7
b = 2	01*	011	0111 1001	01111 10101	011111 101101 110001	0111111 1011101 1101001
3	—	—	—	20211	—	2202101
4	—	132*	3021*	—	213312 310221 330201	3203211
5	13*	—	3032	—	—	—
6	—	253*	—	41532*	325523 420432 530421*	—
7	—	—	—	—	—	—
8	25*	374*	—	—	437734 640632	6417532*
9	—	—	—	62853*	—	—
10	—	495*	6174*	—	549945 631764	—
11	37*	—	—	—	—	—
12	—	5(11)6*	—	83(11)74*	65(11)(11)56	962(11)853*
13	—	—	—	—	951(10)74*	—
14	49*	6(13)7*	—	—	76(13)(13)67	—
15	—	—	92(11)6*	(10)4(14)95*	—	—

Then, we obtain the following list of the numbers $\nu = \nu(b,n)$, $\nu_r = \nu_{reg}(b,n)$ and $\nu_{nr} = \nu_{non\text{-}reg}(b,n)$:

	n = 2			n = 3			n = 4			n = 5			n = 6			n = 7		
b	ν	ν_r	ν_{nr}	ν	ν_r	ν_{nr}	ν	ν_r	ν_{nr}	ν	ν_r	ν_{nr}	ν	ν_r	ν_{nr}	ν	ν_r	ν_{nr}
2	1	1	0	1	0	1	2	0	2	2	0	2	3	0	3	3	0	3
3	0	0	0	0	0	0	0	0	0	1	0	1	0	0	0	1	0	1
4	0	0	0	1	1	0	1	1	0	0	0	0	3	0	3	1	0	1
5	1	1	0	0	0	0	1	0	1	0	0	0	0	0	0	0	0	0
6	0	0	0	1	1	0	0	0	0	1	1	0	3	1	2	0	0	0
7	0	0	0	0	0	0	0	0	0	0	0	0	0	0	0	0	0	0
8	1	1	0	1	1	0	0	0	0	0	0	0	2	0	2	1	1	0
9	0	0	0	0	0	0	0	0	0	1	1	0	0	0	0	0	0	0
10	0	0	0	1	1	0	1	1	0	0	0	0	2	0	2	0	0	0
11	1	1	0	0	0	0	0	0	0	0	0	0	0	0	0	0	0	0
12	0	0	0	1	1	0	0	0	0	1	1	0	1	0	1	1	1	0
13	0	0	0	0	0	0	0	0	0	0	0	0	1	1	0	0	0	0
14	1	1	0	1	1	0	0	0	0	0	0	0	1	0	1	0	0	0
15	0	0	0	0	0	0	1	1	0	1	1	0	0	0	0	0	0	0

Now, we have the following:

Questions: (1) Are there any formulas for $\nu(b,n)$, $\nu_{reg}(b,n)$ and $\nu_{non\text{-}reg}(b,n)$ in terms of b and n?

(2) Are there any formulas for b-adic n-digit regular or non-regular Kaprekar constants in terms of b and n?

Known results: There are some known results that answer some parts of the questions above as follows:

(1) In the case where $n = 2$, by the results on the two-digit Kaprekar transformation given by Young [4] (cf. [1], Theorem 3.1), we see that for any integer $b \geq 2$, there exists a b-adic two-digit non-trivial Kaprekar constant if and only if $b + 1$ is divisible by three.

Since there is no two-digit non-regular Kaprekar constant by definition, we see immediately that for any integer $b \geq 2$,
$$v_{\text{non-reg}}(b,2) = 0 \quad \text{and} \quad v(b,2) = v_{\text{reg}}(b,2).$$

In this article, we shall prove in Theorem 4(1) and Corollary 3(1) that any two-digit regular Kaprekar constant is of the form:
$$(m(2m+1))_{3m+2}$$

with any integers $m \geq 0$ and:
$$v_{\text{reg}}(b,2) = \begin{cases} 1 & \text{if } 3 \mid (b+1), \\ 0 & \text{otherwise.} \end{cases}$$

(2) In the case where $n = 3$, Eldridge and Sagong [5] proved that any three-digit non-trivial Kaprekar constant is of the form:
$$(m(2m+1)(m+1))_{2m+2}$$

with any integers $m \geq 0$ and that for any integer $b \geq 2$,
$$v(b,3) = \begin{cases} 1 & \text{if } b \text{ is even,} \\ 0 & \text{if } b \text{ is odd.} \end{cases}$$

In particular, we see immediately that:
$$v_{\text{reg}}(b,3) = \begin{cases} 1 & \text{if } b \geq 4 \text{ is even,} \\ 0 & \text{if } b = 2 \text{ or } b \geq 3 \text{ is odd,} \end{cases}$$

and:
$$v_{\text{non-reg}}(b,3) = \begin{cases} 1 & \text{if } b = 2, \\ 0 & \text{if } b \geq 3. \end{cases}$$

(3) In the case where $n = 4$, Hasse and Prichett [6] obtained a formula:
$$((3m+3)m(4m+3)(2m+2))_{5m+5}$$

for $(5m+5)$-adic four-digit non-trivial Kaprekar constants with any integer $m \geq 0$. This implies that if $b \geq 5$ and $5 \mid b$, then $v_{\text{reg}}(b,4) \geq 1$.

In this article, we shall prove in Theorem 4(2) and Corollary 3(2) that any four-digit regular Kaprekar constant is equal to $(3021)_4$ or given by the above formula obtained by Hasse and Prichett with $m \geq 1$ and that for any integer $b \geq 2$,
$$v_{\text{reg}}(b,4) = \begin{cases} 1 & \text{if } b = 4 \text{ or, } b \geq 10 \text{ and } 5 \mid b, \\ 0 & \text{otherwise.} \end{cases}$$

(4) In the case where $n = 5$, Prichett [7] obtained a formula:
$$((2m+2)m(3m+2)(2m+1)(m+1))_{3m+3}$$

for $(3m + 3)$-adic five-digit non-trivial Kaprekar constants with any integers $m \geq 0$. This implies that if $b \geq 6$ and $3 \mid b$, then $\nu_{reg}(b, 5) \geq 1$.

In this article, we shall prove in Theorem 3(1) and Corollary 3(3) that any five-digit regular Kaprekar constant is given by the above formula obtained by Prichett with $m \geq 1$ and that for any integer $b \geq 2$,

$$\nu_{reg}(b, 5) = \begin{cases} 1 & \text{if } b \geq 6 \text{ and } 3 \mid b, \\ 0 & \text{otherwise.} \end{cases}$$

(5) In the case where $b = 2$, the first author [1] showed that for any $n \geq 2$, all two-adic n-digit non-trivial Kaprekar constants are of the form:

$$(\overbrace{1\cdots1}^{k-1}0\overbrace{1\cdots1}^{n-2k}0\overbrace{0\cdots0}^{k-1}1)_2$$

with all integers $1 \leq k \leq \left[\dfrac{n}{2}\right]$ and $\nu(2, n) = \left[\dfrac{n}{2}\right]$. In particular, we see immediately that:

$$\nu_{reg}(2, n) = \begin{cases} 1 & \text{if } n = 2, \\ 0 & \text{if } n \geq 3 \end{cases}$$

and:

$$\nu_{\text{non-reg}}(2, n) = \begin{cases} 0 & \text{if } n = 2, \\ \left[\dfrac{n}{2}\right] & \text{if } n \geq 3. \end{cases}$$

(6) In the case where $b = 3$, the authors [8] showed that for any $n \geq 2$, all three-adic n-digit non-trivial Kaprekar constants are of the form:

$$(\overbrace{2\cdots2}^{k}\overbrace{1\cdots1}^{\ell-k-1}0\overbrace{2\cdots2}^{\ell-k}\overbrace{1\cdots1}^{\ell-k}0\overbrace{0\cdots0}^{k-1}1)_3.$$

with all pairs (k, ℓ) of integers satisfying $0 < k < \ell$ and $n = 3\ell - k$, and:

$$\nu(3, n) = \left[\dfrac{1}{6}\left(n - \dfrac{1 + 3(-1)^n}{2}\right)\right].$$

In particular, we see immediately that:

$$\nu_{reg}(3, n) = 0, \quad \nu_{\text{non-reg}}(3, n) = \nu(3, n).$$

We have the impression that the behavior of the values of $\nu(b, n)$, $\nu_{reg}(b, n)$ and $\nu_{\text{non-reg}}(b, n)$ in the list in Example 2 is not only complicated, but also suggestive of some general rules. It seems that it is very hard to obtain general results without observing any case-by-case results. The aim of this article is to see formulas for b-adic n-digit regular and non-regular Kaprekar constants and to study the properties of $\nu_{reg}(b, n)$ and $\nu_{\text{non-reg}}(b, n)$ towards answers to the questions above.

Firstly, we see formulas for Kaprekar constants in the following:

Theorem 1. *Let $m \geq 0$ and $n \geq 2$ be any integers. We put:*

$$b(m, n) = \begin{cases} 3m + 2 & \text{if } n = 2, \\ 2^{\frac{n-4}{2}}(4m + 3) + m + 2 & \text{if } n \text{ is even and } n \geq 4, \\ \dfrac{n+1}{2}(m + 1) & \text{if } n \text{ is odd.} \end{cases}$$

(1) We assume that n is even and define the $b(m,n)$-adic n-digit integer:

$$K(m,n) = \begin{cases} (m(2m+1))_{b(m,2)} & \text{if } n = 2, \\ ((3m+3)m(4m+3)(2m+2))_{b(m,4)} & \text{if } n = 4, \\ (a_{n-1}a_{n-2}\cdots a_i \cdots a_{\frac{n}{2}+1}a_{\frac{n}{2}}a_{\frac{n}{2}-1}\cdots a_j \cdots a_1 a_0)_{b(m,n)} & \text{if } n \geq 6, \end{cases}$$

where we put:

$$a_{n-1} = 2^{\frac{n-4}{2}}(4m+3) - m,$$

$$a_i = (2^{\frac{n-4}{2}} - 2^{n-i-2})(4m+3) + m + 1 \quad \text{for } n-2 \geq i \geq \frac{n}{2}+1,$$

$$a_{\frac{n}{2}} = m,$$

$$a_j = 2^{j-1}(4m+3) \quad \text{for } \frac{n}{2}-1 \geq j \geq 1,$$

$$a_0 = 2m+2.$$

Then, $K(m,n)$ is a non-trivial Kaprekar constant, which is regular if and only if $n = 2$ or $m \geq 1$.

(2) We assume that n is odd and define the $b(m,n)$-adic n-digit integer:

$$L(m,n) = \begin{cases} (m(2m+1)(m+1))_{b(m,3)} & \text{if } n = 3, \\ (b_{n-1}\cdots b_i \cdots b_{\frac{n+3}{2}}b_{\frac{n+1}{2}}b_{\frac{n-1}{2}}b_{\frac{n-3}{2}}\cdots b_j \cdots b_1 b_0)_{b(m,n)} & \text{if } n \geq 5, \end{cases}$$

where we put:

$$b_i = \left(i - \frac{n-1}{2}\right)(m+1) \quad \text{for } n-1 \geq i \geq \frac{n+3}{2},$$

$$b_{\frac{n+1}{2}} = m,$$

$$b_{\frac{n-1}{2}} = \frac{n+1}{2}(m+1) - 1,$$

$$b_j = (j+1)(m+1) - 1 \left(= b_{\frac{n+1}{2}+j} - 1\right) \quad \text{for } \frac{n-3}{2} \geq j \geq 1,$$

$$b_0 = m+1.$$

Then, $L(m,n)$ is a non-trivial Kaprekar constant, which is regular if and only if $m \geq 1$.

Remark 1. (1) We can see that for any integer $n \geq 2$, the sequence:

$$b(n) := \{b(m,n) \mid m = 0, 1, 2, \ldots\}$$

consisting of bases defined in Theorem 1 is the arithmetic progression with the common difference:

$$\begin{cases} 3 & \text{if } n = 2, \\ 2^{\frac{n}{2}} + 1 & \text{if } n \text{ is even and } n \geq 4, \\ \dfrac{n+1}{2} & \text{if } n \text{ is odd} \end{cases}$$

and the first term:

$$\begin{cases} 2 & \text{if } n = 2, \\ 3 \times 2^{\frac{n-4}{2}} + 2 & \text{if } n \text{ is even and } n \geq 4, \\ \dfrac{n+1}{2} & \text{if } n \text{ is odd}. \end{cases}$$

(2) *As we have seen in the known results above, the regular Kaprekar constants $K(m,4)$, $L(m,3)$, and $L(m,5)$ have already been obtained by Hasse and Prichett [6], Eldridge and Sagong [5], and Prichett [7], respectively.*

Definition 3. (1) *We call the double series:*

$$K := \{K(m,n) \mid m = 1,2,3,\ldots, n = 2,4,6,\ldots\},$$
$$L := \{L(m,n) \mid m = 1,2,3,\ldots, n = 3,5,7,\ldots\}$$

the systems of regular Kaprekar constants.

(2) *Let $n \geq 2$ be any integer. We call the sequence:*

$$K(n) := \{K(m,n) \mid m = 1,2,3,\ldots\} \quad \text{with even } n, \text{ or}$$
$$L(n) := \{L(m,n) \mid m = 1,2,3,\ldots\} \quad \text{with odd } n$$

the progression of n-digit regular Kaprekar constants with arithmetic progression $b(n) \setminus \{b(0,n)\}$ of bases.

By Theorem 1, we see that the formulas for the numbers $a_{n-1},\ldots,a_0$ (resp. $b_{n-1},\ldots,b_0$) of digits of members in $K(n)$ (resp. $L(n)$) are given by polynomials in m of degree one. This implies that they can be regarded as arithmetic progressions indexed by $m = 1,2,3,\ldots$, as well as the arithmetic progression $b(n) \setminus \{b(0,n)\}$ of bases.

(3) *Let $m \geq 1$ be any integer. We call the sequences:*

$$K[m] := \{K(m,n) \mid n = 2,4,6,\ldots\}$$
$$(\text{resp. } L[m] := \{L(m,n) \mid n = 3,5,7,\ldots\})$$

the m-th chain of regular Kaprekar constants in the system K (resp. L) with ascending even (resp. odd) digits.

Example 3. (1) *Here are examples of some members $K(m,n)$ in the progressions $K(n)$ and the chains $K[m]$ of regular Kaprekar constants with $1 \leq m \leq 5$ and $n = 2,4,6$.*

	$K(2)$	$K(4)$	$K(6)$
$K[1]$	$(13)_5$	$(6174)_{10}$	$((13)91(14)74)_{17}$
$K[2]$	$(25)_8$	$(92(11)6)_{15}$	$((20)(14)2(22)(11)6)_{26}$
$K[3]$	$(37)_{11}$	$((12)3(15)8)_{20}$	$((27)(19)3(30)(15)8)_{35}$
$K[4]$	$(49)_{14}$	$((15)4(19)(10))_{25}$	$((34)(24)4(38)(19)(10))_{44}$
$K[5]$	$(5(11))_{17}$	$((18)5(23)(12))_{30}$	$((41)(29)5(46)(23)(12))_{53}$

(2) *Here are examples of some members $L(m,n)$ in the progressions $L(n)$ and the chains $L[m]$ of regular Kaprekar constants with $1 \leq m \leq 5$ and $n = 3,5,7$.*

	$L(3)$	$L(5)$	$L(7)$
$L[1]$	$(132)_4$	$(41532)_6$	$(6417532)_8$
$L[2]$	$(253)_6$	$(62853)_9$	$(962(11)853)_{12}$
$L[3]$	$(374)_8$	$(83(11)74)_{12}$	$((12)83(15)(11)74)_{16}$
$L[4]$	$(495)_{10}$	$((10)4(14)95)_{15}$	$((15)(10)4(19)(14)95)_{20}$
$L[5]$	$(5(11)6)_{12}$	$((12)5(17)(11)6)_{18}$	$((18)(12)5(23)(17)(11)6)_{24}$

Remark 2. *By the cases where $n = 4$ and $n = 6$ in the lists in Examples 2 and 3, we see that the progressions $K(n)$ and $L(n)$ of n-digit regular Kaprekar constants may not consist of all n-digit regular Kaprekar constants in general. Actually, for any $n \geq 2$, it seems that it is very hard to obtain formulas for all n-digit regular Kaprekar constants. In Section 2, we obtain some partial results on them with specified n.*

As a corollary of Theorem 1, we immediately obtain some results on the positivity of the numbers $\nu_{\text{reg}}(b, n)$ of all b-adic n-digit regular Kaprekar constants as in the following:

Corollary 1. (1) *Let $n \geq 2$ and $b \geq 2$ be any integers. If n and b satisfy one of the following conditions:*

(i) $n = 2$ and $b = 3m + 2$ with $m \geq 1$,

(ii) n is even, $n \geq 4$ and $b = 2^{\frac{n-4}{2}}(4m + 3) + m + 2$ with $m \geq 1$,

(iii) n is odd and $b = \dfrac{n+1}{2}(m+1)$ with $m \geq 1$,

then:
$$\nu_{\text{reg}}(b, n) \geq 1.$$

(2) *If an integer $b \geq 4$ is not a prime number, then for any non-trivial divisor d of b,*
$$\nu_{\text{reg}}(b, 2d - 1) \geq 1.$$

Therefore, the number of all b-adic odd-digit regular Kaprekar constants is greater than or equal to the number of all non-trivial divisors of b.

Secondly, we obtain formulas for non-regular Kaprekar constants by means of double series of regular Kaprekar constants obtained in Theorem 1 in the following:

Theorem 2. *Let the notation be as in Theorem 1.*
(1) *We assume that $m \equiv 1 \pmod 3$ and $n \equiv 0 \pmod 4$, and put:*
$$\beta_{m,n} = \frac{b(m,n) - 1}{3}.$$

For any integer $r \geq 2$, we denote by $K(m, n, r)$ the $b(m, n)$-adic $(n + 2r)$-digit integer:

$$\left((3m+3) \overbrace{\beta_{m,4} \cdots \beta_{m,4}}^{r} \, m(4m+3) \overbrace{(2\beta_{m,4}) \cdots (2\beta_{m,4})}^{r} (2m+2) \right)_{b(m,4)}$$

in the case where $n = 4$, and:

$$\left(a_{n-1} \cdots a_{\frac{n}{2}+1} \overbrace{\beta_{m,n} \cdots \beta_{m,n}}^{r} a_{\frac{n}{2}} a_{\frac{n}{2}-1} \overbrace{(2\beta_{m,n}) \cdots (2\beta_{m,n})}^{r} a_{\frac{n}{2}-2} \cdots a_0 \right)_{b(m,n)}$$

in the case where $n \geq 8$. Then, $K(m, n, r)$ is a non-regular Kaprekar constant.

(2) *We assume that $m = 1$, $n \equiv 3 \pmod 6$ and $n \geq 9$. For any integer $r \geq 2$, we denote by $L(1, n, r)$ the $b(1, n)(= n + 1)$-adic $(n + 2r)$-digit integer:*

$$\left(b_{n-1} \cdots b_{\frac{2n}{3}} \overbrace{\frac{n}{3} \cdots \frac{n}{3}}^{r} b_{\frac{2n}{3}-1} \cdots b_{\frac{n}{3}} \overbrace{\frac{2n}{3} \cdots \frac{2n}{3}}^{r} b_{\frac{n}{3}-1} \cdots b_0 \right)_{b(1,n)}$$

$$\left(= T_{(b(1,n),n)} \left(n \cdots \frac{2n+3}{3} \overbrace{\frac{2n}{3} \cdots \frac{2n}{3}}^{r} \frac{2n}{3} \cdots \frac{n+3}{3} \overbrace{\frac{n}{3} \cdots \frac{n}{3}}^{r} \frac{n}{3} \cdots 1 \right) \right).$$

Then, $L(1,n,r)$ is a non-regular Kaprekar constant.

Example 4. (1) *Here is an example of the non-regular constant $K(m,n,r)$ obtained in Theorem 2(1) in the case where $m = 4$, $n = 8$, and $r = 2$.*

(m,n)	$(4,8)$
$b(m,n)$	82
$K(m,n)$	$((72)(62)(43)4(76)(38)(19)(10))_{82}$
$\beta_{m,n}, 2\beta_{m,n}$	27, 54
$K(m,n,r)$	$((72)(62)(43)(27)(27)4(76)(54)(54)(38)(19)(10))_{82}$

(2) *Here is an example of the non-regular constant $L(1,n,r)$ obtained in Theorem 2(2) in the case where $n = 9$ and $r = 4$.*

$(1,n)$	$(1,9)$
$b(1,n)$	10
$L(1,n)$	$(864197532)_{10}$
$\frac{n}{3}, \frac{2n}{3}$	3, 6
$L(1,n,r)$	$(86433331976666532)_{10}$

As a corollary of Theorem 2, we immediately obtain the following result on the positivity of the numbers $\nu_{\text{reg}}(b,n)$ of all b-adic n-digit non-regular Kaprekar constants:

Corollary 2. *For any integers $m \geq 1$ and $n \geq 4$ satisfying:*

$$m \equiv 1 \pmod 3 \text{ and } n \equiv 0 \pmod 4$$

or:

$$m = 1, \ n \equiv 3 \pmod 6 \text{ and } n \geq 9,$$

and for any integer $r \geq 2$, we see that:

$$\nu_{\text{non-reg}}(b(m,n), n+2r) \geq 1.$$

In Section 1, we shall prove Theorems 1 and 2 and Corollaries 1 and 2. In Section 2.1, we shall obtain some formulas for all n-digit regular Kaprekar constants in Theorem 3 for $n = 5, 7, 9, 11$ and Theorem 4 for $n = 2, 4, 6, 8$. Moreover, we shall see some conditional results on formulas for n-digit regular Kaprekar constants in Proposition 1 for $n = 13, 15, 17$. Then, we shall see in Section 2.2 some observations on the values of $\nu_{\text{reg}}(b,n)$. We think that this article is fit for the Special Issue "Number Theory and Symmetry," since Kaprekar constants have the symmetric property that they are fixed points for recursive number theoretical functions $T_{(b,n)}$.

2. Proofs of Theorems and Corollaries in the Introduction

In this section, we prove Theorem 1 and Corollary 1 on regular Kaprekar constants and Theorem 2 and Corollary 2 on non-regular Kaprekar constants, respectively.

2.1. A Proof of Theorem 1

(1) Let the notation be as in Part (1) of Theorem 1. Here, we omit proving the Parts (i)–(iii), since they can be checked by direct calculations.

(iv) In the case where $n \geq 8$ is even, let:

$$K(m,n) = (a_{n-1}a_{n-2} \cdots a_i \cdots a_{\frac{n}{2}+1} a_{\frac{n}{2}} a_{\frac{n}{2}-1} \cdots a_j \cdots a_1 a_0)_{b(m,n)}$$

be the $b(m,n)$-adic n-digit integer defined in the assertion of Theorem 1(1). Let $c_{n-1} \geq \cdots \geq c_1 \geq c_0$ be the rearrangement of the numbers $a_0, \ldots, a_{n-1}$ of all digits of $K(m,n)$ in descending order. Then, the relation between $a_0, \ldots, a_{n-1}$ and $c_0, \ldots, c_{n-1}$ is given as in the following:

Lemma 1. *In the situation above, we see that:*

$$c_{n-1} = a_{\frac{n}{2}-1}, \quad c_{n-2} = a_{n-1},$$
$$c_i = a_{i+1}, \quad c_{n-i-1} = a_{n-i-2} \quad \text{for } n-3 \geq i \geq \frac{n}{2},$$
$$c_1 = a_0, \quad c_0 = a_{\frac{n}{2}}.$$

Proof. Since for any $n-3 \geq i \geq \frac{n}{2}$,

$$a_{i+1} = (2^{\frac{n-4}{2}} - 2^{n-i-3})(4m+3) + m + 1,$$
$$a_{n-i-2} = 2^{n-i-3}(4m+3),$$

we see easily that:

$$a_{n-2} > a_{n-3} > \cdots > a_{\frac{n}{2}+1}$$

and:

$$a_{\frac{n}{2}-2} > a_{\frac{n}{2}-3} > \cdots > a_1.$$

Moreover,

$$a_{\frac{n}{2}-1} = 2^{\frac{n-4}{2}}(4m+3)$$
$$\geq 2^{\frac{n-4}{2}}(4m+3) - m = a_{n-1}$$
$$> 2^{\frac{n-4}{2}}(4m+3) - (3m+2) = a_{n-2},$$
$$a_{\frac{n}{2}+1} - a_{\frac{n}{2}-2} = (2^{\frac{n-4}{2}} - 2^{\frac{n}{2}-3})(4m+3) + (m+1) - 2^{\frac{n}{2}-3}(4m+3)$$
$$= m+1 > 0$$

and:

$$a_1 = 4m+3 > a_0 = 2m+2 > a_{\frac{n}{2}} = m.$$

Therefore, the lemma is proven. □

We put:

$$T_{(b(m,n),n)}(K(m,n)) = (a'_{n-1} \cdots a'_1 a'_0)_{b(m,n)}$$

with integers $0 \leq a'_0, a'_1, \ldots, a'_{n-1} \leq b(m,n) - 1$. By Ref. [1] (Theorem 1.1 (6)) and Lemma 1, we then see that:

$$a'_{n-1} = c_{n-1} - c_0 = a_{\frac{n}{2}-1} - a_{\frac{n}{2}} = 2^{\frac{n-4}{2}}(4m+3) - m = a_{n-1},$$

$$a'_{n-2} = c_{n-2} - c_1 = a_{n-1} - a_0 = 2^{\frac{n-4}{2}}(4m+3) - m - (2m+2)$$
$$= (2^{\frac{n-4}{2}} - 1)(4m+3) + m + 1 = a_{n-2},$$

$$a'_{\frac{n}{2}} = c_{\frac{n}{2}} - c_{\frac{n}{2}-1} - 1 = a_{\frac{n}{2}+1} - a_{\frac{n}{2}-2} - 1$$
$$= (2^{\frac{n-4}{2}} - 2^{\frac{n}{2}-3})(4m+3) + (m+1) - 2^{\frac{n}{2}-3}(4m+3) - 1$$
$$= m = a_{\frac{n}{2}},$$

$$a'_{\frac{n}{2}-1} = b(m,n) - 1 - (c_{\frac{n}{2}} - c_{\frac{n}{2}-1})$$
$$= 2^{\frac{n-4}{2}}(4m+3) + m + 2 - 1 - (m+1)$$
$$= 2^{\frac{n-4}{2}}(4m+3) = a_{\frac{n}{2}-1},$$

$$a'_1 = b(m,n) - 1 - (c_{n-2} - c_1)$$
$$= 2^{\frac{n-4}{2}}(4m+3) + m + 2 - 1 - ((2^{\frac{n-4}{2}} - 1)(4m+3) + m + 1)$$
$$= 4m + 3 = a_1,$$

$$a'_0 = b(m,n) - (c_{n-1} - c_0)$$
$$= 2^{\frac{n-4}{2}}(4m+3) + m + 2 - (2^{\frac{n-4}{2}}(4m+3) - m)$$
$$= 2m + 2 = a_0.$$

Moreover, we see that for any $n - 3 \geq i \geq \frac{n}{2} + 1$,

$$a'_i = c_i - c_{n-i-1} = a_{i+1} - a_{n-i-2}$$
$$= (2^{\frac{n-4}{2}} - 2^{n-i-3})(4m+3) + m + 1 - 2^{n-i-3}(4m+3)$$
$$= (2^{\frac{n-4}{2}} - 2^{n-i-2})(4m+3) + m + 1 = a_i,$$

$$a'_{n-i-1} = b(m,n) - 1 - a'_i$$
$$= 2^{\frac{n-4}{2}}(4m+3) + m + 2 - 1 - ((2^{\frac{n-4}{2}} - 2^{n-i-2})(4m+3) + m + 1)$$
$$= 2^{n-i-2}(4m+3) = a_{n-i-1}.$$

Therefore, we see that:

$$T_{(b(m,n),n)}(K(m,n)) = (a'_{n-1} \cdots a'_1 a'_0)_{b(m,n)}$$
$$= (a_{n-1} \cdots a_1 a_0)_{b(m,n)}$$
$$= K(m,n),$$

i.e., $K(m,n)$ is a non-trivial Kaprekar constant, which is regular if and only if $m \geq 1$, which implies that $a_{\frac{n}{2}-1} \neq a_{n-1}$.

(2) Let the notation be as in Part (2) of Theorem 1.

As we have seen in the known results (2) and (4) in the Introduction, the cases where $n = 3$ and $n = 5$ have already been proven by Eldridge and Sagong [5] and Prichett [7], respectively. Therefore, it suffices to prove Part (2) in the case where $n \geq 7$.

For any odd integer $n \geq 7$, let:

$$L(m,n) = (b_{n-1} \cdots b_i \cdots b_{\frac{n+3}{2}} b_{\frac{n+1}{2}} b_{\frac{n-1}{2}} b_{\frac{n-3}{2}} \cdots b_j \cdots b_1 b_0)_{b(m,n)}$$

be the $b(m,n)$-adic n-digit integer defined in the assertion of Theorem 1(2). Let $c_{n-1} \geq \cdots \geq c_1 \geq c_0$ be the rearrangement of the numbers $b_0, \ldots, b_{n-1}$ of all digits of $L(m,n)$ in descending order. Then, the relation between $b_0, b_1, \ldots, b_{n-1}$ and $c_0, c_1, \ldots, c_{n-1}$ is given as in the following:

Lemma 2. *In the situation above, we see that:*

$$c_{n-1} = b_{\frac{n-1}{2}},$$
$$c_{2i-1} = b_{\frac{n-1}{2}+i}, \quad c_{2i-2} = b_{i-1} \quad \text{for} \quad \frac{n-1}{2} \geq i \geq 2,$$
$$c_1 = b_0, \quad c_0 = b_{\frac{n+1}{2}}.$$

Proof. By the definition of the numbers of all digits of $L(m,n)$ in Theorem 1(2), we see immediately that:

$$c_{n-1} = \frac{n+1}{2}(m+1) - 1 = b_{\frac{n-1}{2}},$$
$$c_{2i-1} = i(m+1) = b_{\frac{n-1}{2}+i}, \quad c_{2i-2} = i(m+1) - 1 = b_{i-1} \quad \text{for} \quad \frac{n-1}{2} \geq i \geq 2,$$
$$c_1 = m+1 = b_0, \quad c_0 = m = b_{\frac{n+1}{2}}.$$

Therefore, the lemma is proven. □

Then, we can prove Part (2) in the case where $n \geq 7$ by the same argument as in the proof of Theorem 1(1)(iv). Therefore, we omit the details of the calculations here.

2.2. A Proof of Corollary 1

(1) In Cases (i) and (ii), we have the $b(m,n)$-adic n-digit regular Kaprekar constant $K(m,n)$ by Theorem 1(1). On the other hand, in Case (iii), we have the $b(m,n)$-adic n-digit regular Kaprekar constant $L(m,n)$ by Theorem 1(2). Therefore, we see that for any integers $b \geq 2$ and $n \geq 2$ satisfying Condition (i), (ii), or (iii),

$$\nu_{\text{reg}}(b,n) \geq 1,$$

and Part (1) is proven.

(2) For any integer $b \geq 4$ that is not a prime number, let d be any non-trivial divisor of b, i.e., d is a divisor of b satisfying $1 < d < b$. We put:

$$m_d = \frac{b}{d} - 1, \quad n_d = 2d - 1.$$

Since $m_d \geq 1$ is an integer and $n_d \geq 3$ is an odd integer satisfying $b(m_d, n_d) = b$, by Theorem 1(2), we have the b-adic n_d-digit regular Kaprekar constant $L(m_d, n_d)$. Therefore, we see that:

$$\nu_{\text{reg}}(b, n_d) \geq 1.$$

Moreover, since $n_d \neq n_{d'}$ for any non-trivial divisors $d \neq d'$ of b, we see that $L(m_d, n_d) \neq L(m_{d'}, n_{d'})$. Therefore, the number of all b-adic odd-digit regular Kaprekar constants is greater than or equal to the number of all non-trivial divisors of b, and Part (2) is proven.

2.3. A Proof of Theorem 2

(1) We assume that:
$$m \equiv 1 \pmod{3} \quad \text{and} \quad n \equiv 0 \pmod{4}.$$

(a) In the case where $n = 4$, $b(m, 4) = 5m + 5$, and:

$$\beta_{m,4} = \frac{b(m,4) - 1}{3} = \frac{5m + 4}{3}$$

which is an integer, since the assumption $m \equiv 1 \pmod{3}$ implies that:

$$b(m, 4) \equiv 2m - 1 \equiv 1 \pmod{3}.$$

Then, for any $r \geq 2$, the $b(m, 4)$-adic $(2r + 4)$-digit integer obtained by rearranging of the numbers of all digits of $K(m, 4, r)$ in descending order is:

$$\left((4m + 3) \overbrace{(2\beta_{m,4}) \cdots (2\beta_{m,4})}^{r} (3m + 3)(2m + 2) \overbrace{\beta_{m,4} \cdots \beta_{m,4}}^{r} m \right)_{b(m,4)}.$$

By Ref. [1] (Theorem 1.1 (6)) and the case where $n = 4$ in Theorem 1 (1), we then see that:

$$T_{(b(m,4), 2r+4)}(K(m, 4, r)) = K(m, 4, r),$$

since $b(m, 4) - 1 - \beta_{m,4} = 2\beta_{m,4}$. Therefore, $K(m, 4, r)$ is a non-regular Kaprekar constant.

(b) In the case where $n \geq 8$, $b(m, n) = 2^{\frac{n-4}{2}}(4m + 3) + m + 2$, and

$$\beta_{m,m} = \frac{b(m, n) - 1}{3} = \frac{1}{3}\left(2^{\frac{n-4}{2}}(4m + 3) + m + 1\right)$$

which is an integer, since $n \equiv 0 \pmod{4}$ implies that $\frac{n-4}{2}$ is even and $m \equiv 1 \pmod{3}$ implies that:

$$b(m, n) \equiv (-1)^{\frac{n-4}{2}} m + m - 1 \equiv 1 \pmod{3}.$$

Let the notation be as in Theorem 1 (1). Since, $n \geq 8$, we see that:

$$a_{\frac{n}{2}-2} - \beta_{m,n} = 2^{\frac{n}{2}-3}(4m+3) - \frac{1}{3}\left(2^{\frac{n-4}{2}}(4m+3) + m + 1\right)$$

$$= \left(\frac{2^{\frac{n}{2}}}{6} - \frac{1}{3}\right) m + \frac{2^{\frac{n}{2}}}{8} - \frac{1}{3} > 0,$$

$$\beta_{m,n} - a_{\frac{n}{2}-3} = \frac{1}{3}\left(2^{\frac{n-4}{2}}(4m+3) + m + 1\right) - 2^{\frac{n}{2}-4}(4m+3)$$

$$= \left(\frac{2^{\frac{n}{2}}}{12} + \frac{1}{3}\right) m + \frac{2^{\frac{n}{2}}}{16} + \frac{1}{3} > 0,$$

$$a_{\frac{n}{2}+2} - 2\beta_{m,n} = \left(2^{\frac{n}{2}-2} - 2^{\frac{n}{2}-4}\right)(4m+3) + m + 1$$

$$- \frac{2}{3}\left(2^{\frac{n-4}{2}}(4m+3) + m + 1\right)$$

$$= \left(\frac{2^{\frac{n}{2}}}{12} + \frac{1}{3}\right) m + \frac{2^{\frac{n}{2}}}{16} + \frac{1}{3} > 0,$$

$$2\beta_{m,n} - a_{\frac{n}{2}+1} = \frac{2}{3}\left(2^{\frac{n-4}{2}}(4m+3) + m + 1\right)$$
$$- \left(\left(2^{\frac{n}{2}-2} - 2^{\frac{n}{2}-3}\right)(4m+3) + m + 1\right)$$
$$= \left(\frac{2^{\frac{n}{2}}}{6} - \frac{1}{3}\right) m + \frac{2^{\frac{n}{2}}}{8} - \frac{1}{3} > 0.$$

By Ref. [1] (Theorem 1.1 (6)) and Lemma 1, we then see that:

$$T_{(b(m,n),n+2r)}(K(m,n,r)) = K(m,n,r),$$

since $b(m,n) - 1 - \beta_{m,n} = 2\beta_{m,n}$. Therefore, $K(m,n,r)$ is a $b(m,n)$-adic $(n+2r)$-digit non-regular Kaprekar constant for any $r \geq 2$.

By (a) and (b) above, Part (1) of Theorem 2 is proven.

(2) We assume that:
$$m = 1, \ n \equiv 3 \pmod{6} \text{ and } n \geq 9.$$

Let the notation be as in Theorem 1 (2). By the definition in *loc. cit.*, the $b(1,n)(= n+1)$-adic n-digit integer obtained by rearranging of the numbers of all digits $b_0, b_1, \ldots, b_{n-1}$ of $L(1,n)$ in descending order is:

$$(n\,(n-1)\cdots 3\,2\,1)_{b(1,n)}$$

given by all integers from 1–n. By Ref. [1] (Theorem 1.1 (8)) and Theorem 1 (2), we then see that:

$$T_{(b(1,n),n+2r)}(L(1,n,r))$$
$$= T_{(b(1,n),n)}\left(n \cdots \frac{2n+3}{3} \overbrace{\frac{2n}{3} \cdots \frac{2n}{3}}^{r} \frac{2n}{3} \cdots \frac{n+3}{3} \overbrace{\frac{n}{3} \cdots \frac{n}{3}}^{r} \frac{n}{3} \cdots 1\right)$$
$$= L(1,n,r),$$

since $n \geq 9$ and $b(1,n) - 1 - \left(\frac{2n}{3} - \frac{n}{3}\right) = \frac{2n}{3}$. Therefore, $L(1,n,r)$ is a $b(1,n)$-adic $(n+2r)$-digit non-regular Kaprekar constant for any $r \geq 2$, and Part (2) of Theorem 2 is proven.

Remark 3. *Although we omit the proof here, we can see that for any integer $m \geq 2$ and odd integer $n \geq 3$, it is impossible to construct any $b(m,n)$-adic $(n+2r)$-digit non-regular Kaprekar constant by adding $\beta_{m,n}$'s and $(2\beta_{m,n})$'s to the $b(m,n)$-adic expression of the $b(m,n)$-adic n-digit regular Kaprekar constant $L(m,n)$, as well as in Part (1) of Theorem 2.*

2.4. A Proof of Corollary 2

We assume that:
$$m \equiv 1 \pmod{3} \text{ and } n \equiv 0 \pmod{4}$$

(resp.
$$m = 1, \ n \equiv 3 \pmod{6} \text{ and } n \geq 9).$$

By Theorem 2, for any integer $r \geq 2$, we then have the $b(m,n)$-adic $(n+2r)$-digit non-regular Kaprekar constant $K(m,n,r)$ (resp. $L(1,n,r)$). Therefore, we see that:

$$\nu_{\text{non-reg}}(b, n+2r) \geq 1,$$

and Corollary 2 is proven.

3. On n-Digit Regular Kaprekar Constants with Specified n

3.1. Some Formulas for All n-Digit Regular Kaprekar Constants with Specified n

Let $K(n)$ and $L(n)$ be the progressions of n-digit regular Kaprekar constants defined in Definition 3(2) for even and odd positive integers n, respectively. On the other hand, it seems that it is very hard to obtain formulas for *all* n-digit regular Kaprekar constants. In this subsection, we shall obtain partial results on such formulas by case-by-case arguments.

Firstly, we shall see formulas for all n-digit regular Kaprekar constants in the cases where $n = 5, 7, 9, 11$ in Theorem 3. Note that, in the case where $n = 3$, Eldridge and Sagong [5] already proved that a three-digit integer x is a regular Kaprekar constant if and only if $x \in L(3)$, i.e., x is of the form:

$$(m(2m+1)(m+1))_{2m+2}$$

with $m \geq 1$.

Although one can obtain a similar result for each odd integer $n \geq 13$, the authors would not like to do tedious calculations for solving simultaneous equations obtained by the uniqueness of b-adic expressions of any positive integer for any integer $b \geq 2$.

Theorem 3. (1) *A five-digit integer x is a regular Kaprekar constant if and only if $x \in L(5)$, i.e., x is of the form:*

$$((2m+2)m(3m+2)(2m+1)(m+1))_{3m+3}$$

with $m \geq 1$.

(2) A seven-digit integer x is a regular Kaprekar constant if and only if $x \in L(7)$, i.e., x is of the form:

$$((3m+3)(2m+2)m(4m+3)(3m+2)(2m+1)(m+1))_{4m+4}$$

with $m \geq 1$.

(3) For any integer $b \geq 2$, a b-adic nine-digit integer x is a regular Kaprekar constant if and only if x is of the form:

$$((b-m-1)(b-2m-2)(b-3m-3)m(b-1)(b-m-2)(3m+2)(2m+1)(m+1))_b,$$

where the base b is in the range $5m + 4 < b < 6m + 5$ with $m \geq 1$.

In particular, when $b = 5m + 5$, x is a member of $L(9)$.

(4) An 11-digit integer x is a regular Kaprekar constant if and only if $x \in L(11)$, i.e., x is of the form:

$$((5m+5)(4m+4)(3m+3)(2m+2)m(6m+5)$$
$$(5m+4)(4m+3)(3m+2)(2m+1)(m+1))_{6m+6}$$

with $m \geq 1$.

Proof. By Theorem 1, it suffices to show that any regular Kaprekar constant in each case is of the form stated in the assertion. In the following, let $b \geq 2$ be any integer.

(1) For any b-adic five-digit regular Kaprekar constant x, we denote by $(c_4 c_3 c_2 c_1 c_0)_b$ with:

$$b - 1 \geq c_4 > c_3 > c_2 > c_1 > c_0 \geq 0$$

the rearrangement in descending order of the numbers of all digits of x. By Ref. [1] (Theorem 1.1 (7)),

$$x = T_{(b,5)}((c_4 c_3 c_2 c_1 c_0)_b)$$
$$= ((c_4 - c_0)(c_3 - c_1 - 1)(b-1)(b-1-(c_3-c_1))(b-(c_4-c_0)))_b.$$

We see the following magnitude relations among the numbers of all digits of x:

$$b - 1 \geq c_4 - c_0 > c_3 - c_1 - 1,$$
$$b - 1 > b - 1 - (c_3 - c_1) > b - (c_4 - c_0).$$

Then, we obtain the following:

Lemma 3.

$$b - 1 = c_4, \quad c_4 - c_0 = c_3, \quad b - 1 - (c_3 - c_1) = c_2,$$
$$b - (c_4 - c_0) = c_1 \quad \text{and} \quad c_3 - c_1 - 1 = c_0.$$

Proof. Since c_4 is the maximum number among all digits of x,

$$b - 1 = c_4.$$

This implies that:

$$c_4 - c_0 = b - 1 - c_0 \quad \text{and} \quad b - (c_4 - c_0) = c_0 + 1.$$

Since c_1 is the second smallest number among all digits of x, we then see that:

$$b - (c_4 - c_0) = c_1.$$

This implies that:

$$c_3 - c_1 - 1 = c_0$$

by the two inequalities above. Moreover, we see that:

$$b - 1 - (c_3 - c_1) = b - 2 - c_0$$
$$< b - 1 - c_0 = c_4 - c_0,$$

which implies that:

$$c_4 - c_0 = c_3 \quad \text{and} \quad b - 1 - (c_3 - c_1) = c_2$$

as desired. □

We then see that the following equality holds:

$$((c_4 - c_0)(c_3 - c_1 - 1)(b-1)(b-1-(c_3-c_1))(b-(c_4-c_0)))_b$$
$$= (c_3 c_0 c_4 c_2 c_1)_b.$$

This implies that $b = 3c_0 + 3$ and:

$$c_4 = 3c_0 + 2, \quad c_3 = 2c_0 + 2, \quad c_2 = 2c_0 + 1, \quad c_1 = c_0 + 1.$$

Putting $m = c_0 \geq 0$, we then see that:

$$x = ((2m+2)m(3m+2)(2m+1)(m+1))_{3m+3}.$$

If $m = 0$, then we see a contradiction that $x = (20211)_3$ is not regular. Therefore, $m \geq 1$, and Part (1) is proven.

(2) For any b-adic seven-digit regular Kaprekar constant x, we denote by $(c_6c_5c_4c_3c_2c_1c_0)_b$ with:

$$b - 1 \geq c_6 > c_5 > c_4 > c_3 > c_2 > c_1 > c_0 \geq 0$$

the rearrangement in descending order of the numbers of all digits of x. By the same argument as in the proof of Part (1), we then see that one of the following two equalities holds:

$$((c_6 - c_0)(c_5 - c_1)(c_4 - c_2 - 1)(b - 1)(b - 1 - (c_4 - c_2))$$
$$(b - 1 - (c_5 - c_1))(b - (c_6 - c_0)))_b$$

$$= \begin{cases} (c_5c_2c_0c_6c_4c_3c_1)_b & \cdots \text{(i)} \\ (c_5c_3c_0c_6c_4c_2c_1)_b & \cdots \text{(ii)} \end{cases}$$

The equality (i) implies a contradiction that $c_2 = -\dfrac{1}{2}$.

The equality (ii) implies that $b = 4c_0 + 4$ and:

$$c_6 = 4c_0 + 3, \quad c_5 = 3c_0 + 3, \quad c_4 = 3c_0 + 2,$$
$$c_3 = 2c_0 + 2, \quad c_2 = 2c_0 + 1, \quad c_1 = c_0 + 1.$$

Putting $m = c_0 \geq 0$, we then see that:

$$x = ((3m + 3)(2m + 2)m(4m + 3)(3m + 2)(2m + 1)(m + 1))_{4m+4}.$$

If $m = 0$, then we see a contradiction that $x = (3203211)_4$ is not regular. Therefore, $m \geq 1$, and Part (2) is proven.

(3) For any b-adic nine-digit regular Kaprekar constant x, we denote by $(c_8c_7c_6c_5c_4c_3c_2c_1c_0)_b$ with:

$$b - 1 \geq c_8 > c_7 > c_6 > c_5 > c_4 > c_3 > c_2 > c_1 > c_0 \geq 0$$

the rearrangement in descending order of the numbers of all digits of x. By the same argument as in the proof of Part (1), we then see that one of the following six equalities holds:

$$((c_8 - c_0)(c_7 - c_1)(c_6 - c_2)(c_5 - c_3 - 1)(b - 1)(b - 1 - (c_5 - c_3))$$
$$(b - 1 - (c_6 - c_2))(b - 1 - (c_7 - c_1))(b - (c_8 - c_0)))_b$$

$$= \begin{cases} (c_7c_5c_4c_0c_8c_6c_3c_2c_1)_b & \cdots \text{(i)} \\ (c_7c_5c_3c_0c_8c_6c_4c_2c_1)_b & \cdots \text{(ii)} \\ (c_7c_5c_2c_0c_8c_6c_4c_3c_1)_b & \cdots \text{(iii)} \\ (c_7c_4c_3c_0c_8c_6c_5c_2c_1)_b & \cdots \text{(iv)} \\ (c_7c_4c_2c_0c_8c_6c_5c_3c_1)_b & \cdots \text{(v)} \\ (c_7c_3c_2c_0c_8c_6c_5c_4c_1)_b & \cdots \text{(vi)} \end{cases}$$

The equalities (i) and (v) imply a contradiction that $c_4 = c_3$.
The equalities (iii), (iv), and (vi) imply a contradiction that $c_5 = c_4$.
The equality (ii) implies that $b = c_3 + 3c_0 + 3$ and:

$$c_8 = c_3 + 3c_0 + 2, \quad c_7 = c_3 + 2c_0 + 2, \quad c_6 = c_3 + 2c_0 + 1,$$
$$c_5 = c_3 + c_0 + 1, \quad c_4 = 3c_0 + 2, \quad c_2 = 2c_0 + 1, \quad c_1 = c_0 + 1.$$

Putting $m = c_0 \geq 0$, we then see that x is equal to:

$$((b-m-1)(b-2m-2)(b-3m-3)m(b-1)(b-m-2)(3m+2)(2m+1)(m+1))_b,$$

where the base b is in the range $5m+4 < b < 6m+5$, since:

$$c_4 = 3m+2 > c_3 = b-3m-3 > c_2 = 2m+1.$$

If $m = 0$, then we see a contradiction that b is in the range $4 < b < 5$. Therefore, $m \geq 1$, and Part (3) is proven.

(4) For any b-adic 11-digit regular Kaprekar constant x, we denote by $(c_{10}c_9c_8c_7c_6c_5c_4c_3c_2c_1c_0)_b$ with:

$$b-1 \geq c_{10} > c_9 > c_8 > c_7 > c_6 > c_5 > c_4 > c_3 > c_2 > c_1 > c_0 \geq 0$$

the rearrangement in descending order of the numbers of all digits of x. By the same argument as in the proof of Part (1), we then see that one of the following twenty equalities holds:

$$((c_{10}-c_0)(c_9-c_1)(c_8-c_2)(c_7-c_3)(c_6-c_4-1)(b-1)(b-1-(c_6-c_4))$$
$$(b-1-(c_7-c_3))(b-1-(c_8-c_2))(b-1-(c_9-c_1))(b-(c_{10}-c_0)))_b$$

$$= \begin{cases}
(c_9c_7c_6c_5c_0c_{10}c_8c_4c_3c_2c_1)_b & \cdots \text{(i)} \\
(c_9c_7c_6c_4c_0c_{10}c_8c_5c_3c_2c_1)_b & \cdots \text{(ii)} \\
(c_9c_7c_6c_3c_0c_{10}c_8c_5c_4c_2c_1)_b & \cdots \text{(iii)} \\
(c_9c_7c_6c_2c_0c_{10}c_8c_5c_4c_3c_1)_b & \cdots \text{(iv)} \\
(c_9c_7c_5c_4c_0c_{10}c_8c_6c_3c_2c_1)_b & \cdots \text{(v)} \\
(c_9c_7c_5c_3c_0c_{10}c_8c_6c_4c_2c_1)_b & \cdots \text{(vi)} \\
(c_9c_7c_5c_2c_0c_{10}c_8c_6c_4c_3c_1)_b & \cdots \text{(vii)} \\
(c_9c_7c_4c_3c_0c_{10}c_8c_6c_5c_2c_1)_b & \cdots \text{(viii)} \\
(c_9c_7c_4c_2c_0c_{10}c_8c_6c_5c_3c_1)_b & \cdots \text{(ix)} \\
(c_9c_7c_3c_2c_0c_{10}c_8c_6c_5c_4c_1)_b & \cdots \text{(x)} \\
(c_9c_4c_3c_2c_0c_{10}c_8c_7c_6c_5c_1)_b & \cdots \text{(xi)} \\
(c_9c_5c_3c_2c_0c_{10}c_8c_7c_6c_4c_1)_b & \cdots \text{(xii)} \\
(c_9c_5c_4c_2c_0c_{10}c_8c_7c_6c_3c_1)_b & \cdots \text{(xiii)} \\
(c_9c_5c_4c_3c_0c_{10}c_8c_7c_6c_2c_1)_b & \cdots \text{(xiv)} \\
(c_9c_6c_3c_2c_0c_{10}c_8c_7c_5c_4c_1)_b & \cdots \text{(xv)} \\
(c_9c_6c_4c_2c_0c_{10}c_8c_7c_5c_3c_1)_b & \cdots \text{(xvi)} \\
(c_9c_6c_4c_3c_0c_{10}c_8c_7c_5c_2c_1)_b & \cdots \text{(xvii)} \\
(c_9c_6c_5c_2c_0c_{10}c_8c_7c_4c_3c_1)_b & \cdots \text{(xviii)} \\
(c_9c_6c_5c_3c_0c_{10}c_8c_7c_4c_2c_1)_b & \cdots \text{(xix)} \\
(c_9c_6c_5c_4c_0c_{10}c_8c_7c_3c_2c_1)_b & \cdots \text{(xx)}
\end{cases}$$

The equality (i) implies a contradiction that $c_5 \leq c_4$.
The equalities (ii), (x), (xii), and (xiii) imply a contradiction that $c_{10} = c_9$.
The equalities (iii), (iv), (vii), (xi), and (xviii) imply a contradiction that $c_7 < c_6$.
The equality (v) implies a contradiction that $c_6 < c_5$.
The equalities (viii) and (xvi) imply a contradiction that $c_7 = c_6$.

The equality (ix) implies that:

$$c_7 = c_2 + 2c_0 + 1, \quad c_6 = 4c_0 + 2, \quad c_3 = 2c_0 + 1,$$

which yields a contradiction that $c_2 > 2c_0 + 1 > c_2$.

The equality (xiv) implies a contradiction that $c_6 = c_5$.
The equality (xv) implies a contradiction that $c_8 = c_7$.
The equality (xvii) implies that:

$$c_7 = 2c_3, \quad c_6 = 3c_3 - 2c_0 - 1, \quad c_2 = 2c_0 + 1,$$

which implies a contradiction that $c_3 > 2c_0 + 1 > c_3$.

The equality (xix) implies a contradiction that $c_7 = 4c_0 + \dfrac{8}{3}$.

The equality (xx) implies a contradiction that $c_8 < c_7$.

The equality (vi) implies that $b = 6c_0 + 6$ and:

$$c_{10} = 6c_0 + 5, \ c_9 = 5c_0 + 5, \ c_8 = 5c_0 + 4, \ c_7 = 4c_0 + 4, \ c_6 = 4c_0 + 3,$$
$$c_5 = 3c_0 + 3, \ c_4 = 3c_0 + 2, \ c_3 = 2c_0 + 2, \ c_2 = 2c_0 + 1, \ c_1 = c + 1.$$

Putting $m = c_0 \geq 0$, we then see that:

$$x = ((5m+5)(4m+4)(3m+3)(3m+2)m(6m+5)$$
$$(5m+4)(4m+3)(3m+2)(2m+1)(m+1))_{6m+6}.$$

If $m = 0$, then we see a contradiction that $x = (54320543211)_6$ is not regular. Therefore, $m \geq 1$, and Part (4) is proven. □

Secondly, we see formulas for all n-digit regular Kaprekar constants in the cases where $n = 2, 4, 6, 8$ in Theorem 4. Although one can obtain a similar result for each even integer $n \geq 10$, the authors would not like to do tedious calculations for solving simultaneous equations obtained by the uniqueness of b-adic expressions of any positive integer for any integer $b \geq 2$.

Note that we shall need more calculations of solving simultaneous equations in the proof for even cases in Theorem 4 than odd cases in Theorem 3, because, in the case where $n \geq 2$ is even, the Kaprekar transformation $T_{(b,n)}$ may not necessarily give us the maximum number $b - 1$ among the numbers of all digits.

Theorem 4. (1) *A two-digit integer x is a regular Kaprekar constant if and only if $x \in K(2) \cup \{(01)_2\}$, i.e, x is of the form:*

$$(m(2m+1))_{3m+2}$$

with $m \geq 0$.

(2) *A four-digit integer x is a regular Kaprekar constant if and only if $x = (3021)_4$ or $x \in K(4)$, i.e., x is of the form:*

$$((3m+3)m(4m+3)(2m+2))_{5m+5}$$

with $m \geq 1$.

(3) A six-digit integer x is a regular Kaprekar constant if and only if x is equal to:

$(530421)_6$,
$((9m+6)(5m+3)(3m+1)(2m+7)(10m+6)(6m+4))_{15m+10}$,
$((5m+4)(3m+2)m(6m+4)(4m+3)(2m+2))_{7m+6}$ or
$((7m+6)(5m+4)m(8m+6)(4m+1)(2m+2))_{9m+8}$ ($\in K(6)$)

with $m \geq 1$.

(4) An eight-digit integer x is a regular Kaprekar constant if and only if x is equal to:

$(97508421)_{10}$, $(75306421)_8$,
$((11m+7)(7m+4)(5m+3)(3m+1)(14m+8)$
$\qquad (12m+7)(10m+6)(6m+4))_{17m+11}$,
$((15m+9)(9m+5)(7m+4)(3m+1)(18m+10)$
$\qquad (14m+8)(12m+7)(6m+4))_{21m+13}$,
$((13m+10)(11m+8)(7m+5)m(14m+10)$
$\qquad (8m+6)(4m+3)(2m+2))_{15m+12}$ or
$((15m+12)(13m+10)(9m+7)m(16m+12)$
$\qquad (8m+6)(4m+3)(2m+2))_{17m+14}$ ($\in K(8)$)

with $m \geq 1$.

Proof. (1) For any b-adic two-digit regular Kaprekar constant x, we denote by $x = (c_1 c_0)_b$ with $b-1 \geq c_1 > c_0 \geq 0$ the rearrangement in descending order of numbers of all digits of x. By Ref. [1] (Theorem 1.1 (2)),

$$x = T_{(b,2)}((c_1 c_0)_b) = ((c_1 - c_0 - 1)(b - (c_1 - c_0)))_b.$$

We then see that one of the following two equalities holds:

$$((c_1 - c_0 - 1)(b - (c_1 - c_0)))_b = \begin{cases} (c_1 c_0)_b & \cdots \text{(i)} \\ (c_0 c_1)_b & \cdots \text{(ii)} \end{cases}$$

The equality (i) implies a contradiction that $c_0 = -1$.
The equality (ii) implies that:

$$c_1 = \frac{2b-1}{3} \quad \text{and} \quad c_0 = \frac{b-2}{3}.$$

Putting $m = c_0 \geq 0$, we then see that:

$$b = 3m + 2 \quad \text{and} \quad c_1 = 2m + 1$$

as desired.

(2) For any b-adic four-digit regular Kaprekar constant x, we denote by $(c_3 c_2 c_1 c_0)_b$ with $b - 1 \geq c_3 > c_2 > c_1 > c_0 \geq 0$ the rearrangement in descending order of the numbers of all digits of x. By Ref. [1] (Theorem 1.1 (6)),

$$x = T_{(b,4)}((c_3 c_2 c_1 c_0)_b)$$
$$= ((c_3 - c_0)(c_2 - c_1 - 1)(b - 1 - (c_2 - c_1)))(b - (c_3 - c_0)))_b.$$

Since:
$$c_3 - c_0 > c_2 - c_1 - 1 \quad \text{and} \quad b - 1 - (c_2 - c_1) > b - (c_3 - c_0),$$

we see that one of the following six equalities holds:

$$((c_3 - c_0)(c_2 - c_1 - 1)(b - 1 - (c_2 - c_1))(b - (c_3 - c_0)))_b$$

$$= \begin{cases} (c_3 c_2 c_1 c_0)_b & \cdots \text{(i)} \\ (c_3 c_1 c_2 c_0)_b & \cdots \text{(ii)} \\ (c_3 c_0 c_2 c_1)_b & \cdots \text{(iii)} \\ (c_1 c_0 c_3 c_2)_b & \cdots \text{(iv)} \\ (c_2 c_0 c_3 c_1)_b & \cdots \text{(v)} \\ (c_2 c_1 c_3 c_0)_b & \cdots \text{(vi)} \end{cases}$$

The equalities (i), (ii), and (vi) imply a contradiction that $c_3 = b$.
The equality (iii) implies that $x = (3021)_4$.
The equality (iv) implies a contradiction that $c_3 < c_2$.
The equality (v) implies that $b = 5c_0 + 5$ and:

$$c_3 = 4c_0 + 3, \quad c_2 = 3c_0 + 3, \quad c_1 = 2c_0 + 2.$$

Putting $m = c_0 \geq 0$, we then see that:

$$x = ((3m+3)m(4m+3)(2m+2))_{5m+5}.$$

If $m = 0$, then we see a contradiction that $x = (3032)_5$ is not regular. Therefore, $m \geq 1$, and Part (2) is proven.

(3) For any b-adic six-digit regular Kaprekar constant x, we denote by $(c_5 c_4 c_3 c_2 c_1 c_0)_b$ with:

$$b - 1 \geq c_5 > c_4 > c_3 > c_2 > c_1 > c_0 \geq 0$$

the rearrangement in descending order of the numbers of all digits of x. By Ref. [1] (Theorem 1.1 (6)),

$$x = T_{(b,6)}((c_5 c_4 c_3 c_2 c_1 c_0)_b)$$
$$= ((c_5 - c_0)(c_4 - c_1)(c_3 - c_2 - 1)(b - 1 - (c_3 - c_2))$$
$$(b - 1 - (c_4 - c_1))(b - (c_5 - c_0)))_b.$$

Since $c_5 - c_0 > c_4 - c_1 > c_3 - c_2 - 1$ and:

$$b - 1 - (c_3 - c_2) > b - 1 - (c_4 - c_1) > b - (c_5 - c_0),$$

we see that $c_3 - c_2 - 1 = c_0$ or $b - (c_5 - c_0) = c_0$. The equality $b - (c_5 - c_0) = c_0$ implies a contradiction that $b = c_5$, and the equality $c_4 - c_1 = c_4$ implies a contradiction that $c_1 = 0 > c_0$. Therefore, we see that one of the following nine equalities holds:

$$((c_5 - c_0)(c_4 - c_1)(c_3 - c_2 - 1)(b - 1 - (c_3 - c_2))$$
$$(b - 1 - (c_4 - c_1))(b - (c_5 - c_0)))_b$$

$$= \begin{cases} (c_5 c_3 c_0 c_4 c_2 c_1)_b & \cdots \text{(i)} \\ (c_5 c_2 c_0 c_4 c_3 c_1)_b & \cdots \text{(ii)} \\ (c_5 c_1 c_0 c_4 c_3 c_2)_b & \cdots \text{(iii)} \\ (c_2 c_1 c_0 c_5 c_4 c_3)_b & \cdots \text{(iv)} \\ (c_3 c_1 c_0 c_5 c_4 c_2)_b & \cdots \text{(v)} \\ (c_3 c_2 c_0 c_5 c_4 c_1)_b & \cdots \text{(vi)} \\ (c_4 c_1 c_0 c_5 c_3 c_2)_b & \cdots \text{(vii)} \\ (c_4 c_2 c_0 c_5 c_3 c_1)_b & \cdots \text{(viii)} \\ (c_4 c_3 c_0 c_5 c_2 c_1)_b & \cdots \text{(ix)} \end{cases}$$

The equality (i) implies that $x = (530421)_6$.
The equality (ii) and (iii) imply a contradiction that $c_2 = c_1$.
The equality (iv) implies that $c_2 = c_0 + 1$, which contradicts the condition that $c_2 > c_1 > c_0$.
The equality (vi) implies a contradiction that $c_2 = c_0$.
The equality (vii) implies a contradiction that $x = (420432)_6$ is not regular.
The equality (v) implies that $b = 5c_0 + 5$ and:

$$c_5 = 4c_0 + 3, \quad c_4 = \frac{10c_0 + 8}{3}, \quad c_3 = 3c_0 + 3,$$
$$c_2 = 2c_0 + 2, \quad c_1 = \frac{5c_0 + 4}{3}.$$

Putting $c_0 = 3m + 1$ with $m \geq 0$, we then see that:

$$x = ((9m + 6)(5m + 3)(3m + 1)(12m + 7)(10m + 6)(6m + 4))_{15m+10}.$$

If $m = 0$, then we see a contradiction that $x = (631764)_{10}$ is not regular. Therefore, $m \geq 1$.
The equality (viii) implies that $b = 7c_0 + 6$ and:

$$c_5 = 6c_0 + 4, \quad c_4 = 5c_0 + 4, \quad c_3 = 4c_0 + 3,$$
$$c_2 = 3c_0 + 2, \quad c_1 = 2c_0 + 2.$$

Putting $m = c_0 \geq 0$, we then see that:

$$x = ((5m + 4)(3m + 2)m(6m + 4)(4m + 3)(2m + 2))_{7m+6}.$$

If $m = 0$, then we see a contradiction that $x = (420432)_6$ is not regular. Therefore, $m \geq 1$.
The equality (ix) implies that $b = 9c_0 + 8$ and:

$$c_5 = 8c_0 + 6, \quad c_4 = 7c_0 + 6, \quad c_3 = 5c_0 + 4,$$
$$c_2 = 4c_0 + 3, \quad c_1 = 2c_0 + 2.$$

Putting $m = c_0 \geq 0$, we then see that:

$$x = ((7m + 6)(5m + 4)m(8m + 6)(4m + 3)(2m + 2))_{9m+8}.$$

If $m = 0$, then we see a contradiction that $x = (640632)_8$ is not regular. Therefore, $m \geq 1$, and Part (3) is proven.

(4) For any b-adic eight-digit regular Kaprekar constant x, we denote by $(c_7c_6c_5c_4c_3c_2c_1c_0)_b$ with:

$$b - 1 \geq c_7 > c_6 > c_5 > c_4 > c_3 > c_2 > c_1 > c_0 \geq 0$$

the rearrangement in descending order of the numbers of all digits of x. By Ref. [1] (Theorem 1.1 (6)),

$$\begin{aligned} x &= T_{(b,8)}((c_7c_6c_5c_4c_3c_2c_1c_0)_b) \\ &= ((c_7 - c_0)(c_6 - c_1)(c_5 - c_2)(c_4 - c_3 - 1)(b - 1 - (c_4 - c_3)) \\ &\quad (b - 1 - (c_5 - c_2))(b - 1 - (c_6 - c_1))(b - (c_7 - c_0)))_b. \end{aligned}$$

Since $c_7 - c_0 > c_6 - c_1 > c_5 - c_2 > c_4 - c_3 - 1$ and:

$$b - 1 - (c_4 - c_3) > b - 1 - (c_5 - c_2) > b - 1 - (c_6 - c_1) > b - (c_7 - c_0),$$

we see that $c_4 - c_3 - 1 = c_0$ or $b - (c_7 - c_0) = c_0$. The equality $b - (c_7 - c_0) = c_0$ implies a contradiction that $b = c_7$, and the equality $c_6 - c_1 = c_6$ implies a contradiction that $c_1 = 0 > c_0$. Therefore, we see that one of the following thirty equalities holds:

$$((c_7 - c_0)(c_6 - c_1)(c_5 - c_2)(c_4 - c_3 - 1)(b - 1 - (c_4 - c_3))$$
$$(b - 1 - (c_5 - c_2))(b - 1 - (c_6 - c_1))(b - (c_7 - c_0)))_b$$

$$= \begin{cases} (c_7 c_5 c_4 c_0 c_6 c_3 c_2 c_1)_b & \cdots \text{(i)} \\ (c_7 c_5 c_3 c_0 c_6 c_4 c_2 c_1)_b & \cdots \text{(ii)} \\ (c_7 c_5 c_2 c_0 c_6 c_4 c_3 c_1)_b & \cdots \text{(iii)} \\ (c_7 c_5 c_1 c_0 c_6 c_4 c_3 c_2)_b & \cdots \text{(iv)} \\ (c_7 c_4 c_3 c_0 c_6 c_5 c_2 c_1)_b & \cdots \text{(v)} \\ (c_7 c_4 c_2 c_0 c_6 c_5 c_3 c_1)_b & \cdots \text{(vi)} \\ (c_7 c_4 c_1 c_0 c_6 c_5 c_3 c_2)_b & \cdots \text{(vii)} \\ (c_7 c_3 c_2 c_0 c_6 c_5 c_4 c_1)_b & \cdots \text{(viii)} \\ (c_7 c_3 c_1 c_0 c_6 c_5 c_4 c_2)_b & \cdots \text{(ix)} \\ (c_7 c_2 c_1 c_0 c_6 c_5 c_4 c_3)_b & \cdots \text{(x)} \\ (c_3 c_2 c_1 c_0 c_7 c_6 c_5 c_4)_b & \cdots \text{(xi)} \\ (c_4 c_2 c_1 c_0 c_7 c_6 c_5 c_3)_b & \cdots \text{(xii)} \\ (c_4 c_3 c_1 c_0 c_7 c_6 c_5 c_2)_b & \cdots \text{(xiii)} \\ (c_4 c_3 c_2 c_0 c_7 c_6 c_5 c_1)_b & \cdots \text{(xiv)} \\ (c_5 c_2 c_1 c_0 c_7 c_6 c_4 c_3)_b & \cdots \text{(xv)} \\ (c_5 c_3 c_1 c_0 c_7 c_6 c_4 c_2)_b & \cdots \text{(xvi)} \\ (c_5 c_3 c_2 c_0 c_7 c_6 c_4 c_1)_b & \cdots \text{(xvii)} \\ (c_5 c_4 c_1 c_0 c_7 c_6 c_3 c_2)_b & \cdots \text{(xviii)} \\ (c_5 c_4 c_2 c_0 c_7 c_6 c_3 c_1)_b & \cdots \text{(xix)} \\ (c_5 c_4 c_3 c_0 c_7 c_6 c_2 c_1)_b & \cdots \text{(xx)} \\ (c_6 c_2 c_1 c_0 c_7 c_5 c_4 c_3)_b & \cdots \text{(xxi)} \\ (c_6 c_3 c_1 c_0 c_7 c_5 c_4 c_2)_b & \cdots \text{(xxii)} \\ (c_6 c_3 c_2 c_0 c_7 c_5 c_4 c_1)_b & \cdots \text{(xxiii)} \\ (c_6 c_4 c_1 c_0 c_7 c_5 c_3 c_2)_b & \cdots \text{(xxiv)} \\ (c_6 c_4 c_2 c_0 c_7 c_5 c_3 c_1)_b & \cdots \text{(xxv)} \\ (c_6 c_4 c_3 c_0 c_7 c_5 c_2 c_1)_b & \cdots \text{(xxvi)} \\ (c_6 c_5 c_1 c_0 c_7 c_4 c_3 c_2)_b & \cdots \text{(xxvii)} \\ (c_6 c_5 c_2 c_0 c_7 c_4 c_3 c_1)_b & \cdots \text{(xxviii)} \\ (c_6 c_5 c_3 c_0 c_7 c_4 c_2 c_1)_b & \cdots \text{(xxix)} \\ (c_6 c_5 c_4 c_0 c_7 c_3 c_2 c_1)_b & \cdots \text{(xxx)} \end{cases}$$

The equality (i) implies that $x = (97508421)_{10}$.
The equality (ii) implies that $x = (75306421)_8$.
The equality (iii) implies a contradiction that $c_6 = c_4$.
The equality (iv) implies a contradiction that $c_5 = c_3$.
The equalities (v), (x), (xv), and (xxi) imply a contradiction that $c_6 = c_5$.
The equality (vi) implies a contradiction that $c_2 = \dfrac{5}{3}$.
The equality (vii) implies a contradiction that $c_7 < c_6$.

The equalities (viii) and (ix) imply a contradiction that $c_3 = c_1$.
The equalities (xi), (xii), (xiii), and (xiv) imply a contradiction that $c_2 = c_1$.
The equality (xvii) implies a contradiction that $c_1 = c_0 = -2$.
The equality (xviii) implies a contradiction that $b = 5c_0 + \frac{14}{3}$.
The equality (xix) implies a contradiction that $b = 2c_2 - \frac{2}{3}$.
The equality (xx) implies a contradiction that $c_7 = c_5$.
The equality (xxii) implies a contradiction that $4 > c_1 > 3$.
The equality (xxiv) implies a contradiction that $b = 2c_1 + \frac{7}{3}$.
The equality (xxv) implies a contradiction that $c_5 = 6c_1 + \frac{14}{3}$.
The equality (xxvi) implies a contradiction that $c_4 = c_1$.
The equality (xxvii) implies a contradiction that $c_0 = -1$.
The equality (xxviii) implies a contradiction that $c_7 = c_4$.
The equality (xvi) implies that $b = \frac{17c_0 + 16}{3}$ and:

$$c_7 = \frac{14c_0 + 10}{3}, \quad c_6 = 4c_0 + 3, \quad c_5 = \frac{11c_0 + 10}{3}, \quad c_4 = \frac{10c_0 + 8}{3},$$
$$c_3 = \frac{7c_0 + 5}{3}, \quad c_2 = 2c_0 + 2, \quad c_1 = \frac{5c_0 + 4}{3}.$$

Putting $c_0 = 3m + 1$ with $m \geq 0$, we then see that:

$$x = ((11m + 7)(7m + 4)(5m + 3)(3m + 1)$$
$$(4m + 8)(12m + 7)(10m + 6)(6m + 4))_{17m+11}.$$

If $m = 0$, then we see a contradiction that $x = (74318764)_{11}$ is not regular. Therefore, $m \geq 1$.
The equality (xxiii) implies that $b = 7c_0 + 6$ and:

$$c_7 = 6c_0 + 4, \quad c_6 = 5c_0 + 4, \quad c_5 = \frac{14c_0 + 10}{3}, \quad c_4 = 4c_0 + 3,$$
$$c_3 = 3c_0 + 2, \quad c_2 = \frac{7c_0 + 5}{3}, \quad c_1 = 2c_0 + 2.$$

Putting $c_0 = 3m + 1$ with $m \geq 0$, we then see that:

$$x = ((15m + 9)(9m + 5)(7m + 4)(3m + 1)$$
$$(18m + 10)(14m + 8)(12m + 7)(6m + 4))_{21m+13}.$$

If $m = 0$, then we see a contradiction that $x = (9541(10)874)_{13}$ is not regular. Therefore, $m \geq 1$.
The equality (xxix) implies that $b = 15c_0 + 12$ and:

$$c_7 = 14c_0 + 10, \quad c_6 = 13c_0 + 10, \quad c_5 = 11c_0 + 8, \quad c_4 = 8c_0 + 6,$$
$$c_3 = 7c_0 + 5, \quad c_2 = 4c_0 + 3, \quad c_1 = 2c_0 + 2.$$

Putting $m = c_0 \geq 0$, we then see that:

$$x = ((13m + 10)(11m + 8)(7m + 5)m$$
$$(14m + 10)(8m + 6)(4m + 3)(2m + 2))_{15m+12}.$$

If $m = 0$, then we see a contradiction that $x = ((10)850(10)632)_{12}$ is not regular. Therefore, $m \geq 1$.

The equality (xxx) implies that $b = 17c_0 + 14$ and:

$$c_7 = 16c_0 + 12, \quad c_6 = 15c_0 + 12, \quad c_5 = 13c_0 + 10, \quad c_4 = 9c_0 + 7,$$
$$c_3 = 8c_0 + 6, \quad c_2 = 4c_0 + 3, \quad c_1 = 2c_0 + 2.$$

Putting $m = c_0 \geq 0$, we then see that:

$$x = ((15m+12)(13m+10)(9m+7)m$$
$$(16m+12)(8m+6)(4m+3)(2m+2))_{17m+14}.$$

If $m = 0$, then we see a contradiction that $x = ((12)(10)70(12)632)_{14}$ is not regular. Therefore, $m \geq 1$, and Part (4) is proven. □

We shall also obtain some conditional results on formulas for n-digit regular Kaprekar constants in the following proposition for which we omit the proof because one can prove them by the same arguments as in the proof of Theorem 3:

Proposition 1. *Let the notation be as in Theorem 3. For any integer $b \geq 2$, we see the following:*
(1) *A b-adic 13-digit integer $x = (a_{12} \cdots a_0)_b$ with $0 \leq a_0, \ldots, a_{12} \leq b-1$ satisfying the condition:*

$$a_{11} > a_4 > a_{10} > a_3 > a_9 > a_2 > a_8 > a_1$$

is a regular Kaprekar constant if and only if $x \in L(13)$ with $b \in b(13)$, i.e., x is of the form:

$$((6m+6)(5m+5)(4m+4)(3m+3)(2m+2)m$$
$$(7m+6)(6m+5)(5m+4)(4m+3)(3m+2)(2m+1)(m+1))_{7m+7}$$

with $m \geq 1$.
(2) *A b-adic 15-digit integer $x = (a_{14} \cdots a_0)_b$ with $0 \leq a_0, \ldots, a_{14} \leq b-1$ satisfying the condition:*

$$a_{13} > a_5 > a_{12} > a_4 > a_{11} > a_3 > a_{10} > a_2 > a_9 > a_1$$

is a regular Kaprekar constant if and only if x is of the form:

$$((b-m_1-1)(b-2m_1-2)(b-3m_1-3)(b-2m_1-m_2-2)$$
$$(b-3m_1-m_2-3)m_2m_1(b-1)(b-m_1-2)(b-m_2-1)$$
$$(3m_1+m_2+2)(2m_1+m_2+1)(3m_1+2)(2m_1+1)(m_1+1))_b,$$

where $m_1 \geq 1$, m_2 is in the range:

$$2m_1 + 1 < m_2 < 3m_1 + 2$$

and b is in the range:

$$6m_1 + m_2 + 5 < b < 5m_1 + 2m_2 + 4.$$

(3) *A b-adic 17-digit integer $x = (a_{16} \cdots a_0)_b$ with $0 \leq a_0, \ldots, a_{16} \leq b-1$ satisfying the condition:*

$$a_{15} > a_6 > a_{14} > a_5 > a_{13} > a_4 > a_{12} > a_3 > a_{11} > a_2 > a_{10} > a_1$$

is a regular Kaprekar constant if and only if x is of the form:

$$((b-m-1)(b-2m-2)(b-3m-3)$$
$$\left(\frac{3b-7m-7}{4}\right)\left(\frac{3b-11m-11}{4}\right)\left(\frac{b-3m-2}{2}\right)\left(\frac{b-m-1}{4}\right)$$
$$m(b-1)(b-m-2)$$
$$\left(\frac{3b+m-3}{4}\right)\left(\frac{b+3m+1}{2}\right)\left(\frac{b+11m+7}{4}\right)\left(\frac{b+7m+3}{4}\right)$$
$$(3m+2)(2m+1)(m+1))_b,$$

where b satisfies the conditions:

$$9m+7 < b < 11m+9 \quad \text{and} \quad b \equiv m+1 \pmod{4}$$

with $m \geq 1$.

3.2. Some Observations on $\nu_{\text{reg}}(b,n)$ with Specified n

As a corollary to Theorems 3 and 4, we can make some observations on the numbers $\nu_{\text{reg}}(b,n)$ of all b-adic n-digit regular Kaprekar constants for $n = 2, 4, 5, 6, 7, 8, 9, 11$ as in the following:

Corollary 3. *Let $b \geq 2$ be any integer. Then, we see the following:*

(1) $\nu_{\text{reg}}(b,2) = \begin{cases} 1 & \text{if } 3 \mid (b+1), \\ 0 & \text{otherwise.} \end{cases}$

(2) $\nu_{\text{reg}}(b,4) = \begin{cases} 1 & \text{if } b = 4 \text{ or, } b \geq 10 \text{ and } 5 \mid b, \\ 0 & \text{otherwise.} \end{cases}$

(3) $\nu_{\text{reg}}(b,5) = \begin{cases} 1 & \text{if } b \geq 6 \text{ and } 3 \mid b, \\ 0 & \text{otherwise.} \end{cases}$

(4) $\nu_{\text{reg}}(b,6) = \begin{cases} 2 & \text{if } b \in (A_1 \cap A_2) \cup (A_2 \cap A_3), \\ 1 & \text{otherwise,} \\ 0 & \text{if } b \neq 6 \text{ and } b \notin A_1 \cup A_2 \cup A_3, \end{cases}$

where the sets A_1, A_2, and A_3 are defined as:

$$A_1 = \{b \in \mathbb{Z} \mid b \geq 25 \text{ and } b \equiv 10 \pmod{15}\},$$
$$A_2 = \{b \in \mathbb{Z} \mid b \geq 13 \text{ and } b \equiv 6 \pmod{7}\},$$
$$A_3 = \{b \in \mathbb{Z} \mid b \geq 17 \text{ and } b \equiv 8 \pmod{9}\}.$$

(5) $\nu_{\text{reg}}(b,7) = \begin{cases} 1 & \text{if } b \geq 8 \text{ and } 4 \mid b, \\ 0 & \text{otherwise.} \end{cases}$

(6) $\nu_{\text{reg}}(b,8) = \begin{cases} 2 & \text{if } b \in (B_1 \cap B_2) \cup (B_1 \cap B_3) \cup (B_2 \cap B_3) \cup (B_3 \cap B_4), \\ 1 & \text{otherwise,} \\ 0 & \text{if } b \neq 8, 10 \text{ and } b \notin B_1 \cup B_2 \cup B_3 \cup B_4. \end{cases}$

where the sets B_1, B_2, B_3, and B_4 are defined as:

$$B_1 = \{b \in \mathbb{Z} \mid b \geq 28 \text{ and } b \equiv 11 \pmod{17}\},$$
$$B_2 = \{b \in \mathbb{Z} \mid b \geq 34 \text{ and } b \equiv 13 \pmod{21}\},$$
$$B_3 = \{b \in \mathbb{Z} \mid b \geq 27 \text{ and } b \equiv 12 \pmod{15}\},$$
$$B_4 = \{b \in \mathbb{Z} \mid b \geq 31 \text{ and } b \equiv 14 \pmod{17}\}.$$

(7) $\quad v_{\text{reg}}(b, 9) = \begin{cases} \left[\dfrac{b}{30}\right] + 1 & \text{if } b \equiv 10, 15, 16, 20, 21, 22, 25, 26, 27, 28 \pmod{30}, \\ \left[\dfrac{b}{30}\right] & \text{otherwise.} \end{cases}$

(8) $\quad v_{\text{reg}}(b, 11) = \begin{cases} 1 & \text{if } b \geq 12 \text{ and } 6 \mid b, \\ 0 & \text{otherwise.} \end{cases}$

Remark 4. (1) *The intersections of the sets A_1, A_2, and A_3 in Corollary 3 (4) are the following:*

$$A_1 \cap A_2 = \{b \in \mathbb{Z} \mid b \geq 55 \text{ and } b \equiv 55 \pmod{105}\},$$
$$A_2 \cap A_3 = \{b \in \mathbb{Z} \mid b \geq 62 \text{ and } b \equiv 62 \pmod{63}\},$$
$$A_1 \cap A_3 = \varnothing.$$

(2) *The intersections of the sets B_1, B_2, B_3, and B_4 in Corollary 3 (6) are the following:*

$$B_1 \cap B_2 = \{b \in \mathbb{Z} \mid b \geq 181 \text{ and } b \equiv 181 \pmod{357}\},$$
$$B_1 \cap B_3 = \{b \in \mathbb{Z} \mid b \geq 147 \text{ and } b \equiv 147 \pmod{255}\},$$
$$B_2 \cap B_4 = \{b \in \mathbb{Z} \mid b \geq 286 \text{ and } b \equiv 286 \pmod{357}\},$$
$$B_3 \cap B_4 = \{b \in \mathbb{Z} \mid b \geq 255 \text{ and } b \equiv 255 \pmod{255}\},$$
$$B_1 \cap B_4 = B_2 \cap B_3 = \varnothing.$$

Remark 5. *We can see that Corollary 3 (1)–(5) matches the values of v_r in the list in Example 2.*

Proof. We see immediately that Parts (1)–(6) and (8) are implied by the respective formulas obtained in Theorem 3 (1), (2), (4) and Theorem 4 for the respective digits n, since these formulas give distinct n-digit regular Kaprekar constants for distinct positive integers m, and we see that:

$$A_1 \cap A_3 = B_1 \cap B_4 = B_2 \cap B_3 = \varnothing$$

as mentioned in Remark 4.

Now, we prove Part (7) for the case where $n = 9$. Since the formula obtained in Theorem 3 (3) gives distinct b-adic nine-digit regular Kaprekar constants for distinct pairs (b, m) of suitable integers b and m, we see that:

$$v_{\text{reg}}(b, 9) = \sharp \left\{ m \in \mathbb{Z} \;\middle|\; m \geq 1, \; \frac{b-5}{6} < m < \frac{b-4}{5} \right\},$$

where the symbol $\sharp$ stands for the number of all elements in the set.

For any integer $b' \geq 0$, we then see that:

$$v_{\text{reg}}(b,9) = \begin{cases} b' & \text{if } 30b'+2 \leq b \leq 30b'+9, \\ b'+1 & \text{if } b = 30b'+10, \\ b' & \text{if } 30b'+11 \leq b \leq 30b'+14, \\ b'+1 & \text{if } 30b'+15 \leq b \leq 30b'+16, \\ b' & \text{if } 30b'+17 \leq b \leq 30b'+19, \\ b'+1 & \text{if } 30b'+20 \leq b \leq 30b'+22, \\ b' & \text{if } 30b'+23 \leq b \leq 30b'+24, \\ b'+1 & \text{if } 30b'+25 \leq b \leq 30b'+28, \\ b' & \text{if } b = 30b'+29, \\ b'+1 & \text{if } 30b'+30 \leq b \leq 30b'+31. \end{cases}$$

Therefore, Part (7) is proven. □

Moreover, as a corollary to Proposition 1, we can obtain lower bounds for $v_{\text{reg}}(b,n)$ with $n = 13, 15, 17$ as in the following:

Corollary 4. *Let $b \geq 2$ be any integer. Then, we have the following estimations:*

(1) $v_{\text{reg}}(b,13) \geq 1$ if $b \geq 14$ and $7 \mid b$.

(2) $v_{\text{reg}}(b,15) \geq \sum_{\frac{b-7}{9} \leq m \leq \frac{b-8}{8}} (b-8m-7) + \sum_{\frac{b-5}{11} \leq m \leq \frac{b-8}{9}} \left(m - \left[\frac{b-9m}{2}\right] + 3\right)$, where the symbol m in the sums stands for positive integers.

(3) $v_{\text{reg}}(b,17) \geq \sharp\left\{k \in \mathbb{Z} \,\bigg|\, k \geq 2, b \equiv k \pmod 4, 0 \leq \frac{b-9k}{4} \leq \left[\frac{k}{2}\right] - 1\right\}$.

Proof. (1) We see immediately that Part (1) is implied by the conditional formula obtained in Proposition 1(1), since the formula gives distinct $(7m+7)$-adic 13-digit regular Kaprekar constants for distinct positive integers m.

(2) Since the conditional formula obtained in Proposition 1(2) gives distinct b-adic 15-digit regular Kaprekar constants for distinct triples (b, m_1, m_2) of suitable integers b, m_1, and m_2, we see that:

$$v_{\text{reg}}(b,15) \geq \sharp\{(m_1, m_2) \in \mathbb{Z} \times \mathbb{Z} \mid m_1 \geq 1, \ 2m_1 + 1 < m_2 < 3m_1 + 2,$$
$$6m_1 + m_2 + 5 < b < 5m_1 + 2m_2 + 4\}.$$

For any integer $m_1 \geq 1$, the list of m_2 and b satisfying the conditions:

$$2m_1 + 1 < m_2 < 3m_1 + 2, \quad 6m_1 + m_2 + 5 < b < 5m_1 + 2m_2 + 4$$

is the following:

m_2	b
$2m_1 + 2$	$8m_1 + 8, \ldots, 9m_1 + 7$
$2m_1 + 3$	$8m_1 + 9, \ldots, 9m_1 + 8, \ 9m_1 + 9$
$\vdots$	$\vdots \quad \ddots \quad \vdots \quad \vdots \quad \ddots$
$3m_1 + 1$	$9m_1 + 7, \ldots, 10m_1 + 6, 10m_1 + 7, \ldots, 11m_1 + 5$

Since the number of b's appearing in the list above is equal to:

$$\begin{cases} (b+1) - (8m_1 + 8) & \text{if } 8m_1 + 8 \leq b \leq 9m_1 + 7, \\ (m_1 - 1) - \left[\dfrac{b - (9m_1 + 8)}{2}\right] & \text{if } 9m_1 + 8 \leq b \leq 11m_1 + 5, \end{cases}$$

the right-hand side in the inequality above is equal to:

$$\sum_{\frac{b-7}{9} \leq m \leq \frac{b-8}{8}} (b - 8m - 7) + \sum_{\frac{b-5}{11} \leq m \leq \frac{b-8}{9}} \left(m - \left[\dfrac{b - 9m}{2}\right] + 3\right),$$

where the symbol m in the sums stands for positive integers. Therefore, Part (2) is proven.

(3) Since the conditional formula obtained in Proposition 1(3) gives distinct b-adic 17-digit regular Kaprekar constants for distinct pairs (b, m) of suitable integers b and m, we see that:

$$\nu_{reg}(b, 17) \geq \sharp\{m \in \mathbb{Z} \mid m \geq 1, 9m + 7 < b < 11m + 9, b \equiv m + 1 \pmod{4}\}.$$

For any integer $m \geq 1$, the first term and the final term in the range $9m + 7 < b < 11m + 9$ of the arithmetic progression with the common difference of four, which are congruent to $m + 1$ modulo four, are $9m + 9$ and $(9m + 9) + 4\left(\left[\dfrac{m+1}{2}\right] - 1\right)$, respectively. Putting $k = m + 1$, we then see that:

$$\sharp\{m \in \mathbb{Z} \mid m \geq 1, 9m + 7 < b < 11m + 9, b \equiv m + 1 \pmod{4}\}$$
$$= \sharp\left\{k \in \mathbb{Z} \mid k \geq 2, b \equiv k \pmod{4}, 0 \leq \dfrac{b - 9k}{4} \leq \left[\dfrac{k}{2}\right] - 1\right\},$$

and Part (3) is proven. □

Author Contributions: Conceptualization, A.Y.; investigation, A.Y. and Y.M.; writing, original draft, A.Y.

Funding: This research did not receive any specific grant from funding agencies in the public, commercial, or not-for-profit sectors.

Acknowledgments: The A.Y. is very grateful to the Y.M., who was one of his students at Soka University, for giving some interesting talks about formulas for Kaprekar constants in seminars held in 2017 at Soka University.

Conflicts of Interest: The authors declare no conflict of interest.

Errata of [1]: Since the reference [1] is very important to readers of this article, we would like to describe the errata of [1] here:

p. 263, ℓ. 32, $N(b, 2)$ and $\ell(b, 2) \to N(b, 5)$ and $\ell(b, 5)$
p. 266, ℓ.7, 14, 16, 18, 19, 20, 21, 23, 24: $(c0)_2 \to (c0)_b$
p. 266, ℓ.16: $((c - 1)(b - c))_2 \to ((c - 1)(b - c))_b$
p. 266, ℓ.14, 19: $((\delta_1(c) - 1)(b - \delta_1(c)))_2 \to ((\delta_1(c) - 1)(b - \delta_1(c)))_b$
p. 266, ℓ.21: $(c - 1)(b - c))_2 \to ((c - 1)(b - c))_b$
p. 266, ℓ.24: $((\delta_{v_2(b+1) - v_2 + 1}(c) - 1)(b - \delta_{v_2(b+1) - v_2 + 1}(c)))_2$
  $\to ((\delta_{v_2(b+1) - v_2(c) + 1}(c) - 1)(b - \delta_{v_2(b+1) - v_2(c) + 1}(c)))_b$
p. 267, ℓ.2, 3: $(c0)_2 \to (c0)_b$
p. 269, ℓ.11: $n \geq 7$ and $\to n \geq 7$; n is odd and
p. 269, ℓ.12: $c_{\frac{n}{2} - 2} \to c_{\frac{n-1}{2} - 2}$
p.280, ℓ.16: Delete the sentence "A.L. Ludington, A bound on Kaprekar constants, J. Reine Angew. Math. 310 (1979) 196–203."

References

1. Yamagami, A. On 2-adic Kaprekar constants and 2-digit Kaprekar distances. *J. Number Theory* **2018**, *185*, 257–280. [CrossRef]
2. Kaprekar, D.R. Another solitaire game. *Scr. Math.* **1949**, *15*, 244–245.
3. Kaprekar, D.R. An interesting property of the number 6174. *Scr. Math.* **1955**, *21*, 304.
4. Young, A.L. A variation on the two-digit Kaprekar routine. *Fibonacci Q.* **1993**, *31*, 138–145.
5. Eldridge, K.E.; Sagong, S. The determination of Kaprekar convergence and loop convergence of all three-digit numbers. *Am. Math. Mon.* **1988**, *95*, 105–112. [CrossRef]
6. Hasse, H.; Prichett, G.D. The determination of all four-digit Kaprekar constants. *J. Reine Angew. Math.* **1978**, *299/300*, 113–124.
7. Prichett, G.D. Terminating cycles for iterated difference values of five digit integers. *J. Reine Angew. Math.* **1978**, *303/304*, 379–388.
8. Yamagami, A.; Matsui, Y. On 3-adic Kaprekar loops. *JP J. Algebra Number Theory Appl.* **2018**, *40*, 957–1028. [CrossRef]

© 2019 by the authors. Licensee MDPI, Basel, Switzerland. This article is an open access article distributed under the terms and conditions of the Creative Commons Attribution (CC BY) license (http://creativecommons.org/licenses/by/4.0/).

Article

Asymptotic Semicircular Laws Induced by p-Adic Number Fields $\mathbb{Q}_p$ and C^*-Algebras over Primes p

Ilwoo Cho

Department of Mathematics & Statistics, Saint Ambrose University, 421 Ambrose Hall, 518 W. Locust St., Davenport, IA 52803, USA; choilwoo@sau.edu

Received: 11 April 2019; Accepted: 4 June 2019; Published: 20 June 2019

Abstract: In this paper, we study asymptotic semicircular laws induced both by arbitrarily fixed C^*-probability spaces, and p-adic number fields $\{\mathbb{Q}_p\}_{p \in \mathcal{P}}$, as $p \to \infty$ in the set $\mathcal{P}$ of all primes.

Keywords: free probability; p-adic number fields $\mathbb{Q}_p$; Banach $*$-probability spaces; C^*-algebras; semicircular elements; the semicircular law; asymptotic semicircular laws

1. Introduction

The main purposes of this paper are (i) to establish *tensor product C^*-probability spaces*

$$(A \otimes_{\mathbb{C}} \mathfrak{S}_p, \psi \otimes \varphi_j^p)$$

induced both by arbitrary unital C^*-probability spaces (A, ψ), and by analytic structures $(\mathfrak{S}_p, \varphi_j^p)$ acting on *p-adic number fields* $\mathbb{Q}_p$ for all primes p in the set $\mathcal{P}$ of all *primes*, where $j \in \mathbb{Z}$, (ii) to consider free-probabilistic structures of (i) affected both by the free probability on (A, ψ), and by the number theory on $\mathbb{Q}_p$ for all $p \in \mathcal{P}$, (iii) to study *asymptotic behaviors* on the structures of (i) as $p \to \infty$ in $\mathcal{P}$, based on the results of (ii), and (iv), and then investigate *asymptotic semicircular laws* from the free-distributional data of (iii).

Our main results illustrate cross-connections among *number theory, representation theory, operator theory, operator algebra theory*, and *stochastic analysis*, via *free probability theory*.

1.1. Preview and Motivation

Relations between primes and *operators* have been studied in various different approaches. In [1], we studied how primes act on *operator algebras* induced by *dynamical systems* on *p-adic*, and *Adelic* objects. Meanwhile, in [2], primes are acting as *linear functionals* on *arithmetic functions*, characterized by *Krein-space operators*.

For number theory and free probability theory, see [3–22], respectively.

In [23], *weighted-semicircular elements*, and *semicircular elements* induced by *p-adic number fields* $\mathbb{Q}_p$ are considered by the author and Jorgensen, for each $p \in \mathcal{P}$, statistically. In [24], the author extended the constructions of *weighted-semicircular elements* of [23] under *free product* of [15,22]. The main results of [24] demonstrate that the (weighted-)semicircular law(s) of [23] is (are) well-determined free-probability-theoretically. As an application, the *free stochastic calculus* was considered in [6].

Independent from the above series of works, we considered *asymptotic semicircular laws* induced by $\{\mathbb{Q}_p\}_{p \in \mathcal{P}}$ in [1]. The constructions of [1] are highly motivated by those of [6,23,24], but they are totally different not only conceptually, but also theoretically. Thus, even though the main results of [1] seem similar to those of [6,24], they indicate-and-emphasize "asymptotic" semicircularity induced by $\{\mathbb{Q}_p\}_{p \in \mathcal{P}}$, as $p \to \infty$. For example, they show that our analyses on $\{\mathbb{Q}_p\}_{p \in \mathcal{P}}$ not only provide natural semicircularity but also asymptotic semicircularity under free probability theory.

In this paper, we study *asymptotic-semicircular laws* over "both" primes and *unital C*-probability spaces*. Since we generalize the asymptotic semicircularity of [25] up to C*-algebra-tensor, the patterns and results of this paper would be similar to those of [25], but generalize-or-universalize them.

1.2. Overview

In Section 2, fundamental concepts and backgrounds are introduced. In Sections 3–6, suitable free-probabilistic models are considered, where they contain *p*-adic number-theoretic information, for our purposes.

In Section 7, we establish-and-study C*-probability spaces containing both analytic data from $\mathbb{Q}_p$, and free-probabilistic information of fixed unital C*-probability spaces. Then, our free-probabilistic structure $\mathfrak{LS}_A$, a free product Banach *-probability space, is constructed, and the free probability on $\mathfrak{LS}_A$ is investigated in Section 8.

In Section 9, asymptotic behaviors on $\mathfrak{LS}_A$ are considered over $\mathcal{P}$, and they analyze the asymptotic semicircular laws on $\mathfrak{LS}_A$ over $\mathcal{P}$ in Section 10.

2. Preliminaries

In this section, we briefly mention backgrounds of our proceeding works.

2.1. Free Probability

See [15,22] (and the cited papers therein) for basic free probability theory. Roughly speaking, *free probability* is the noncommutative operator-algebraic extension of measure theory (containing probability theory) and statistical analysis. As an independent branch of operator algebra theory, it is applied not only to mathematical analysis (e.g., [5,12–14,26]), but also to related fields (e.g., [18,27–31]).

Here, combinatorial free probability is used (e.g., [15–17]). In the text, *free moments*, *free cumulants*, and the *free product of *-probability spaces* are considered without detailed introduction.

2.2. Analysis on $\mathbb{Q}_p$

For *p-adic analysis* and *Adelic analysis*, see [21,22]. We use definitions, concepts, and notations from there. Let $p \in \mathcal{P}$ be a prime, and let $\mathbb{Q}$ be the set of all *rational numbers*. Define a *non-Archimedean norm* $|.|_p$, called the *p-norm on $\mathbb{Q}$* by

$$|x|_p = \left| p^k \frac{a}{b} \right|_p = \frac{1}{p^k},$$

for all $x = p^k \frac{a}{b} \in \mathbb{Q}$, where $k, a \in \mathbb{Z}$, and $b \in \mathbb{Z} \setminus \{0\}$.

The normed space $\mathbb{Q}_p$ is the maximal *p*-norm closures in $\mathbb{Q}$, i.e., the set $\mathbb{Q}_p$ forms a *Banach space*, for $p \in \mathcal{P}$ (e.g., [22]). Each element x of $\mathbb{Q}_p$ is uniquely expressed by

$$x = \sum_{k=-N}^{\infty} x_k p^k, \ x_k \in \{0, 1, ..., p-1\},$$

for $N \in \mathbb{N}$, decomposed by

$$x = \sum_{l=-N}^{-1} x_l p^l + \sum_{k=0}^{\infty} x_k p^k.$$

If $x = \sum_{k=0}^{\infty} x_k p^k$ in $\mathbb{Q}_p$, then x is said to be a *p-adic integer*, and it satisfies $|x|_p \leq 1$. Thus, one can define the *unit disk* $\mathbb{Z}_p$ of $\mathbb{Q}_p$,

$$\mathbb{Z}_p = \{x \in \mathbb{Q}_p : |x|_p \leq 1\}.$$

For the *p-adic addition* and the *p-adic multiplication* in the sense of [22], the algebraic structure $\mathbb{Q}_p$ forms a *field*, and hence, $\mathbb{Q}_p$ is a *Banach field*.

Note that $\mathbb{Q}_p$ is also a *measure space*,

$$\mathbb{Q}_p = \left(\mathbb{Q}_p, \sigma(\mathbb{Q}_p), \mu_p \right),$$

equipped with the σ-algebra $\sigma(\mathbb{Q}_p)$ of $\mathbb{Q}_p$, and a left-and-right additive invariant *Haar measure* on μ_p, satisfying

$$\mu_p(\mathbb{Z}_p) = 1.$$

If we take
$$U_k = p^k \mathbb{Z}_p = \{p^k x \in \mathbb{Q}_p : x \in \mathbb{Z}_p\}, \tag{1}$$
in $\sigma(\mathbb{Q}_p)$, for all $k \in \mathbb{Z}$, then these subsets U_k's of (1) satisfy

$$\mathbb{Q}_p = \bigcup_{k \in \mathbb{Z}} U_k,$$

and
$$\mu_p(U_k) = \tfrac{1}{p^k} = \mu_p(x + U_k), \tag{2}$$

for all $x \in \mathbb{Q}_p$, and
$$\cdots \subset U_2 \subset U_1 \subset U_0 = \mathbb{Z}_p \subset U_{-1} \subset U_{-2} \subset \cdots,$$

i.e., the family $\{U_k\}_{k \in \mathbb{Z}}$ of (1) is a *topological basis element of* $\mathbb{Q}_p$ (e.g., [22]).

Define subsets $\partial_k \in \sigma(\mathbb{Q}_p)$ by
$$\partial_k = U_k \setminus U_{k+1}, \tag{3}$$
for all $k \in \mathbb{Z}$.

Such μ_p-measurable subsets ∂_k of (3) are called the *k-th boundaries (of U_k)* in $\mathbb{Q}_p$, for all $k \in \mathbb{Z}$. By (2) and (3),

$$\mathbb{Q}_p = \bigsqcup_{k \in \mathbb{Z}} \partial_k, \tag{4}$$

$$\mu_p(\partial_k) = \mu_p(U_k) - \mu_p(U_{k+1}) = \tfrac{1}{p^k} - \tfrac{1}{p^{k+1}},$$

where $\sqcup$ is the *disjoint union*, for all $k \in \mathbb{Z}$,

Let $\mathcal{M}_p$ be an algebraic *algebra*,
$$\mathcal{M}_p = \mathbb{C}\left[\{\chi_S : S \in \sigma(\mathbb{Q}_p)\}\right], \tag{5a}$$

where χ_S are the usual *characteristic functions* of μ_p-measurable subsets S of $\mathbb{Q}_p$. Thus, $f \in \mathcal{M}_p$, if and only if

$$f = \sum_{S \in \sigma(\mathbb{Q}_p)} t_S \chi_S; t_S \in \mathbb{C}, \tag{5b}$$

where $\sum$ is the *finite sum*. Note that the algebra $\mathcal{M}_p$ of (5a) is a *-algebra over $\mathbb{C}$, with its well-defined *adjoint*,

$$\left(\sum_{S \in \sigma(G_p)} t_S \chi_S\right)^* \stackrel{def}{=} \sum_{S \in \sigma(G_p)} \overline{t_S}\, \chi_S,$$

for $t_S \in \mathbb{C}$ with their *conjugates* $\overline{t_S}$ in $\mathbb{C}$.

If $f \in \mathcal{M}_p$ is given as in (5b), then one defines the *integral of f* by

$$\int_{\mathbb{Q}_p} f\, d\mu_p = \sum_{S \in \sigma(\mathbb{Q}_p)} t_S\, \mu_p(S). \tag{6a}$$

Remark that, by (5a), the integral (6a) is unbounded on $\mathcal{M}_p$, i.e.,

$$\int_{\mathbb{Q}_p} \chi_{\mathbb{Q}_p} d\mu_p = \mu_p(\mathbb{Q}_p) = \infty, \tag{6b}$$

by (2).

Note that, by (4), for each $S \in \sigma(\mathbb{Q}_p)$, there exists a corresponding subset Λ_S of $\mathbb{Z}$,

$$\Lambda_S = \{j \in \mathbb{Z} : S \cap \partial_j \neq \varnothing\}, \tag{7}$$

satisfying

$$\int_{\mathbb{Q}_p} \chi_S \, d\mu_p = \int_{\mathbb{Q}_p} \sum_{j \in \Lambda_S} \chi_{S \cap \partial_j} \, d\mu_p$$
$$= \sum_{j \in \Lambda_S} \mu_p(S \cap \partial_j)$$

by (6a)

$$\leq \sum_{j \in \Lambda_S} \mu_p(\partial_j) = \sum_{j \in \Lambda_S} \left(\frac{1}{p^j} - \frac{1}{p^{j+1}} \right), \tag{8}$$

by (4), for the set Λ_S of (7).

Remark again that the right-hand side of (8) can be ∞; for instance, $\Lambda_{\mathbb{Q}_p} = \mathbb{Z}$, e.g., see (4), (6a) and (6b). By (8), one obtains the following proposition.

Proposition 1. *Let $S \in \sigma(\mathbb{Q}_p)$, and let $\chi_S \in \mathcal{M}_p$. Then, there exists $r_j \in \mathbb{R}$, such that*

$$0 \leq r_j = \frac{\mu_p(S \cap \partial_j)}{\mu_p(\partial_j)} \leq 1, \forall j \in \Lambda_S;$$
$$\tag{9}$$
$$\int_{\mathbb{Q}_p} \chi_S \, d\mu_p = \sum_{j \in \Lambda_S} r_j \left(\frac{1}{p^j} - \frac{1}{p^{j+1}} \right).$$

3. Statistical Models on $\mathcal{M}_p$

In this section, fix $p \in \mathcal{P}$, and let $\mathbb{Q}_p$ be the p-adic number field, and let $\mathcal{M}_p$ be the $*$-algebra (5a). We here establish a suitable statistical model on $\mathcal{M}_p$ with free-probabilistic language.

Let U_k be the basis elements (1), and ∂_k, their boundaries (3) of $\mathbb{Q}_p$, i.e.,

$$U_k = p^k \mathbb{Z}_p,$$

for all $k \in \mathbb{Z}$, and

$$\partial_k = U_k \setminus U_{k+1}; k \in \mathbb{Z}. \tag{10}$$

Define a linear functional $\varphi_p : \mathcal{M}_p \to \mathbb{C}$ by the *integration* (6a), i.e.,

$$\varphi_p(f) = \int_{\mathbb{Q}_p} f \, d\mu_p, \tag{11}$$

for all $f \in \mathcal{M}_p$.

Then, by (9), one obtains that $\varphi_p\left(\chi_{U_j}\right) = \frac{1}{p^j}$, and $\varphi_p\left(\chi_{\partial_j}\right) = \frac{1}{p^j} - \frac{1}{p^{j+1}}$, since $\Lambda_{U_j} = \{k \in \mathbb{Z} : k \geq j\}$, and $\Lambda_{\partial_j} = \{j\}$, for all $j \in \mathbb{Z}$, where Λ_S are in the sense of (7) for all $S \in \sigma(\mathbb{Q}_p)$.

Definition 1. *The pair $(\mathcal{M}_p, \varphi_p)$ is called the p-adic (unbounded-)measure space for $p \in \mathcal{P}$, where φ_p is the linear functional (11) on $\mathcal{M}_p$.*

Let ∂_k be the k-th boundaries (10) of $\mathbb{Q}_p$, for all $k \in \mathbb{Z}$. Then, for $k_1, k_2 \in \mathbb{Z}$, one obtains that

$$\chi_{\partial_{k_1}} \chi_{\partial_{k_2}} = \chi_{\partial_{k_1} \cap \partial_{k_2}} = \delta_{k_1, k_2} \chi_{\partial_{k_1}},$$

and hence,

$$\varphi_p\left(\chi_{\partial_{k_1}} \chi_{\partial_{k_2}}\right) = \delta_{k_1, k_2} \varphi_p\left(\chi_{\partial_{k_1}}\right)$$
$$= \delta_{k_1, k_2} \left(\frac{1}{p^{k_1}} - \frac{1}{p^{k_1+1}} \right). \tag{12}$$

Proposition 2. *Let $(j_1, ..., j_N) \in \mathbb{Z}^N$, for $N \in \mathbb{N}$. Then,*

$$\prod_{l=1}^{N} \chi_{\partial_{j_l}} = \delta_{(j_1,...,j_N)} \chi_{\partial_{j_1}} \text{ in } \mathcal{M}_p,$$

and hence,

$$\varphi_p \left(\prod_{l=1}^{N} \chi_{\partial_{j_l}} \right) = \delta_{(j_1,...,j_N)} \left(\frac{1}{p^{j_1}} - \frac{1}{p^{j_1+1}} \right), \quad (13)$$

where

$$\delta_{(j_1,...,j_N)} = \left(\prod_{l=1}^{N-1} \delta_{j_l, j_{l+1}} \right) \left(\delta_{j_N, j_1} \right).$$

Proof. The computation (13) is shown by the induction on (12). □

Recall that, for any $S \in \sigma(\mathbb{Q}_p)$,

$$\varphi_p(\chi_S) = \sum_{j \in \Lambda_S} r_j \left(\frac{1}{p^j} - \frac{1}{p^{j+1}} \right), \quad (14)$$

for some $0 \leq r_j \leq 1$, for $j \in \Lambda_S$, by (9). Thus, by (14), if $S_1, S_2 \in \sigma(\mathbb{Q}_p)$, then

$$\begin{aligned}
\chi_{S_1} \chi_{S_2} &= \left(\sum_{k \in \Lambda_{S_1}} \chi_{S_1 \cap \partial_k} \right) \left(\sum_{j \in \Lambda_{S_2}} \chi_{S_2 \cap \partial_j} \right) \\
&= \sum_{(k,j) \in \Lambda_{S_1} \times \Lambda_{S_2}} \left(\chi_{S_1 \cap \partial_k} \chi_{S_2 \cap \partial_j} \right) \\
&= \sum_{(k,j) \in \Lambda_{S_1} \times \Lambda_{S_2}} \delta_{k,j} \chi_{(S_1 \cap S_2) \cap \partial_j} \\
&= \sum_{j \in \Lambda_{S_1, S_2}} \chi_{(S_1 \cap S_2) \cap \partial_j},
\end{aligned} \quad (15)$$

where

$$\Lambda_{S_1, S_2} = \Lambda_{S_1} \cap \Lambda_{S_2},$$

by (4).

Proposition 3. *Let $S_l \in \sigma(\mathbb{Q}_p)$, and let $\chi_{S_l} \in (\mathcal{M}_p, \varphi_p)$, for $l = 1, ..., N$, for $N \in \mathbb{N}$. Let*

$$\Lambda_{S_1,...,S_N} = \bigcap_{l=1}^{N} \Lambda_{S_l} \text{ in } \mathbb{Z},$$

where Λ_{S_l} are in the sense of (7), for $l = 1, ..., N$. Then, there exists $r_j \in \mathbb{R}$, such that

$$0 \leq r_j \leq 1 \text{ in } \mathbb{R},$$

for all $j \in \Lambda_{S_1,...,S_N}$, and

$$\varphi_p \left(\prod_{l=1}^{N} \chi_{S_l} \right) = \sum_{j \in \Lambda_{S_1,...,S_N}} r_j \left(\frac{1}{p^j} - \frac{1}{p^{j+1}} \right). \quad (16)$$

Proof. The proof of (16) is done by the induction on (15), and by (13). □

4. Representation of $(\mathcal{M}_p, \varphi_p)$

Fix a prime $p \in \mathcal{P}$. Let $(\mathcal{M}_p, \varphi_p)$ be the p-adic measure space. By understanding $\mathbb{Q}_p$ as a measure space, construct the L^2-space,

$$H_p \stackrel{def}{=} L^2\left(\mathbb{Q}_p, \sigma(\mathbb{Q}_p), \mu_p\right) = L^2\left(\mathbb{Q}_p\right), \tag{17}$$

over $\mathbb{C}$. Then, this *Hilbert space* H_p of (17) consists of all square-integrable elements of $\mathcal{M}_p$, equipped with its *inner product* $<,>_2$,

$$\langle f_1, f_2 \rangle_2 \stackrel{def}{=} \int_{\mathbb{Q}_p} f_1 f_2^* \, d\mu_p, \tag{18a}$$

for all $f_1, f_2 \in H_p$. Naturally, H_p is has its L^2-*norm* $\|.\|_2$ on $\mathcal{M}_p$,

$$\|f\|_2 \stackrel{def}{=} \sqrt{\langle f, f \rangle_2}, \tag{18b}$$

for all $f \in H_p$, where $<,>_2$ is the inner product (18a) on H_p.

Definition 2. *The Hilbert space H_p of (17) is called the p-adic Hilbert space.*

Our $*$-algebra $\mathcal{M}_p$ acts on the p-adic Hilbert space H_p, via an action α^p,

$$\alpha^p(f)(h) = fh, \text{ for all } h \in H_p, \tag{19a}$$

for all $f \in \mathcal{M}_p$. i.e., the morphism α^p of (19a) is a $*$-homomorphism from $\mathcal{M}_p$ to the *operator algebra* $B(H_p)$, consisting of all Hilbert-space operators on H_p. For instance,

$$\alpha^p\left(\chi_{\mathbb{Q}_p}\right)\left(\sum_{S \in \sigma(\mathbb{Q}_p)} t_S \chi_S\right) = \sum_{S \in \sigma(\mathbb{Q}_p)} t_S \chi_{\mathbb{Q}_p \cap S}$$

$$= \sum_{S \in \sigma(\mathbb{Q}_p)} t_S \chi_S, \tag{19b}$$

for all $h = \sum_{S \in \sigma(\mathbb{Q}_p)} t_S \chi_S \in H_p$, with $\|h\|_2 < \infty$, for $\chi_{\mathbb{Q}_p} \in \mathcal{M}_p$, even though $\chi_{\mathbb{Q}_p} \notin H_p$.

Indeed, It is not difficult to check that

$$\alpha^p(f_1 f_2) = \alpha^p(f_1)\alpha^p(f_2) \text{ on } H_p, \forall f_1, f_2 \in \mathcal{M}_p,$$

$$(\alpha^p(f))^* = \alpha(f^*) \text{ on } H_p, \forall f \in \mathcal{M}_p. \tag{20a}$$

Notation 1. Denote $\alpha^p(f)$ by α_f^p, for all $f \in \mathcal{M}_p$. In addition, for convenience, denote $\alpha_{\chi_S}^p$ simply by α_S^p, for all $S \in \sigma(\mathbb{Q}_p)$.

Note that, by (19b), one can have a well-defined operator $\alpha_{\mathbb{Q}_p}^p = \alpha_{\chi_{\mathbb{Q}_p}}^p$ in $B(H_p)$, and it satisfies that

$$\alpha_{\mathbb{Q}_p}^p(h) = h = 1_{H_p}(h), \forall h \in H_p, \tag{20b}$$

where $1_{H_p} \in B(H_p)$ is the identity operator on H_p.

Proposition 4. *The pair (H_p, α^p) is a Hilbert-space representation of $\mathcal{M}_p$.*

Proof. It suffices to show that α^p is an algebra-action of $\mathcal{M}_p$ on H_p. However, this morphism α^p is a $*$-homomorphism from $\mathcal{M}_p$ into $B(H_p)$, by (20a). □

Definition 3. *The Hilbert-space representation (H_p, α^p) is called the p-adic representation of $\mathcal{M}_p$.*

Depending on the p-adic representation (H_p, α^p) of $\mathcal{M}_p$, one can define the C^*-subalgebra M_p of $B(H_p)$ as follows.

Definition 4. *Let M_p be the operator-norm closure of $\mathcal{M}_p$,*

$$M_p \stackrel{def}{=} \overline{\alpha^p(\mathcal{M}_p)} = \overline{\mathbb{C}\left[\alpha_f^p : f \in \mathcal{M}_p\right]} \qquad (21)$$

in $B(H_p)$, where $\overline{X}$ are the operator-norm closures of subsets X of $B(H_p)$. This C^*-algebra M_p is said to be the p-adic C^*-algebra of $(\mathcal{M}_p, \varphi_p)$.

By (21), the p-adic C^*-algebra M_p is a unital C^*-algebra contains its *unity* (or the unit, or the multiplication-identity) $1_{H_p} = \alpha_{\mathbb{Q}_p}^p$, by (20b).

5. Statistics on M_p

In this section, fix $p \in \mathcal{P}$, and let M_p be the corresponding p-adic C^*-algebra of (21). Define a linear functional $\varphi_j^p : M_p \to \mathbb{C}$ by

$$\varphi_j^p(a) \stackrel{def}{=} \left\langle a(\chi_{\partial_j}), \chi_{\partial_j} \right\rangle_2, \forall a \in M_p, \qquad (22a)$$

for $\chi_{\partial_j} \in H_p$, where $\langle,\rangle_2$ is the inner product (4.2) on the p-adic Hilbert space H_p of (4.1), and ∂_j are the boundaries (3.1) of $\mathbb{Q}_p$, for all $j \in \mathbb{Z}$. It is not hard to check such a linear functional φ_j^p on M_p is bounded, since

$$\begin{aligned}\varphi_j^p\left(\alpha_S^p\right) &= \left\langle \alpha_S^p\left(\chi_{\partial_j}\right), \chi_{\partial_j}\right\rangle_2 = \left\langle \chi_S \chi_{\partial_j}, \chi_{\partial_j}\right\rangle_2 \\ &= \left\langle \chi_{S \cap \partial_j}, \chi_{\partial_j}\right\rangle_2 = \int_{\mathbb{Q}_p} \chi_{S \cap \partial_j} d\mu_p \\ &\leq \int_{\mathbb{Q}_p} \chi_{\partial_j} d\mu_p = \mu_p\left(\partial_j\right) = \frac{1}{p^j} - \frac{1}{p^{j+1}},\end{aligned} \qquad (22b)$$

for all $S \in \sigma(\mathbb{Q}_p)$, for any fixed $j \in \mathbb{Z}$.

Definition 5. *Let φ_j^p be bounded linear functionals (22a) on the p-adic C^*-algebra M_p, for all $j \in \mathbb{Z}$. Then, the pairs $\left(M_p, \varphi_j^p\right)$ are said to be the j-th p-adic C^*-measure spaces, for all $j \in \mathbb{Z}$.*

Thus, one can get the system

$$\{(M_p, \varphi_j^p) : j \in \mathbb{Z}\}$$

of the j-th p-adic C^*-measure spaces (M_p, φ_j^p)'s.

Note that, for any fixed $j \in \mathbb{Z}$, and (M_p, φ_j^p), the unity

$$1_{M_p} \stackrel{denote}{=} 1_{H_p} = \alpha_{\mathbb{Q}_p}^p \text{ of } M_p$$

satisfies that

$$\begin{aligned}\varphi_j^p\left(1_{M_p}\right) &= \left\langle \chi_{\mathbb{Q}_p \cap \partial_j}, \chi_{\partial_j}\right\rangle_2 \\ &= \left\|\chi_{\partial_j}\right\|^2 = \frac{1}{p^j} - \frac{1}{p^{j+1}}.\end{aligned} \qquad (23)$$

Thus, the j-th p-adic C^*-measure space (M_p, φ_j^p) is a bounded-measure space, but not a probability space, in general.

Proposition 5. *Let $S \in \sigma\left(\mathbb{Q}_p\right)$, and $\alpha_S^p \in \left(M_p, \varphi_j^p\right)$, for a fixed $j \in \mathbb{Z}$. Then, there exists $r_S \in \mathbb{R}$, such that*

$$0 \leq r_S \leq 1 \text{ in } \mathbb{R},$$

and

$$\varphi_j^p\left(\left(\alpha_S^p\right)^n\right) = r_S\left(\frac{1}{p^j} - \frac{1}{p^{j+1}}\right); n \in \mathbb{N}. \tag{24}$$

Proof. Remark that the element α_S^p is a projection in M_p, in the sense that:

$$\left(\alpha_S^p\right)^* = \alpha_{(\chi_S^*)}^p = \alpha_S^p = \alpha_{(\chi_S \cap \chi_S)}^p = \left(\alpha_S^p\right)^2, \text{ in } M_p,$$

and hence,

$$\left(\alpha_S^p\right)^n = \alpha_S^p,$$

for all $n \in \mathbb{N}$. Thus, we obtain the formula (24) by (22b). □

As a corollary of (24), one obtains that, if ∂_k is a k-th boundaries of $\mathbb{Q}_p$, then

$$\varphi_j^p\left(\left(\alpha_{\partial_k}^p\right)^n\right) = \delta_{j,k}\left(\frac{1}{p^j} - \frac{1}{p^{j+1}}\right), \tag{25}$$

for all $n \in \mathbb{N}$, for $k \in \mathbb{Z}$.

6. The C^*-Subalgebra $\mathfrak{S}_p$ of M_p

Let M_p be the p-adic C^*-algebra for $p \in \mathcal{P}$. Let

$$P_{p,j} = \alpha_{\partial_j}^p \in M_p, \tag{26}$$

for all $j \in \mathbb{Z}$. By (24) and (25), these operators $P_{p,j}$ of (26) are *projections* on the p-adic Hilbert space H_p, in M_p, for all $p \in \mathcal{P}$, $j \in \mathbb{Z}$.

Definition 6. *Let $p \in \mathcal{P}$, and let $\mathfrak{S}_p$ be the C^*-subalgebra*

$$\mathfrak{S}_p = C^*\left(\{P_{p,j}\}_{j \in \mathbb{Z}}\right) = \overline{\mathbb{C}\left[\{P_{p,j}\}_{j \in \mathbb{Z}}\right]} \text{ of } M_p, \tag{27}$$

where $P_{p,j}$ are in the sense of ((26)), for all $j \in \mathbb{Z}$. We call $\mathfrak{S}_p$, the p-adic boundary (C^*-)subalgebra of M_p.

Proposition 6. *If $\mathfrak{S}_p$ is the p-adic boundary subalgebra (27), then*

$$\mathfrak{S}_p \overset{*\text{-iso}}{=} \bigoplus_{j \in \mathbb{Z}} \left(\mathbb{C} \cdot P_{p,j}\right) \overset{*\text{-iso}}{=} \mathbb{C}^{\oplus |\mathbb{Z}|}, \tag{28}$$

in the p-adic C^*-algebra M_p.

Proof. It is enough to show that the generating operators $\{P_{p,j}\}_{j \in \mathbb{Z}}$ of $\mathfrak{S}_p$ are mutually orthogonal from each other. It is not hard to check that

$$P_{p,j_1} P_{p,j_2} = \alpha^p\left(\chi_{\partial_{j_1}^p \cap \partial_{j_2}^p}\right) = \delta_{j_1,j_2} \alpha_{\partial_{j_1}^p}^p = \delta_{j_1,j_2} P_{p,j_1},$$

in $\mathfrak{S}_p$, for all $j_1, j_2 \in \mathbb{Z}$. Therefore, the structure theorem (28) is shown. □

By (27), one can define the measure spaces,

$$\mathfrak{S}_p(j) \overset{\text{denote}}{=} \left(\mathfrak{S}_p, \varphi_j^p\right), \forall j \in \mathbb{Z}, \tag{29}$$

for $p \in \mathcal{P}$, where the linear functionals φ_j^p of (29) are the restrictions $\varphi_j^p|_{\mathfrak{S}_p}$ of (22a), for all $p \in \mathcal{P}$, $j \in \mathbb{Z}$.

7. On the Tensor Product C^*-Probability Spaces $\left(A \otimes_{\mathbb{C}} \mathfrak{S}_p, \psi \otimes \varphi_j^p\right)$

In this section, we define and study our main objects of this paper. Let (A, ψ) be an arbitrary unital C^*-probability space (e.g., [22]), satisfying

$$\psi(1_A) = 1,$$

where 1_A is the unity of a C^*-algebra A. In addition, let

$$\mathfrak{S}_p(j) = \left(\mathfrak{S}_p, \varphi_j^p\right) \tag{30}$$

be the p-adic C^*-measure spaces (29), for all $p \in \mathcal{P}, j \in \mathbb{Z}$.

Fix now a unital C^*-probability space (A, ψ), and $p \in \mathcal{P}, j \in \mathbb{Z}$. Define a tensor product C^*-algebra

$$\mathfrak{S}_p^A \stackrel{def}{=} A \otimes_{\mathbb{C}} \mathfrak{S}_p, \tag{31}$$

and a linear functional ψ_j^p on $\mathfrak{S}_p^A$ by a linear morphism satisfying

$$\psi_j^p\left(a \otimes P_{p,k}\right) = \varphi_j^p\left(\psi(a) P_{p,k}\right), \tag{32}$$

for all $a \in (A, \psi)$, and $k \in \mathbb{Z}$.

Note that, by the structure theorem (28) of the p-adic boundary subalgebra $\mathfrak{S}_p$,

$$\mathfrak{S}_p^A \stackrel{*\text{-iso}}{=} A \otimes_{\mathbb{C}} \left(\mathbb{C}^{\oplus |\mathbb{Z}|}\right) \stackrel{*\text{-iso}}{=} A^{\oplus |\mathbb{Z}|}, \tag{33}$$

by (31).

By (33), one can verify that a morphism ψ_j^p of (32) is indeed a well-defined bounded linear functional on $\mathfrak{S}_p^A$.

Definition 7. *For any arbitrarily fixed $p \in \mathcal{P}, j \in \mathbb{Z}$, let $\mathfrak{S}_p^A$ be the tensor product C^*-algebra (31), and ψ_j^p, the linear functional (32) on $\mathfrak{S}_p^A$. Then, we call $\mathfrak{S}_p^A$, the A-tensor p-adic boundary algebra. The corresponding structure,*

$$\mathfrak{S}_p^A(j) \stackrel{denote}{=} \left(\mathfrak{S}_p^A, \psi_j^p\right) \tag{34}$$

is said to be the j-th p-adic A-(tensor C^*-probability-)space.

Note that, by (22a), (22b) and (32), the j-th p-adic A-space $\mathfrak{S}_p^A(j)$ of (34) is not a "unital" C^*-probability space, even though (A, ψ) is. Indeed, the C^*-algebra $\mathfrak{S}_p^A$ of (31) has its unity $1_A \otimes 1_{M_p}$, satisfying

$$\psi_j^p\left(1_A \otimes 1_{M_p}\right) = \varphi_j^p\left(\psi(1_A) 1_{M_p}\right)$$

$$= 1 \cdot \varphi_j^p(1_{M_p}) = \frac{1}{p^j} - \frac{1}{p^{j+1}},$$

for $j \in \mathbb{Z}$.

Remark that, by (32),

$$\psi_j^p\left(a \otimes P_{p,k}\right) = \psi(a) \, \varphi_j^p\left(P_{p,k}\right), \tag{35a}$$

for all $a \in (A, \psi)$, and $k \in \mathbb{Z}$. Thus, by abusing notation, one may write the definition (32) by

$$\psi_j^p = \psi \otimes \varphi_j^p \text{ on } A \otimes_{\mathbb{C}} \mathfrak{S}_p = \mathfrak{S}_p^A, \tag{35b}$$

in the sense of (35a), for all $p \in \mathcal{P}, j \in \mathbb{Z}$.

Proposition 7. Let $a \in (A, \psi)$, and $P_{p,k}$, the k-th generating projection of $\mathfrak{S}_p$, for all $k \in \mathbb{Z}$, and let $a \otimes P_{p,k}$ be the corresponding free random variable of the j-th p-adic A-space $\mathfrak{S}_p^A(j)$, for $j \in \mathbb{Z}$. Then,

$$\psi_j^p\left(\left(a \otimes P_{p,k}\right)^n\right) = \delta_{j,k}\, \psi(a^n) \left(\frac{1}{p^j} - \frac{1}{p^{j+1}}\right), \tag{36}$$

for all $n \in \mathbb{N}$.

Proof. Let $T_{p,k}^a = a \otimes P_{p,k}$ be a given free random variable of $\mathfrak{S}_p^A(j)$. Then,

$$\left(T_{p,k}^a\right)^n = \left(a \otimes P_{p,k}\right)^n = a^n \otimes P_{p,k} = T_{p,k}^{a^n},$$

and hence

$$\psi_j^p\left(\left(T_{p,k}^a\right)^n\right) = \psi_j^p\left(T_{p,k}^{a^n}\right)$$
$$= \psi(a^n)\, \varphi_j^p\left(P_{p,k}\right) = \psi(a^n) \left(\delta_{j,k}\left(\frac{1}{p^j} - \frac{1}{p^{j+1}}\right)\right)$$

by (35a)

$$= \delta_{j,k}\psi(a^n)\left(\frac{1}{p^j} - \frac{1}{p^{j+1}}\right),$$

for all $n \in \mathbb{N}$. Therefore, the free-distributional data (36) holds. □

Suppose a is a "self-adjoint" free random variable in (A, ψ) in the above proposition. Then, formula (36) completely characterizes the free distribution of $a \otimes P_{p,k}$ in the j-th p-adic A-space $\mathfrak{S}_p^A(j)$ of (34), i.e., the free distribution of $a \otimes P_{p,k}$ is characterized by the sequence,

$$\left(\delta_{j,k}\psi(a^n)\left(\frac{1}{p^j} - \frac{1}{p^{j+1}}\right)\right)_{n=1}^\infty$$

for all $p \in \mathcal{P}$, and $j, k \in \mathbb{Z}$ because $a \otimes P_{p,k}$ is self-adjoint in $\mathfrak{S}_p^A$ too.

It illustrates that the free probability on $\mathfrak{S}_p^A(j)$ is determined both by the free probability on (A, ψ), and by the statistical data on $\mathfrak{S}_p(j)$ of (30) (implying p-adic analytic information), for $p \in \mathcal{P}, j \in \mathbb{Z}$.

Notation. From below, for convenience, let's denote the free random variables $a \otimes P_{p,k}$ of $\mathfrak{S}_p^A(j)$, with $a \in (A, \psi)$ and $k \in \mathbb{Z}$, by $T_{p,k}^a$, i.e.,

$$T_{p,k}^a \stackrel{denote}{=} a \otimes P_{p,k},$$

for all $p \in \mathcal{P}, j \in \mathbb{Z}$.

In the proof of (36), it is observed that

$$\left(T_{p,k}^a\right)^n = T_{p,k}^{a^n} \in \mathfrak{S}_p^A(j) \tag{37}$$

for all $n \in \mathbb{N}$. More generally, the following free-distributional data is obtained.

Theorem 1. Fix $p \in \mathcal{P}$, and $j \in \mathbb{Z}$, and let $\mathfrak{S}_p^A(j)$ be the j-th p-adic A-space (34). Let $T_{p,k_l}^{a_l} \in \mathfrak{S}_p^A(j)$, for $l = 1, ..., N$, for $N \in \mathbb{N}$. Then,

$$\psi_j^p\left(\prod_{l=1}^N \left(T_{p,k_l}^{a_l}\right)^{n_l}\right) = \left(\prod_{l=1}^N \delta_{j,k_l}\right)\left(\frac{1}{p^j} - \frac{1}{p^{j+1}}\right)\psi\left(\prod_{l=1}^N a_l^{n_l}\right), \tag{38}$$

for all $n_1, ..., n_N \in \mathbb{N}$.

Proof. Let $T_{p,k_l}^{a_l} = a_l \otimes P_{p,k_l}$ be free random variables of $\mathfrak{S}_p^A(j)$, for $l = 1, ..., N$. Then, by (37),

$$\left(T_{p,k_l}^{a_l}\right)^{n_l} = T_{p,k_l}^{a_l^{n_l}} \in \mathfrak{S}_p^A(j), \text{ for } n_l \in \mathbb{N},$$

for all $l = 1, ..., N$. Thus,

$$T = \prod_{l=1}^{N} \left(T_{p,k_l}^{a_l}\right)^{n_l} = \left(\prod_{l=1}^{N} a_l^{n_l}\right) \otimes \left(\delta_{j:k_1,\ldots,k_N} P_{p,j}\right)$$

in $\mathfrak{S}_p^A(j)$, with

$$\delta_{j:k_1,\ldots,k_N} = \prod_{l=1}^{N} \delta_{j,k_l} \in \{0,1\}.$$

Therefore,

$$\psi_j^p(T) = \delta_{j:k_1,\ldots,k_N} \psi\left(\prod_{l=1}^{N} a_l^{n_l}\right) \varphi_j^p(P_{p,j})$$

$$= \delta_{j:k_1,\ldots,k_N} \left(\frac{1}{p^j} - \frac{1}{p^{j+1}}\right) \psi\left(\prod_{l=1}^{N} a_l^{n_l}\right),$$

by (35a). Thus, the joint free-distributional data (38) holds. □

Definitely, if $N = 1$ in (38), one obtains the formula (36).

8. On the Banach ∗-Probability Spaces $\mathfrak{LS}_{p,j}^A$

Let (A, ψ) be an arbitrarily fixed unital C^*-probability space, and let $\mathfrak{S}_p(j)$ be in the sense of (30), for all $p \in \mathcal{P}, j \in \mathbb{Z}$. Then, one can construct the tensor product C^*-probability spaces, the j-th p-adic A-space,

$$\mathfrak{S}_p^A(j) = \left(\mathfrak{S}_p^A, \psi_j^p\right) = \left(A \otimes_{\mathbb{C}} \mathfrak{S}_p, \psi \otimes \varphi_j^p\right)$$

of (34), for $p \in \mathcal{P}, j \in \mathbb{Z}$.

Throughout this section, we fix $p \in \mathcal{P}, j \in \mathbb{Z}$, and the corresponding j-th p-adic A-space $\mathfrak{S}_p^A(j)$. In addition, we keep using our notation $T_{p,k}^a$ for the free random variables $a \otimes P_{p,k}$ of $\mathfrak{S}^A(j)$, for all $a \in (A, \psi)$ and $k \in \mathbb{Z}$, where $P_{p,k}$ are the generating projections (26) of the p-adic boundary subalgebra $\mathfrak{S}_p$.

Recall that, by (36) and (38),

$$\psi_j^p\left(T_{p,k}^a\right) = \delta_{j,k} \psi(a) \left(\frac{1}{p^j} - \frac{1}{p^{j+1}}\right), \forall k \in \mathbb{Z}. \tag{39}$$

Now, let ϕ be the *Euler totient function*,

$$\phi : \mathbb{N} \to \mathbb{C},$$

defined by

$$\phi(n) = |\{k \in \mathbb{N} : k \leq n, \gcd(n,k) = 1\}|, \tag{40}$$

for all $n \in \mathbb{N}$, where $|X|$ are the *cardinalities of sets* X, and gcd is the *greatest common divisor*.

By the definition (40),

$$\phi(n) = n \left(\prod_{q \in \mathcal{P}, q | n} \left(1 - \frac{1}{q}\right)\right), \tag{41}$$

for all $n \in \mathbb{N}$, where "$q \mid n$" means "q divides n." Thus,

$$\phi(q) = q - 1 = q\left(1 - \frac{1}{q}\right), \forall q \in \mathcal{P}, \tag{42}$$

by (40) and (41).

By (42), we have

$$\varphi_j^p\left(P_{p,k}\right) = \frac{\delta_{j,k}}{p^j}\left(1 - \frac{1}{p}\right)$$

$$= \frac{\delta_{j,k}\phi(p)}{p^{j+1}},$$

for $P_{p,k} \in \mathfrak{S}_p$, and hence,

$$\psi_j^p\left(T_{p,k}^a\right) = \delta_{j,k}\left(\tfrac{\phi(p)}{p^{j+1}}\right)\psi(a), \qquad (43)$$

for all $T_{p,k}^a \in \mathfrak{S}_p^A(j)$, by (39).

Let's consider the following estimates.

Lemma 1. *Let ϕ be the Euler totient function (40). Then,*

$$\lim_{p\to\infty}\tfrac{\phi(p)}{p^{j+1}} = \begin{cases} 0, & \text{if } j > 0, \\ 1, & \text{if } j = 0, \\ \infty, \text{ Undefined}, & \text{if } j < 0, \end{cases} \qquad (44)$$

for all $j \in \mathbb{Z}$, where "$p \to \infty$" means "p is getting bigger and bigger in $\mathcal{P}$."

Proof. Observe that

$$\lim_{p\to\infty}\tfrac{\phi(p)}{p} = \lim_{p\to\infty}\left(1 - \tfrac{1}{p}\right) = 1,$$

by (42). Thus, one can get that

$$\lim_{p\to\infty}\tfrac{\phi(p)}{p^{j+1}} = \lim_{p\to\infty}\left(\tfrac{\phi(p)}{p}\right)\left(\tfrac{1}{p^j}\right) = \lim_{p\to\infty}\tfrac{1}{p^j},$$

for $j \in \mathbb{Z}$. Thus,

$$\lim_{p\to\infty}\tfrac{\phi(p)}{p^{j+1}} = \lim_{p\to\infty}\tfrac{1}{p^j} = \begin{cases} 0, & \text{if } j > 0, \\ 1, & \text{if } j = 0, \\ \lim_{p\to\infty} p^{|j|} = \infty, & \text{if } j < 0, \end{cases}$$

where $|j|$ are the absolute values of $j \in \mathbb{Z}$. Thus, the estimation (44) holds. □

8.1. Semicircular Elements

Let (B, φ) be an arbitrary *topological $*$-probability space* (C^*-probability space, or W^*-probability space, or Banach $*$-probability space, etc.) equipped with a topological $*$-algebra B (C^*-algebra, resp., W^*-algebra, resp., Banach $*$-algebra), and a linear functional φ on B.

Definition 8. *A self-adjoint operator $a \in B$ is said to be semicircular in (B, φ), if*

$$\varphi(a^n) = \omega_n c_{\frac{n}{2}}; n \in \mathbb{N}, \omega_n = \begin{cases} 1, & \text{if } n \text{ is even}, \\ 0, & \text{if } n \text{ is odd}, \end{cases} \qquad (45)$$

and c_k are the k-th Catalan numbers,

$$c_k = \tfrac{1}{k+1}\binom{2k}{k} = \tfrac{(2k)!}{k!(k+1)!},$$

for all $k \in \mathbb{N}_0 = \mathbb{N} \cup \{0\}$.

By [15–17], if $k_n(...)$ is the *free cumulant on B in terms of φ*, then a self-adjoint operator a is semicircular in (B, φ), if and only if

$$k_n\big(\underbrace{a, a, \ldots, a}_{n\text{-times}}\big) = \begin{cases} 1, & \text{if } n = 2, \\ 0, & \text{otherwise}, \end{cases} \qquad (46)$$

for all $n \in \mathbb{N}$. The above characterization (46) of the semicircularity (45) holds by the *Möbius inversion of* [15]. For example, definition (45) and the characterization (46) give equivalent free distributions, *the semicircular law.*

If a_l are semicircular elements in topological $*$-probability spaces (B_l, φ_l), for $l = 1, 2$, then the free distributions of a_l are completely characterized by the free-moment sequences,

$$\left(\varphi_l(a_l^n)\right)_{n=1}^{\infty}, \text{ for } l = 1, 2,$$

by the self-adjointness of a_1 and a_2; and by (45), one obtains that

$$\begin{aligned}\left(\varphi_1(a_1^n)\right)_{n=1}^{\infty} &= \left(\omega_n c_{\frac{n}{2}}\right)_{n=1}^{\infty} \\ &= (0, c_1, 0, c_2, 0, c_3, \ldots) \\ &= \left(\varphi_2(a_2^n)\right)_{n=1}^{\infty}.\end{aligned}$$

Equivalently, the free distributions of the semicircular elements a_1 and a_2 are characterized by the free-cumulant sequences,

$$\left(k_n^1(a_1, \ldots, a_1)\right)_{n=1}^{\infty} = (0, 1, 0, 0, 0, \ldots) = \left(k_n^2(a_2, \ldots, a_2)\right)_{n=1}^{\infty},$$

by (46), where $k_n^l(\ldots)$ are the free cumulants on B_l in terms of φ_l, for all $l = 1, 2$.

It shows the universality of free distributions of semicircular elements. For example, the free distributions of any semicircular elements are universally characterized by either the free-moment sequence

$$\left(\omega_n c_{\frac{n}{2}}\right)_{n=1}^{\infty}, \tag{47}$$

or the free-cumulant sequence

$$(0, 1, 0, 0, \ldots).$$

Definition 9. *Let a be a semicircular element of a topological $*$-probability space (B, φ). The free distribution of a is called "the" semicircular law.*

8.2. Tensor Product Banach $*$-Algebra $\mathfrak{LS}_p^A$

Let $\mathfrak{S}_p^A(k) = \left(\mathfrak{S}_p^A, \psi_k^p\right)$ be the k-th p-adic A-space (34), for all $p \in \mathcal{P}$, $k \in \mathbb{Z}$. Throughout this section, we fix $p \in \mathcal{P}$, $k \in \mathbb{Z}$, and $\mathfrak{S}_p^A(k)$. In addition, denote $a \otimes P_{p,j}$ by $T_{p,j}^a$ in $\mathfrak{S}_p^A(k)$, for all $a \in (A, \psi)$ and $j \in \mathbb{Z}$.

Define now bounded linear transformations $\mathbf{c}_p^A$ and $\mathbf{a}_p^A$ "acting on the tensor product C^*-algebra $\mathfrak{S}_p^A$," by linear morphisms satisfying,

$$\mathbf{c}_p^A\left(T_{p,j}^a\right) = T_{p,j+1}^a,$$

$$\mathbf{a}_p^A\left(T_{p,j}^a\right) = T_{p,j-1}^a, \tag{48}$$

on $\mathfrak{S}_p$, for all $j \in \mathbb{Z}$.

By the definitions (27) and (31), and by the structure theorem (33), the above linear morphisms $\mathbf{c}_p^A$ and $\mathbf{a}_p^A$ of (48) are well-defined on $\mathfrak{S}_p^A$.

By (48), one can understand $\mathbf{c}_p^A$ and $\mathbf{a}_p^A$ as bounded linear transformations contained in the *operator space* $B(\mathfrak{S}_p^A)$ consisting of all bounded linear operators acting on $\mathfrak{S}_p^A$, by regarding the C^*-algebra $\mathfrak{S}_p^A$ as a *Banach space* equipped with its C^*-norm (e.g., [32]). Under this sense, the operators $\mathbf{c}_p^A$ and $\mathbf{a}_p^A$ of (48) are well-defined *Banach-space operators* on $\mathfrak{S}_p^A$.

Definition 10. *The Banach-space operators* $\mathbf{c}_p^A$ *and* $\mathbf{a}_p^A$ *on* $\mathfrak{S}_p^A$, *in the sense of (48), are called the A-tensor p-creation, respectively, the A-tensor p-annihilation on* $\mathfrak{S}_p^A$. *Define a new Banach-space operator* $\mathbf{1}_p^A$ *by*

$$\mathbf{1}_p^A = \mathbf{c}_p^A + \mathbf{a}_p^A \text{ on } \mathfrak{S}_p^A. \tag{49}$$

We call this operator $\mathbf{1}_p^A$, the A-tensor p-radial operator on $\mathfrak{S}_p^A$.

Let $\mathbf{1}_p^A$ be the A-tensor p-radial operator $\mathbf{c}_p^A + \mathbf{a}_p^A$ of (49) in $B(\mathfrak{S}_p^A)$. Construct a *closed subspace* $\mathfrak{L}_p^A$ of $B(\mathfrak{S}_p^A)$ by

$$\mathfrak{L}_p^A = \overline{\mathbb{C}[\{\mathbf{1}_p^A\}]} \subset B(\mathfrak{S}_p^A), \tag{50}$$

equipped with the inherited *operator-norm* $\|.\|$ from the operator space $B(\mathfrak{S}_p^A)$, defined by

$$\|T\| = \sup\{\|Tx\|_{\mathfrak{S}_p^A} : x \in \mathfrak{S}_p^A \text{ s.t., } \|x\|_{\mathfrak{S}_p^A} = 1\},$$

where $\|.\|_{\mathfrak{S}_p^A}$ is the C^*-norm on the A-tensor p-adic algebra $\mathfrak{S}_p^A$ (e.g., [32]).

By the definition (50), the set $\mathfrak{L}_p^A$ is not only a closed subspace of $B(\mathfrak{S}_p^A)$, but also an algebra over $\mathbb{C}$. Thus, the subspace $\mathfrak{L}_p^A$ is a Banach algebra embedded in $B(\mathfrak{S}_p^A)$.

On the Banach algebra $\mathfrak{L}_p^A$ of (50), define a unary operation (*) by

$$\left(\sum_{k=0}^{\infty} s_k \left(\mathbf{1}_p^A\right)^k\right)^* = \sum_{k=0}^{\infty} \overline{s_k} \left(\mathbf{1}_p^A\right)^k \text{ in } \mathfrak{L}_p^A, \tag{51}$$

where $s_k \in \mathbb{C}$, with their conjugates $\overline{s_k} \in \mathbb{C}$.

Then, the operation (51) is a well-defined *adjoint on* $\mathfrak{L}_p^A$. Thus, equipped with the adjoint (51), this Banach algebra $\mathfrak{L}_p^A$ of (50) forms a *Banach $*$-algebra* in $B(\mathfrak{S}_p^A)$. For example, all elements of $\mathfrak{L}_p^A$ are adjointable (in the sense of [32]) in $B(\mathfrak{S}_p^A)$.

Let $\mathfrak{L}_p^A$ be in the sense of (50). Construct now the tensor product Banach $*$-algebra $\mathfrak{LS}_p^A$ by

$$\mathfrak{LS}_p^A \overset{def}{=} \mathfrak{L}_p^A \otimes_{\mathbb{C}} \mathfrak{S}_p^A = \mathfrak{L}_p^A \otimes_{\mathbb{C}} (A \otimes_{\mathbb{C}} \mathfrak{S}_p), \tag{52}$$

where $\otimes_{\mathbb{C}}$ is the *tensor product* of Banach $*$-algebras. Since $\mathfrak{S}_p^A$ is a C^*-algebra, it is a Banach $*$-algebra too.

Take now a generating element $\left(\mathbf{1}_p^A\right)^n \otimes T_{p,j}^a$, for some $n \in \mathbb{N}_0$, and $j \in \mathbb{Z}$, where $T_{p,j}^a = a \otimes P_{p,j}$ are in the sense of (37) in $\mathfrak{S}_p^A$, with axiomatization:

$$\left(\mathbf{1}_p^A\right)^0 = 1_{\mathfrak{S}_p^A},$$

the *identity operator on* $\mathfrak{S}_p^A$ in $B\left(\mathfrak{S}_p^A\right)$, satisfying

$$1_{\mathfrak{S}_p^A}(T) = T,$$

for all $T \in \mathfrak{S}_p^A$. Define now a bounded linear morphism $E_p^A : \mathfrak{LS}_p^A \to \mathfrak{S}_p^A$ by a linear transformation satisfying that:

$$E_p^A \left(\left(\mathbf{1}_p^A\right)^k \otimes T_{p,j}^a\right) = \frac{1}{\left[\frac{k}{2}\right]+1} \left(\mathbf{1}_p^A\right)^k (T_{p,j}^a), \tag{53}$$

for all $k \in \mathbb{N}_0, j \in \mathbb{Z}$, where $\left[\frac{k}{2}\right]$ is the *minimal integer greater than or equal to* $\frac{k}{2}$, for all $k \in \mathbb{N}_0$, for example,

$$\left[\tfrac{3}{2}\right] = 2 = \left[\tfrac{4}{2}\right].$$

By the cyclicity (50) of the tensor factor $\mathfrak{L}_p^A$ of $\mathfrak{LG}_p^A$, and by the structure theorem (33) of the other tensor factor $\mathfrak{S}_p^A$ of $\mathfrak{LG}_p^A$, the above morphism E_p^A of (53) is a well-defined bounded linear transformation from $\mathfrak{LG}_p^A$ onto $\mathfrak{S}_p^A$.

Now, consider how our A-tensor p-radial operator $1_p^A = c_p^A + a_p^A$ acts on $\mathfrak{S}_p^A$. First, observe that: if c_p^A and a_p^A are the A-tensor p-creation, respectively, the A-tensor p-annihilation on $\mathfrak{S}_p^A$, then

$$c_p^A a_p^A \left(T_{p,j}^a \right) = T_{p,j}^a = a_p^A c_p^A \left(T_{p,j}^a \right),$$

for all $a \in (A, \psi)$, and for all $j \in \mathbb{Z}, p \in \mathcal{P}$, and, hence,

$$c_p^A a_p^A = 1_{\mathfrak{S}_p^A} = a_p^A c_p^A \text{ on } \mathfrak{S}_p^A. \tag{54}$$

Lemma 2. *Let c_p^A, a_p^A be the A-tensor p-creation, respectively, the A-tensor p-annihilation on $\mathfrak{S}_p^A$. Then,*

$$\left(c_p^A \right)^n \left(a_p^A \right)^n = 1_{\mathfrak{S}_p^A} = \left(a_p^A \right)^n \left(c_p^A \right)^n,$$

$$\left(c_p^A \right)^{n_1} \left(a_p^A \right)^{n_2} = \left(a_p^A \right)^{n_2} \left(c_p^A \right)^{n_1}, \tag{55}$$

on $\mathfrak{S}_p^A$, for all $n, n_1, n_2 \in \mathbb{N}$.

Proof. The formulas in (55) hold by induction on (54). □

By (55), one can get that

$$\left(1_p^A \right)^n = \left(c_p^A + a_p^A \right)^n = \sum_{k=0}^n \binom{n}{k} \left(c_p^A \right)^k \left(a_p^A \right)^{n-k}, \tag{56}$$

with identity:

$$\left(c_p^A \right)^0 = 1_{\mathfrak{S}_p^A} = \left(a_p^A \right)^0,$$

for all $n \in \mathbb{N}$, where

$$\binom{n}{k} = \frac{n!}{k!(n-k)!},$$

for all $k \leq n \in \mathbb{N}_0$. By (56), one obtains the following proposition.

Proposition 8. *Let $1_p^A \in \mathfrak{L}_p^A$ be the A-tensor p-radial operator on $\mathfrak{S}_p^A$. Then,*

$$\left(1_p^A \right)^{2m-1} \text{ does not contain } 1_{\mathfrak{S}_p^A}\text{-term, and} \tag{57}$$

$$\left(1_p^A \right)^{2m} \text{ contains its } 1_{\mathfrak{S}_p^A}\text{-term, } \binom{2m}{m} \cdot 1_{\mathfrak{S}_p^A}, \tag{58}$$

for all $m \in \mathbb{N}$.

Proof. The proofs of (57) and (58) are done by straightforward computations of (56) with the help of (55). □

8.3. Free-Probabilistic Information of $Q_{p,j}^a$ in $\mathfrak{LS}_p^A$

Fix $p \in \mathcal{P}$, and a unital C^*-probability space (A, ψ), and let $\mathfrak{LS}_p^A$ be the Banach $*$-algebra (52). Let $E_p^A : \mathfrak{LS}_p^A \to \mathfrak{S}_p^A$ be the linear transformation (53). Throughout this section, let

$$Q_{p,j}^a \stackrel{denote}{=} 1_p^A \otimes T_{p,j}^a \in \mathfrak{LS}_p^A, \tag{59}$$

for all $j \in \mathbb{Z}$, where $T_{p,j}^a = a \otimes P_{p,j} \in \mathfrak{S}_p^A$ are in the sense of (37) generating $\mathfrak{S}_p^A$, for $a \in (A, \psi)$, and $j \in \mathbb{Z}$. Observe that

$$\begin{aligned}\left(Q_{p,j}^a\right)^n &= \left(1_p^A \otimes T_{p,j}^a\right)^n \\ &= \left(1_p^A\right)^n \otimes \left(T_{p,j}^a\right)^n = \left(1_p^A\right)^n \otimes T_{p,j}^{a^n},\end{aligned} \tag{60}$$

by (37), for all $n \in \mathbb{N}$, for all $j \in \mathbb{Z}$.

If $Q_{p,j}^a \in \mathfrak{LS}_p^A$ is in the sense of (59) for $j \in \mathbb{Z}$, then

$$E_p^A\left(\left(Q_{p,j}^a\right)^n\right) = \frac{1}{\left[\frac{n}{2}\right]+1} \left(1_p^A\right)^n \left(T_{p,j}^{a^n}\right), \tag{61}$$

by (53) and (60), for all $n \in \mathbb{N}$.

For any fixed $j \in \mathbb{Z}$, define a linear functional τ_j^p on $\mathfrak{LS}_p^A$ by

$$\tau_j^p = \psi_j^p \circ E_p^A \text{ on } \mathfrak{LS}_p^A, \tag{62}$$

where $\psi_j^p = \psi \otimes \varphi_j^p$ is a linear functional (35a), or (35b) on $\mathfrak{S}_p^A$.

By the linearity of both ψ_j^p and E_p^A, the morphism τ_j^p of (62) is a well-defined linear functional on $\mathfrak{LS}_p^A$ for $j \in \mathbb{Z}$. Thus, the pair $\left(\mathfrak{LS}_p^A, \tau_j^p\right)$ forms a Banach $*$-probability space (e.g., [22]).

Definition 11. *The Banach $*$-probability spaces*

$$\mathfrak{LS}_{p,j}^A \stackrel{denote}{=} \left(\mathfrak{LS}_p^A, \tau_j^p\right) \tag{63}$$

are called the A-tensor j-th p-adic (free-)filters, for all $p \in \mathcal{P}, j \in \mathbb{Z}$, where τ_j^p are in the sense of (62).

By (61) and (62), if $Q_{p,j}^a$ is in the sense of (59) in $\mathfrak{LS}_{p,j}^A$, then

$$\tau_j^p\left(\left(Q_{p,j}^a\right)^n\right) = \frac{1}{\left[\frac{n}{2}\right]+1} \psi_j^p\left(\left(1_p^A\right)^n \left(T_{p,j}^{a^n}\right)\right), \tag{64}$$

for all $n \in \mathbb{N}$.

Theorem 2. *Let $Q_{p,k}^a = 1_p^A \otimes T_{p,k}^a = 1_p^A \otimes \left(a \otimes P_{p,k}\right)$ be a free random variable (59) of the A-tensor j-th p-adic filter $\mathfrak{LS}_{p,j}^A$ of (63), for $p \in \mathcal{P}, j \in \mathbb{Z}$, for all $k \in \mathbb{Z}$. Then,*

$$\tau_j^p\left(\left(Q_{p,k}^a\right)^n\right) = \delta_{j,k} \omega_n \psi(a^n) c_{\frac{n}{2}}\left(\frac{\phi(p)}{p^{j+1}}\right), \tag{65}$$

where ω_n are in the sense of (45), for all $n \in \mathbb{N}$.

Proof. Let $Q_{p,j}^a$ be in the sense of (59) in $\mathfrak{LS}_{p,j}^A$, for the fixed $p \in \mathcal{P}$ and $j \in \mathbb{Z}$. Then,

$$\tau_j^p\left(\left(Q_{p,j}^a\right)^{2n-1}\right) = \psi_j^p\left(E_p^A\left(\left(Q_{p,j}^a\right)^{2n-1}\right)\right)$$

by (62)

$$= \left(\frac{1}{\left[\frac{2n-1}{2}\right]+1}\right) \psi_j^p\left(\left(1_p^A\right)^{2n-1} \left(T_{p,j}^{a^{2n-1}}\right)\right)$$

by (64)

$$= \left(\frac{1}{[\frac{2n-1}{2}]+1}\right) \psi_j^p \left(\left(\sum_{k=0}^{n} \binom{2n-1}{k} (c_p^A)^k (a_p^A)^{2n-1-k}\right) \left(T_{p,j}^{a^{2n-1}}\right)\right)$$

by (56)

$$= 0,$$

by (57), for all $n \in \mathbb{N}$.

Observe now that, for any $n \in \mathbb{N}$,

$$\tau_j^p \left(\left(Q_{p,j}^a\right)^{2n}\right) = \left(\frac{1}{[\frac{2n}{2}]+1}\right) \psi_j^p \left((1_p^A)^{2n} \left(T_{p,j}^{a^{2n}}\right)\right)$$

by (64)

$$= \left(\frac{1}{n+1}\right) \psi_j^p \left(\left(\sum_{k=0}^{2n} \binom{2n}{k} (c_p^A)^k (a_p^A)^{2n-k}\right) \left(T_{p,j}^{a^{2n}}\right)\right)$$

by (56)

$$= \left(\frac{1}{n+1}\right) \psi_j^p \left(\left(\binom{2n}{n}\right) T_{p,j}^{a^{2n}} + [\text{Rest terms}]\right)$$

by (58)

$$= \frac{1}{n+1} \binom{2n}{n} \psi_j^p \left(T_{p,j}^{a^{2n}}\right) = \frac{1}{n+1} \binom{2n}{n} \psi(a^{2n}) \left(\frac{\phi(p)}{p^{j+1}}\right)$$

by (39) and (43)

$$= c_n \psi(a^{2n}) \left(\frac{\phi(p)}{p^{j+1}}\right),$$

where c_n are the n-th Catalan numbers.

If $k \neq j$ in $\mathbb{Z}$, and if $Q_{p,k}^a$ are in the sense of (59) in $\mathfrak{LG}_{p,j}^A$, then

$$\tau_j^p \left(\left(Q_{p,k}^a\right)^n\right) = 0,$$

for all $n \in \mathbb{N}$, by the definition (22a) of the linear functional φ_j^p on $\mathfrak{G}_p$, inducing the linear functional $\psi_j^p = \psi \otimes \varphi_j^p$ on the tensor factor $\mathfrak{G}_p^A$ of $\mathfrak{LG}_{p,j}^A$.

Therefore, the free-distributional data (65) holds true. □

Note that, if a is self-adjoint in (A, ψ), then the generating operators $Q_{p,k}^a$ of the A-tensor j-th p-adic filter $\mathfrak{LG}_{p,j}^A$ are self-adjoint in $\mathfrak{LG}_p^A$, since

$$\left(Q_{p,k}^a\right)^* = \left(1_p^A \otimes T_{p,k}^a\right)^* = (1_p^A)^* \otimes \left(T_{p,k}^a\right)^*$$
$$= 1_p^A \otimes T_{p,k}^{a^*} = Q_{p,k}^a,$$

for all $k \in \mathbb{Z}$, for $p \in \mathcal{P}, j \in \mathbb{Z}$, by (51).

Thus, if a is a self-adjoint free random variable of (A, ψ), then the above formula (65) fully characterizes the free distributions (up to τ_j^p) of the generating operators $Q_{p,k}^a$ of $\mathfrak{LG}_p^A$, for all $k, j \in \mathbb{Z}$, for $p \in \mathcal{P}$.

The free-distributional data (65) can be refined as follows: if $p \in \mathcal{P}, j \in \mathbb{Z}$, and if $\mathfrak{LG}_{p,j}^A$ is the corresponding A-tensor j-th p-adic filter (63), then

$$\tau_j^p \left(\left(Q_{p,j}^a\right)^n\right) = \omega_n c_{\frac{n}{2}} \psi(a^n) \left(\frac{\phi(p)}{p^{j+1}}\right), \quad (66)$$

for all $n \in \mathbb{N}$, and

$$\tau_j^p\left(\left(Q_{p,k}^a\right)^n\right) = 0, \tag{67}$$

for all $n \in \mathbb{N}$, whenever $k \neq j$ in $\mathbb{Z}$, for all $n \in \mathbb{N}$.

Before we focus on non-zero free-distributional data (66) of $Q_{p,j}^a$, let's conclude the following result for $\{Q_{p,k}^a\}_{k \neq j \in \mathbb{Z}}$.

Corollary 1. *Let $p \in \mathcal{P}$, $j \in \mathbb{Z}$, and let $\mathfrak{LS}_{p,j}^A$ be the A-tensor j-th p-adic filter (63). Then, the generating operators*

$$Q_{p,k}^a = 1_p^A \otimes T_{p,j}^a = 1_p^A \otimes \left(a \otimes P_{p,j}\right) \in \mathfrak{LS}_{p,j}^A$$

have the zero free distribution, whenever $k \neq j$ in $\mathbb{Z}$.

Proof. It is proven by (65) and (67). □

By the above corollary, we now restrict our interests to the "j-th" generating operators $Q_{p,j}^a$ of (59) in the A-tensor "j-th" p-adic filter $\mathfrak{LS}_{p,j}^A$, for all $p \in \mathcal{P}$, $j \in \mathbb{Z}$, having non-zero free distributions determined by (66).

9. On the Free Product Banach ∗-Probability Space $\mathfrak{LS}_A$

Throughout this section, let (A, ψ) be a fixed unital C^*-probability space, and let

$$\mathfrak{LS}_{p,j}^A = \left(\mathfrak{LS}_p^A, \tau_j^p\right) \tag{68}$$

be A-tensor j-th p-adic filters, where

$$\mathfrak{LS}_p^A = \mathfrak{L}_p^A \otimes_{\mathbb{C}} \mathfrak{S}_p^A = \mathfrak{L}_p^A \otimes_{\mathbb{C}} \left(A \otimes_{\mathbb{C}} \mathfrak{S}_p\right),$$

are in the sense of (52), and τ_j^p are the linear functionals (62) on $\mathfrak{LS}_p^A$, for all $p \in \mathcal{P}$, $j \in \mathbb{Z}$.

Let $Q_{p,k}^a = 1_p^A \otimes T_{p,k}^a = 1_p^A \otimes \left(a \otimes P_{p,k}\right)$ be the generating elements (59) of $\mathfrak{LS}_{p,j}^A$ of (68), for $a \in (A, \psi)$, $p \in \mathcal{P}$, and $k, j \in \mathbb{Z}$. Then, these operators $Q_{p,k}^a$ of $\mathfrak{LS}_{p,j}^A$ have their free-distributional data,

$$\tau_j^p\left(\left(Q_{p,k}^a\right)^n\right) = \delta_{j,k} \omega_n \psi(a^n) c_{\frac{n}{2}} \left(\frac{\phi(p)}{p^{j+1}}\right), \tag{69}$$

for all $n \in \mathbb{N}$, by (65).

By (66) and (67), we here concentrate on the "j-th" generating operators of $\mathfrak{LS}_{p,j}^A$ having non-zero free distributions (69) for all $j \in \mathbb{Z}$, for all $p \in \mathcal{P}$.

9.1. Free Product Banach ∗-Probability Space $(\mathfrak{LS}_A, \tau)$

By (68), we have the family

$$\left\{\mathfrak{LS}_{p,j}^A : p \in \mathcal{P}, j \in \mathbb{Z}\right\}$$

of Banach ∗-probability spaces, consisting of the A-tensor j-th p-adic filters $\mathfrak{LS}_{p,j}^A$.

Define the *free product Banach ∗-probability space*,

$$(\mathfrak{LS}_A, \tau) \stackrel{\text{def}}{=} \underset{p \in \mathcal{P}, j \in \mathbb{Z}}{\star} \mathfrak{LS}_{p,j}^A,$$

$$= \left(\underset{p \in \mathcal{P}, j \in \mathbb{Z}}{\star} \mathfrak{LS}_p^A, \underset{p \in \mathcal{P}, j \in \mathbb{Z}}{\star} \tau_j^p\right) \tag{70}$$

in the sense of [15,22].

By (70), the A-tensor j-th p-adic filters $\mathfrak{LG}_{p,j}$ of (68) are the *free blocks* of the Banach $*$-probability space $(\mathfrak{LG}_A, \tau)$ of (70).

All operators of the Banach $*$-algebra $\mathfrak{LG}_A$ in (70) are the Banach-topology limits of linear combinations of noncommutative free reduced words (under operator-multiplication) in

$$\bigsqcup_{p \in \mathcal{P}, j \in \mathbb{Z}} \mathfrak{LG}_{p,j}^A.$$

More precisely, since each free block $\mathfrak{LG}_{p,j}^A$ is generated by $\{Q_{p,k}^a\}_{a \in A, k \in \mathbb{Z}}$, for all $p \in \mathcal{P}$, $j \in \mathbb{Z}$, all elements of $\mathfrak{LG}_A$ are the Banach-topology limits of linear combinations of free words in

$$\bigsqcup_{p \in \mathcal{P}, j \in \mathbb{Z}} \{Q_{p,k}^a \in \mathfrak{LG}_{p,j} : a \in A, k \in \mathbb{Z}\}.$$

In particular, all noncommutative free words have their unique free "reduced" words (as operators of $\mathfrak{LG}_A$ under operator-multiplication) formed by

$$\prod_{l=1}^{N} \left(Q_{p_l,k_l}^{a_l}\right)^{n_l}, \text{ where } Q_{p_l,k_l}^{a_l} \in \mathfrak{LG}_{p_l,j_l}^A$$

in $\mathfrak{LG}_A$, for all $a_1, ..., a_N \in (A, \psi)$, and $n_1, ..., n_N \in \mathbb{N}$, where either the N-tuple

$$(p_1, ..., p_N), \text{ or } (j_1, ..., j_N)$$

is alternating in $\mathcal{P}$, respectively, in $\mathbb{Z}$, in the sense that:

$$p_1 \neq p_2, p_2 \neq p_3, ..., p_{N-1} \neq p_N \text{ in } \mathcal{P},$$

respectively,

$$j_1 \neq j_2, j_2 \neq j_3, ..., j_{N-1} \neq j_N \text{ in } \mathbb{Z}$$

(e.g., see [22]).

For example, a 5-tuple

$$(2, 2, 3, 7, 2)$$

is not alternating in $\mathcal{P}$, while a 5-tuple

$$(2, 3, 2, 7, 2)$$

is alternating in $\mathcal{P}$, etc.

By (70), if $Q_{p,j}^a$ are the j-th a-tensor generating operators of a free block $\mathfrak{LG}_{p,j}^A$ of the Banach $*$-probability space $(\mathfrak{LG}_A, \tau)$, for all $j \in \mathbb{Z}$, for $p \in \mathcal{P}$, $j \in \mathbb{Z}$, then $\left(Q_{p,j}^a\right)^n$ are contained in the same free block $\mathfrak{LG}_{p,j}^A$ of $(\mathfrak{LG}_A, \tau)$, and, hence, they are free reduced words with their lengths-1, for all $n \in \mathbb{N}$. Therefore, we have

$$\begin{aligned} \tau\left(\left(Q_{p,j}^a\right)^n\right) &= \tau_j^p\left(\left(Q_{p,j}^a\right)^n\right) \\ &= \omega_n c_{\frac{n}{2}} \psi(a^n) \left(\frac{\phi(p)}{p^{j+1}}\right), \end{aligned} \quad (71)$$

for all $n \in \mathbb{N}$, by (69).

Definition 12. *The Banach $*$-probability space $\mathfrak{LG}_A \overset{denote}{=} (\mathfrak{LG}_A, \tau)$ of (70) is called the A-tensor (free-)Adelic filterization of $\{\mathfrak{LG}_{p,j}^A\}_{p \in \mathcal{P}, j \in \mathbb{Z}}$.*

As we discussed at the beginning of Section 9, we now focus on studying free random variables of the A-tensor Adelic filterization $\mathfrak{LS}_A$ of (70) having "non-zero" free distributions.

Define a subset $\mathcal{U}$ of $\mathfrak{LS}_A$ by

$$\mathcal{U} = \left\{ Q_{p,j}^{1_A} \in \mathfrak{LS}_{p,j}^A \,|\, \forall p \in \mathcal{P}, j \in \mathbb{Z} \right\} \tag{72}$$

in $\mathfrak{LS}_A$, where 1_A is the unity of A, and $Q_{p,j}^{1_A}$ are the "j-th" 1_A-tensor generating operators of $\mathfrak{LS}_A$, in the free blocks $\mathfrak{LS}_{p,j}^A$, for all $p \in \mathcal{P}, j \in \mathbb{Z}$.

Then, the elements $Q_{p,j}^{1_A}$ of $\mathcal{U}$ have their non-zero free distributions,

$$\left(\omega_n c_{\frac{n}{2}} \psi(1_A^n) \left(\frac{\phi(p)}{p^{j+1}} \right) \right)_{n=1}^{\infty} = \left(\omega_n c_{\frac{n}{2}} \left(\frac{\phi(p)}{p^{j+1}} \right) \right)_{n=1}^{\infty},$$

by (71), since

$$\psi(1_A^n) = \psi(1_A) = 1,$$

for all $n \in \mathbb{N}$. Now, define a Cartesian product set

$$\mathcal{U}_A \stackrel{def}{=} A \times \mathcal{U}, \tag{73a}$$

set-theoretically, where $\mathcal{U}$ is in the sense of (72).

Define a function $\Omega: \mathcal{U}_A \to \mathfrak{LS}_A$ by

$$\Omega\left((a, Q_{p,j}^{1_A}) \right) \stackrel{def}{=} Q_{p,j}^a \text{ in } \mathfrak{LS}_A, \tag{73b}$$

for all $(a, Q_{p,j}^{1_A}) \in \mathcal{U}_A$, where $\mathcal{U}_A$ is in the sense of (73a).

It is not difficult to check that this function Ω of (73b) is a well-defined injective map. Moreover, it induces all j-th a-tensor generating elements $Q_{p,j}^a$ of $\mathfrak{LS}_{p,j}^a$ in $\mathfrak{LS}_A$, for all $p \in \mathcal{P}$, and $j \in \mathbb{Z}$.

Define a Banach $*$-subalgebra $\mathbb{LS}_A$ of the A-tensor Adelic filterization $\mathfrak{LS}_A$ of (70) by

$$\mathbb{LS}_A \stackrel{def}{=} \overline{\mathbb{C}[\Omega(\mathcal{U}_A)]} \text{ in } \mathfrak{LS}_A, \tag{74a}$$

where $\Omega(\mathcal{U}_A)$ is the subset of $\mathfrak{LS}_A$, induced by (73a) and (73b), and $\overline{Y}$ mean the Banach-topology closures of subsets Y of $\mathfrak{LS}_A$.

Then, this Banach $*$-subalgebra $\mathbb{LS}_A$ of (74a) has a sub-structure,

$$\mathbb{LS}_A \stackrel{denote}{=} (\mathbb{LS}_A, \tau = \tau|_{\mathbb{LS}_A}) \tag{74b}$$

in the A-tensor Adelic filterization $\mathfrak{LS}_A$.

Theorem 3. *Let $\mathbb{LS}_A$ be the Banach $*$-algebra (74a) in the A-tensor Adelic filterization $\mathfrak{LS}_A$. Then,*

$$\mathbb{LS}_A \stackrel{*-iso}{=} \overline{\underset{p \in \mathcal{P}, j \in \mathbb{Z}}{\star} \mathbb{C}\left[\{Q_{p,j}^a : a \in (A, \psi)\} \right]}$$

$$\stackrel{*-iso}{=} \overline{\mathbb{C}\left[\underset{p \in \mathcal{P}, j \in \mathbb{Z}}{\star} \{Q_{p,j}^a : a \in (A, \psi)\} \right]}, \tag{75}$$

where $Q_{p,j}^a \in \Omega(\mathcal{U}_A)$ of (73b). Here, $(\star)$ in the first $*$-isomorphic relation in (75) is the free-probability-theoretic free product determined by the linear functional τ of (70), or of (74b) (e.g., [15,22]), and $(\star)$ in the second $*$-isomorphic relation in (75) is the pure-algebraic free product generating noncommutative free words in $\Omega(\mathcal{U}_A)$.

Proof. Let $\mathbb{LS}_A$ be the Banach $*$-subalgebra (74a) in $\mathfrak{LS}_A$. Then,

$$\mathbb{LS}_A = \overline{\mathbb{C}\left[\{Q_{p,j}^a \in \mathfrak{LS}_{p,j}^A : a \in (A, \psi)\}_{p \in \mathcal{P}, j \in \mathbb{Z}} \right]}$$

by (73a), (73b) and (74a)

$$\overset{*\text{-iso}}{=} \overline{\underset{p\in\mathcal{P},\,j\in\mathbb{Z}}{\star} \mathbb{C}\left[\{Q_{p,j}^{a} : a \in (A,\,\psi)\}\right]}$$

in $\mathfrak{LG}_A$, since all elements $Q_{p,j}^a \in \Omega(\mathcal{U}_A)$ are chosen from mutually distinct free blocks $\mathfrak{LG}_{p,j}^A$ of the A-tensor Adelic filterization $\mathfrak{LG}_A$, and, hence, the operators $\{Q_{p,j}^a, Q_{p,j}^{a^*}\}_{p\in\mathcal{P},\,j\in\mathbb{Z}}$ are free from each other in $\mathfrak{LG}_A$, for any $a \in (A,\,\psi)$, for all $p \in \mathcal{P}, j \in \mathbb{Z}$, moreover,

$$\overset{*\text{-iso}}{=} \overline{\mathbb{C}\left[\underset{p\in\mathcal{P},\,j\in\mathbb{Z}}{\star}\{Q_{p,j}^{a} : a \in (A,\,\psi)\}\right]},$$

because all elements of $\mathbb{LS}_A$ are the (Banach-topology limits of) linear combinations of free words in $\Omega(\mathcal{U}_A)$, by the very above $*$-isomorphic relation. Indeed, for any noncommutative (pure-algebraic) free words in

$$\underset{p\in\mathcal{P},\,j\in\mathbb{Z}}{\cup}\{Q_{p,j}^{a} : a \in (A,\,\psi)\}$$

have their unique free "reduced" words under operator-multiplication on $\mathfrak{LG}_A$, as operators of $\mathbb{LS}_A$. Therefore, the structure theorem (75) holds. □

The above theorem characterizes the free-probabilistic structure of the Banach $*$-algebra $\mathbb{LS}_A$ of (74a) in the A-tensor Adelic filterization $\mathfrak{LG}_A$. This structure theorem (75) demonstrates that the Banach $*$-probability space $(\mathbb{LS}_A, \tau)$ of (74b) is well-determined, having its natural inherited free probability from that on $\mathfrak{LG}_A$.

Definition 13. *Let $(\mathbb{LS}_A, \tau)$ be the Banach $*$-probability space (74b). Then, we call*

$$\mathbb{LS}_A \overset{denote}{=} (\mathbb{LS}_A, \tau),$$

the A-tensor (Adelic) sub-filterization of the A-tensor Adelic filterization $\mathfrak{LG}_A$.

By (69), (71), (72) and (75), one can verify that the free probability on the A-tensor sub-filterization $\mathbb{LS}_A$ provide "possible" non-zero free distributions on the A-tensor Adelic filterization $\mathfrak{LG}_A$, up to free probability on (A, ψ). i.e., if $a \in (A, \psi)$ have their non-zero free distributions, then $Q_{p,j}^a \in \mathbb{LS}_A$ have non-zero free distributions, and, hence, they have their non-zero free distributions on $\mathfrak{LG}_A$.

Theorem 4. *Let $Q_{p,j}^a \in \Omega(\mathcal{U}_A)$ be free random variables of the A-tensor sub-filterization $\mathbb{LS}_A$, for $a \in (A, \psi)$, and $p \in \mathcal{P}$, and $j \in \mathbb{Z}$. Then,*

$$\tau\left(\left(Q_{p,j}^a\right)^n\right) = \omega_n c_{\frac{n}{2}} \psi(a^n) \left(\frac{\phi(p)}{p^{j+1}}\right),$$

(76)

$$\tau\left(\left(\left(Q_{p,j}^a\right)^*\right)^n\right) = \omega_n c_{\frac{n}{2}} \overline{\psi(a^n)} \left(\frac{\phi(p)}{p^{j+1}}\right),$$

for all $n \in \mathbb{N}$.

Proof. The first formula of (76) is shown by (71). Thus, it suffices to prove the second formula of (76) holds. Note that

$$\left(Q_{p,j}^a\right)^* = \left(1_p^A \otimes T_{p,j}^a\right)^* = \left(1_p^A \otimes (a \otimes P_{p,j})\right)^*$$
$$= \left(1_p^A\right)^* \otimes \left(a \otimes P_{p,j}\right)^* = 1_p^A \otimes \left(a^* \otimes P_{p,j}\right),$$

and, hence,

$$\left(Q_{p,j}^a\right)^* = Q_{p,j}^{a^*} \text{ in } \mathbb{LS}_A,$$

(77)

for all $Q^a_{p,j} \in \Omega(\mathcal{U}_A)$. Thus, one has

$$\left(\left(Q^a_{p,j}\right)^*\right)^n = \left(Q^{a^*}_{p,j}\right)^n = Q^{(a^*)^n}_{p,j} = Q^{(a^n)^*}_{p,j} \text{ in } \mathbb{LS}_A,$$

by (77).

Thus, one has

$$\tau\left(\left(\left(Q^a_{p,j}\right)^*\right)^n\right) = \omega_n c_{\frac{n}{2}} \psi\left((a^n)^*\right)\left(\frac{\phi(p)}{p^{j+1}}\right)$$

$$= \omega_n c_{\frac{n}{2}} \overline{\psi(a^n)}\left(\frac{\phi(p)}{p^{j+1}}\right),$$

by (71), for all $n \in \mathbb{N}$. Therefore, the second formula of (76) holds too. □

9.2. Prime-Shifts on $\mathbb{LS}_A$

Let $\mathbb{LS}_A$ be the A-tensor sub-filterization (70) of the A-tensor Adelic filterization $\mathfrak{LS}_A$. In this section, we define a certain ∗-homomorphism on $\mathbb{LS}_A$, and study asymptotic free-distributional data on $\mathbb{LS}_A$ (and hence those on $\mathfrak{LS}_A$) over primes.

Let $\mathcal{P}$ be the set of all primes in $\mathbb{N}$, regarded as a *totally ordered set* (in short, a TOset) for the usual ordering ($\leq$), i.e.,

$$\mathcal{P} = \{q_1 < q_2 < q_3 < q_4 < \cdots\}, \tag{78}$$

with

$$q_1 = 2, q_2 = 3, q_3 = 5, q_4 = 7, q_5 = 11, \ldots, \text{etc.}$$

Define an injective function $h : \mathcal{P} \to \mathcal{P}$ by

$$h(q_k) = q_{k+1}; k \in \mathbb{N}, \tag{79}$$

where q_k are primes of (78), for all $k \in \mathbb{N}$.

Definition 14. *Let h be an injective function (79) on the TOset $\mathcal{P}$ of (78). We call h the shift on $\mathcal{P}$.*

Let h be the shift (79) on the TOset $\mathcal{P}$, and let

$$h^{(n)} \stackrel{\text{def}}{=} \underbrace{h \circ h \circ h \circ \cdots \circ h}_{n\text{-times}}, \text{ on } \mathcal{P}, \tag{80}$$

for all $n \in \mathbb{N}$, where ($\circ$) is the usual functional *composition*.

By the definitions (79) and (80),

$$h^{(n)}(q_k) = q_{k+n}, \tag{81}$$

for all $n \in \mathbb{N}$, in $\mathcal{P}$. For instance, $h^{(3)}(2) = 7$, and $h^{(4)}(5) = 17$, etc.

These injective functions $h^{(n)}$ of (80) are called the *n-shifts* on $\mathcal{P}$, for all $n \in \mathbb{N}$.

For the shift h on $\mathcal{P}$, one can define a ∗-homomorphism π_h on the A-tensor sub-filterization $\mathbb{LS}_A$ by a bounded "multiplicative" linear transformation, satisfying that

$$\pi_h\left(Q^a_{q_k,j}\right) = Q^a_{h(q_k),j} = Q^a_{q_{k+1},j}, \tag{82}$$

for all $Q_{q_k,j} \in \Omega(\mathcal{U}_A)$, for all $q_k \in \mathcal{P}$, for all $j \in \mathbb{Z}$, where h is the shift (79) on $\mathcal{P}$.

By (82), we have

$$\pi_h\left(\prod_{l=1}^{N}\left(Q^{a_l}_{q_{k_l},j_l}\right)^{n_l}\right) = \prod_{l=1}^{N}\left(Q^{a_l}_{h(q_{k_l}),j_l}\right)^{n_l} = \prod_{l=1}^{N}\left(Q^{a_l}_{q_{k_l+1},j_l}\right)^{n_l}, \tag{83}$$

in $\mathbb{LS}_A$, for all $Q^a_{q_{k_l},j_l} \in \Omega(\mathcal{U}_A)$, for $q_{k_l} \in \mathcal{P}$, $j_l \in \mathbb{Z}$, for $l = 1, ..., N$, for $N \in \mathbb{N}$, where $n_1, ..., n_N \in \mathbb{N}$.

Remark 1. *Note that the multiplicative linear transformation π_h of (82) is indeed a $*$-homomorphism satisfying*

$$\pi_h(T^*) = (\pi_h(T))^*,$$

for all $T \in \mathbb{LS}_A$, because

$$\begin{aligned}\pi_h\left(\left(Q^a_{p,j}\right)^*\right) &= \pi_h\left(Q^{a^*}_{p,j}\right) = Q^{a^*}_{h(p),j} \\ &= \left(Q^a_{h(p),j}\right)^* = \left(\pi_h\left(Q^a_{p,j}\right)\right)^*,\end{aligned}$$

for all $Q^a_{p,j} \in \Omega(\mathcal{U}_A)$.

In addition, by (82), we obtain the $*$-homomorphisms,

$$\pi^n_h = \underbrace{\pi_h \pi_h \pi_h \cdots \pi_h}_{n\text{-times}}, \text{ on } \mathbb{LS}_A, \tag{84}$$

the products (or compositions) of the n-copies of the $*$-homomorphism π_h of (82), acting on $\mathbb{LS}_A$. It is not difficult to check that

$$\begin{aligned}\pi^n_h\left(Q^a_{p,j}\right) &= \pi^{n-1}_h\left(Q^a_{h(p),j}\right) = \pi^{n-2}_h\left(Q^a_{h^{(2)}(p),j}\right) \\ &= \cdots = \pi_h\left(Q^a_{h^{(n-1)}(p),j}\right) = Q^a_{h^{(n)}(p),j'}\end{aligned} \tag{85}$$

for all $Q^a_{p,j} \in \Omega(\mathcal{U}_A)$ in $\mathbb{LS}_A$, where $h^{(k)}$ are the k-shifts (80) on $\mathcal{P}$, for all $k \in \mathbb{N}$.

Definition 15. *Let π_h be the $*$-homomorphism (82) on the A-tensor sub-filterization $\mathbb{LS}_A$, and let π^n_h be the products (84) acting on $\mathbb{LS}_A$, for all $n \in \mathbb{N}$, with $\pi^1_h = \pi_h$. Then, we call π^n_h, the n-prime-shift ($*$-homomorphism) on $\mathbb{LS}_A$, for all $n \in \mathbb{N}$. In particular, the 1-prime-shift π_h is simply said to be the prime-shift ($*$-homomorphism) on $\mathbb{LS}_A$.*

Thus, for any $Q^a_{q_k,j} \in \Omega(\mathcal{U}_A)$ in $\mathbb{LS}_A$, for $q_k \in \mathcal{P}$ (in the sense of (78) with $k \in \mathbb{N}$), the n-prime-shift π^n_h satisfies

$$\pi^n_h\left(Q^a_{q_k,j}\right) = Q^a_{h^{(n)}(q_k),j} = Q^a_{q_{k+n},j'} \tag{86}$$

by (81) and (85), and, hence,

$$\pi^n_h\left(\prod_{l=1}^N \left(Q^{a_l}_{q_{k_l},j_l}\right)^{n_l}\right) = \prod_{l=1}^N \left(Q^{a_l}_{q_{k_l+n},j_l}\right)^{n_l}, \tag{87}$$

by (83) and (86), for all $n \in \mathbb{N}$.

By (86) and (87), one may write as follows;

$$\pi^n_h = \pi_{h^{(n)}} \text{ on } \mathbb{LS}_A, \text{ for all } n \in \mathbb{N},$$

where $h^{(n)}$ are the n-shifts (81) on the TOset $\mathcal{P}$.

Consider now the sequence

$$\Pi = \left(\pi^n_h\right)^\infty_{n=1} \tag{88}$$

of the n-prime-shifts on $\mathbb{LS}_A$.

For any fixed $T \in \mathbb{LS}_A$, the sequence Π of (88) induces the sequence of operators,

$$\Pi(T) = \left(\pi^n_h(T)\right)^\infty_{n=1} = \left(\pi_h(T), \pi^2_h(T), \pi^3_h(T), \cdots\right)$$

in $\mathbb{LS}_A$, and this sequence $\Pi(T)$ has its corresponding free-distributional data, represented by the following $\mathbb{C}$-sequence:

$$\tau(\Pi(T)) = \left(\tau\left(\pi_h^n(T)\right)\right)_{n=1}^{\infty}. \tag{89}$$

We are interested in the convergence of the $\mathbb{C}$-sequence $\tau(\Pi(T))$ of (89), as $n \to \infty$.

Either convergent or divergent, the $\mathbb{C}$-sequence $\tau(\Pi(T))$ of (89), induced by any fixed operator $T \in \mathbb{LS}_A$, shows the asymptotic free distributional data of the family $\{\pi_h^n(T)\}_{n=1}^{\infty} \subset \mathbb{LS}_A$, as $n \to \infty$ in $\mathbb{N}$, equivalently, as $q_n \to \infty$ in $\mathcal{P}$.

9.3. Asymptotic Behaviors in $\mathbb{LS}_A$ over $\mathcal{P}$

Recall that, by (44), we have

$$\lim_{p \to \infty} \frac{\phi(p)}{p^{j+1}} = \begin{cases} 0, & \text{if } j > 0, \\ 1, & \text{if } j = 0, \\ \infty, \text{ Undefined}, & \text{if } j < 0, \end{cases} \tag{90}$$

for $j \in \mathbb{Z}$.

Recall also that there are bounded $*$-homomorphisms

$$\Pi = \left(\pi_h^n\right)_{n=1}^{\infty}, \text{ acting on } \mathbb{LS}_A,$$

of (88), where π_h^n are the n-prime shifts of (84), where h is the shift (79) on the TOset $\mathcal{P}$ of (78). Then, these $*$-homomorphisms of Π satisfies

$$\lim_{n \to \infty} \left(\pi_h^n\left(Q_{p,j}^a\right)\right) = \lim_{n \to \infty} \left(Q_{h^{(n)}(p),\,j}^a\right), \tag{91}$$

for all $Q_{p,j}^a \in \Omega(\mathcal{U}_A)$ in $\mathbb{LS}_A$, where $h^{(n)}$ are the n-shifts (80) on $\mathcal{P}$, for all $n \in \mathbb{N}$.

Thus, one can get that: if $\prod_{l=1}^{N}\left(Q_{p_l,j_l}^{a_l}\right)^{n_l}$ is a free reduced words of $\mathbb{LS}_A$ in $\Omega(\mathcal{U}_A)$, then

$$\lim_{n \to \infty} \pi_h^n\left(\prod_{l=1}^{N}\left(Q_{p_l,j_l}^{a_l}\right)^{n_l}\right) = \lim_{n \to \infty}\left(\prod_{l=1}^{N} \pi_h^n\left(\left(Q_{p_l,j_l}^{a_l}\right)^{n_l}\right)\right)$$

$$= \lim_{n \to \infty}\left(\prod_{l=1}^{N}\left(\pi_h^n\left(Q_{p_l,j_l}^{a_l}\right)\right)^{n_l}\right)$$

since π_h^n are $*$-homomorphisms on $\mathbb{LS}_A$

$$= \lim_{n \to \infty}\left(\prod_{l=1}^{N}\left(Q_{h^{(n)}(p_l),j_l}^{a_l}\right)^{n_l}\right)$$

by (91)

$$= \prod_{l=1}^{N}\left(\lim_{n \to \infty}\left(Q_{h^{(n)}(p_l),\,j_l}^{a_l}\right)^{n_l}\right), \tag{92}$$

under the Banach-topology for $\mathbb{LS}_A$, for all $Q_{p_l,j_l}^{a_l} \in \Omega(\mathcal{U}_A)$, for $a_l \in (A, \psi)$, $p_l \in \mathcal{P}$, $j_l \in \mathbb{Z}$, for $l = 1, \dots, N$, for all $N \in \mathbb{N}$.

Notation 2. (in short, **N 2** from below) For convenience, we denote $\lim_{n \to \infty} \pi_h^n$ symbolically by π, for the sequence $\Pi = \left(\pi_h^n\right)_{n=1}^{\infty}$ of (88).

Lemma 3. Let $Q_{p_l,j_l}^{a_l} \in \Omega(\mathcal{U}_A)$ be generators of the A-tensor sub-filterization $\mathbb{LS}_A$, for $l = 1, \dots, N$, for $N \in \mathbb{N}$. In addition, let Π be the sequence (88) acting on $\mathbb{LS}_A$. If π is in the sense of **N 2**, then

$$\pi\left(Q_{p_1,j_1}^{a_1}\right) = \lim_{n \to \infty}\left(Q_{\left(h^{(n)}(p_1)\right),\,j_1}^{a_1}\right),$$

$$(93)$$

$$\pi\left(\prod_{l=1}^{N}\left(Q_{p_l,j_l}^{a_l}\right)^{n_l}\right) = \lim_{n\to\infty}\left(\prod_{l=1}^{N}\left(Q_{h^{(n)}(p_l),j_l}^{a_l}\right)^{n_l}\right),$$

for all $n_1, ..., n_N \in \mathbb{N}$, where $h^{(n)}$ are the n-shifts (80) on $\mathcal{P}$.

Proof. The proof of (93) is done by (91) and (92). □

By abusing notation, one may/can understand the above formula (93) as follows

$$\pi\left(Q_{p_1,j_1}^{a_1}\right) = \lim_{p_1\to\infty} Q_{p_1,j_1}^{a_1}, \tag{94a}$$

$$\pi\left(\prod_{l=1}^{N} Q_{p_l,j_l}^{n_l}\right) = \prod_{l=1}^{N}\left(\lim_{p_l\to\infty}\left(Q_{p_l,j_l}^{n_l}\right)\right),$$

respectively, where "$\lim_{q\to\infty}$" for $q \in \mathcal{P}$ is in the sense of (44).

Such an understanding (94a) of the formula (93) is meaningful by the constructions (80) of n-shifts $h^{(n)}$ on $\mathcal{P}$. For example,

$$\lim_{n\to\infty} h^{(n)}(q) = \lim_{p\to\infty} p, \text{ for } q \in \mathcal{P}, \tag{94b}$$

where the right-hand side of (94b) means that: starting with q, take bigger primes again and again in the TOset $\mathcal{P}$ of (78).

Assumption and Notation: From below, for convenience, the notations in (94a) are used for (93), if there is no confusion.

We now define a new (unbounded) linear functional τ_0 on $\mathbb{LS}_A$ with respect to the linear functional τ of (74a), by

$$\tau_0 \stackrel{\text{def}}{=} \tau \circ \pi \text{ on } \mathbb{LS}_A, \tag{95}$$

where π is in the sense of **N 2**.

Theorem 5. *Let $\mathbb{LS}_A = (\mathbb{LS}_A, \tau)$ be the A-tensor sub-filterization (74b), and let $\tau_0 = \tau \circ \pi$ be the new linear functional (95) on the Banach ∗-algebra $\mathbb{LS}_A$ of (74a). Then, for the generators*

$$\{Q_{p,j}^a\}_{p\in\mathcal{P}} \subset \Omega(\mathcal{U}_A) \text{ of } \mathbb{LS}_A,$$

for an arbitrarily fixed $a \in (A, \psi)$ and $j \in \mathbb{Z}$, we have that

$$\tau_0\left(\left(Q_{p,j}^a\right)^n\right) = \begin{cases} 0, & \text{if } j > 0, \\ \omega_n c_{\frac{n}{2}}\psi(a^n), & \text{if } j = 0, \\ \infty, \text{ Undefined}, & \text{if } j < 0, \end{cases} \tag{96}$$

for all $n \in \mathbb{N}$.

Proof. Let $\{Q_{p,j}^a\}_{p\in\mathcal{P}} \subset \Omega(\mathcal{U}_A)$ in $\mathbb{LS}_A$, for fixed $a \in (A, \psi)$ and $j \in \mathbb{Z}$. Then,

$$\tau_0\left(\left(Q_{p,j}^a\right)^n\right) = (\tau \circ \pi)\left(\left(Q_{p,j}^a\right)^n\right) = \tau\left(\lim_{p\to\infty}\left(Q_{p,j}^a\right)^n\right)$$

by (93) and (94a)

$$= \lim_{p\to\infty} \tau\left(\left(Q_{p,j}^a\right)^n\right)$$

by the boundedness of τ for the (norm, or strong) topology for $\mathbb{LS}_A$

$$= \lim_{p \to \infty} \tau_j^p \left(\left(Q_{p,j}^a \right)^n \right) = \lim_{p \to \infty} \left(\omega_n c_{\frac{n}{2}} \psi(a^n) \left(\frac{\phi(p)}{p^{j+1}} \right) \right)$$

by (70), (75) and (77)

$$= \left(\omega_n c_{\frac{n}{2}} \psi(a^n) \right) \left(\lim_{p \to \infty} \frac{\phi(p)}{p^{j+1}} \right)$$

$$= \begin{cases} 0, & \text{if } j > 0, \\ \omega_n c_{\frac{n}{2}} \psi(a^n), & \text{if } j = 0, \\ \infty, \text{ Undefined}, & \text{if } j < 0, \end{cases}$$

by (90), for each $n \in \mathbb{N}$. Therefore, the free-distributional data (96) holds for τ_0. □

By (96), we obtain the following corollary.

Corollary 2. *Let $Q_{p,0}^{1_A} \in \Omega(\mathcal{U}_A)$ be free random variables of the A-tensor sub-filterization $\mathbb{LS}_A$, for all $p \in \mathcal{P}$, where 1_A is the unity of (A, ψ). Then, the asymptotic free distribution of the family*

$$\mathcal{Q}_0^{1_A} = \{ Q_{p,0}^{1_A} \in \Omega(\mathcal{U}_A) \}_{p \in \mathcal{P}}$$

follows the semicircular law asymptotically as $p \to \infty$ in $\mathcal{P}$.

Proof. Let $\mathcal{Q}_0^{1_A} = \{Q_{p,0}^{1_A}\}_{p \in \mathcal{P}} \subset \Omega(\mathcal{U}_A)$ in $\mathbb{LS}_A$. Then, for the linear functional τ_0 of (95) on $\mathbb{LS}_A$,

$$\tau_0 \left(\left(Q_{p,0}^{1_A} \right)^n \right) = \omega_n c_{\frac{n}{2}},$$

for all $n \in \mathbb{N}$, by (96), since

$$\psi(1_A^n) = \psi(1_A) = 1; n \in \mathbb{N}.$$

If $p \to \infty$ in $\mathcal{P}$, then the asymptotic free distribution of the family $\mathcal{Q}_0^{1_A}$ is the semicircular law by the self-adjointness of all $Q_{p,0}^{1_A}$'s, and by the semicircularity (45) and (47). □

Independent from (96), we obtain the following asymptotic free-distributional data on $\mathbb{LS}_A$.

Theorem 6. *Let $j_1, ..., j_N$ be "mutually distinct" in $\mathbb{Z}$, for $N > 1$ in $\mathbb{N}$, and hence the N-tuple*

$$[j] = (j_1, ..., j_N) \in \mathbb{Z}^N$$

is alternating in $\mathbb{Z}$. In addition, let

$$[a] = (a_1, ..., a_N)$$

be an arbitrarily fixed N-tuple of free random variables $a_1, ..., a_N$ of the unital C^*-probability space (A, ψ), and let's fix

$$[n] = (n_1, ..., n_N) \in \mathbb{N}^N.$$

Now, define a family $\mathcal{T}_{[j]}^{[a],[n]}$ of free reduced words with their lengths-N,

$$\mathcal{T}_{[j]}^{[a],[n]} = \left\{ T = \prod_{l=1}^{N} \left(Q_{p_l, j_l}^{a_l} \right)^{n_l} : p_1, ..., p_N \in \mathcal{P} \right\}, \quad (97)$$

in $\mathbb{LS}_A$, for $Q_{p_l, j_l}^{a_l} \in \Omega(\mathcal{U}_A)$, for all $p_l \in \mathcal{P}$, where $a_l \in [a]$, $j_l \in [j]$, for $l = 1, ..., N$.

For any free reduced words $T \in \mathcal{T}_{[j]}^{[a],[n]}$, if τ_0 is the linear functional (95) on $\mathbb{LS}_A$, then

$$\tau_0(T) = \begin{cases} 0, & \text{if } \sum_{l=1}^{N} j_l > 1 - N, \\ \prod_{l=1}^{N} \left(\omega_{n_l} c_{\frac{n_l}{2}} \psi(a^{n_l}) \right), & \text{if } \sum_{l=1}^{N} j_l = 1 - N, \\ \infty, \text{ Undefined}, & \text{if } \sum_{l=1}^{N} j_l < 1 - N, \end{cases} \quad (98)$$

for all $n \in \mathbb{N}$.

Proof. Let $T \in \mathcal{T}_{[j]}^{[a],[n]}$ be in the sense of (97) in the A-tensor sub-filterization $\mathbb{LS}_A$. Then, these operators T form free reduced words with their lengths-N in $\mathbb{LS}_A$, since $[j]$ is an alternating N-tuple of "mutually distinct" integers. Observe that

$$\tau_0(T) = \tau(\pi(T)) = \tau \left(\prod_{l=1}^{N} \left(\lim_{p_l \to \infty} \left(Q_{p_l, j_l}^{a_l} \right) \right)^{n_l} \right)$$

by (93) and (94a)

$$= \tau \left(\prod_{l=1}^{N} \left(\lim_{p \to \infty} \left(Q_{p, j_l}^{a_l} \right) \right)^{n_l} \right)$$

because

$$\lim_{p \to \infty} p = \lim_{n \to \infty} h^{(n)}(p_l) = \lim_{p_l \to \infty} p_l, \text{ in } \mathcal{P},$$

in the sense of (44), for all $l = 1, ..., N$, and, hence, it goes to

$$= \lim_{p \to \infty} \left(\tau \left(\left(\prod_{l=1}^{N} Q_{p, j_l}^{a_l} \right)^{n_l} \right) \right)$$

by the boundedness of τ for the (norm, or strong) topology for $\mathbb{LS}_A$

$$= \lim_{p \to \infty} \left(\prod_{l=1}^{N} \left(\omega_{n_l} c_{\frac{n_l}{2}} \psi(a_l^{n_l}) \left(\frac{\phi(p)}{p^{j_l+1}} \right) \right) \right)$$

since $[j]$ consists of "mutually-distinct" integers, by the Möbius inversion

$$= \left(\prod_{l=1}^{N} \omega_{n_l} c_{\frac{n_l}{2}} \psi(a_l^{n_l}) \right) \left(\lim_{p \to \infty} \left(\prod_{l=1}^{N} \left(\frac{\phi(p)}{p^{j_l+1}} \right) \right) \right)$$

$$= \left(\prod_{l=1}^{N} \omega_{n_l} c_{\frac{n_l}{2}} \psi(a_l^{n_l}) \right) \left(\lim_{p \to \infty} \left(\frac{\phi(p)}{p^{N+\Sigma_{l=1}^{N} j_l}} \right) \right)$$

$$= \left(\prod_{l=1}^{N} \omega_{n_l} c_{\frac{n_l}{2}} \psi(a_l^{n_l}) \right) \left(\lim_{p \to \infty} \left(\frac{\phi(p)}{p^{(N-1+\Sigma_{l=1}^{N} j_l)+1}} \right) \right)$$

$$= \left(\prod_{l=1}^{N} \omega_{n_l} c_{\frac{n_l}{2}} \psi(a_l^{n_l}) \right) \left(\lim_{p \to \infty} \left(\frac{\phi(p)}{p^{(N-1+\Sigma_{l=1}^{N} j_l)+1}} \right) \right)$$

$$= \begin{cases} 0 & \text{if } N - 1 + \sum_{l=1}^{N} j_l > 0 \\ \prod_{l=1}^{N} \left(\omega_{n_l} c_{\frac{n_l}{2}} \psi(a_l^{n_l}) \right) & \text{if } N - 1 + \sum_{l=1}^{N} j_l = 0 \\ \infty & \text{if } N - 1 + \sum_{l=1}^{N} j_l < 0, \end{cases}$$

by (90), for all $n \in \mathbb{N}$. Therefore, the family $\mathcal{T}_{[j]}^{[a],[n]}$ of (97) satisfies the asymptotic free-distributional data (98) in the A-tensor sub-filterization $\mathbb{LS}_A$ over $\mathcal{P}$. □

The above two theorems illustrate the asymptotic free-probabilistic behaviors on the A-tensor sub-filterization $\mathbb{LS}_A$ over $\mathcal{P}$, by (96) and (98).

As a corollary of (96), we showed that the family

$$\mathcal{Q}_0^{1_A} = \{Q_{p,0}^{1_A}\}_{p \in \mathcal{P}} \subset \mathbb{LS}_A$$

has its asymptotic free distribution, the semicircular law in $\mathbb{LS}_A$, as $p \to \infty$. More generally, the following theorem is obtained.

Theorem 7. *Let a be a self-adjoint free random variable of our unital C^*-probability space (A, ψ). Assume that it satisfies*

(i) $\psi(a) \in \mathbb{R}^{\times} = \mathbb{R} \setminus \{0\}$ in $\mathbb{C}$,
(ii) $\psi(a^{2n}) = \psi(a)^{2n}$, for all $n \in \mathbb{N}$.

Then, the family

$$\mathcal{X}_0^a = \left\{ X_{p,0}^a = \tfrac{1}{\psi(a)} Q_{p,0}^a : p \in \mathcal{P} \right\} \tag{99}$$

follows the asymptotic semicircular law, in $\mathbb{LS}_A$ over $\mathcal{P}$.

Proof. Let $a \in (A, \psi)$ be a self-adjoint free random variable satisfying two conditions (i) and (ii), and let $\mathcal{X}_0^a$ be the family (99) of the A-tensor sub-filterization $\mathbb{LS}_A$. Then, all elements

$$X_{p,0}^a = \tfrac{1}{\psi(a)} Q_{p,0}^a = 1_p^A \otimes \left(\left(\tfrac{1}{\psi(a)} a \right) \otimes P_{p,0} \right) \text{ of } \mathcal{X}_0^a$$

are self-adjoint in $\mathbb{LS}_A$, by the self-adjointness of $Q_{p,0}^a$, and by the condition (i).

For any $X_{p,0}^a \in \mathcal{X}_0^a$, observe that

$$\tau_0 \left(\left(X_{p,0}^a \right)^n \right) = \tfrac{1}{\psi(a)^n} \tau_0 \left(\left(Q_{p,0}^a \right)^n \right)$$

$$= \tfrac{1}{\psi(a)^n} \left(\omega_n c_{\frac{n}{2}} \psi(a^n) \right)$$

by (96)

$$= \left(\omega_n c_{\frac{n}{2}} \left(\tfrac{\psi(a^n)}{\psi(a^n)} \right) \right)$$

by the condition (ii)

$$= \omega_n c_{\frac{n}{2}},$$

for all $n \in \mathbb{N}$. Therefore, the family $\mathcal{X}_0^a$ has its asymptotic semicircular law over $\mathcal{P}$, by (45). □

Similar to the construction of $\mathcal{X}_0^a$ of (99), if we construct the families $\mathcal{X}_j^a$,

$$\mathcal{X}_j^a = \left\{ \tfrac{1}{\psi(a)} Q_{p,j}^a : Q_{p,j}^a \in \Omega(\mathcal{U}_A) \right\}_{p \in \mathcal{P}}, \tag{100}$$

for a fixed $a \in (A, \psi)$ satisfying the conditions (i) and (ii) of the above theorem, and, for a fixed $j \in \mathbb{Z}$, then one obtains the following corollary.

Corollary 3. Fix $a \in (A, \psi)$ satisfying the conditions (i) and (ii) of the above theorem. Let's fix $j \in \mathbb{Z}$, and let $\mathcal{X}_j^a$ be the corresponding family (100) in the A-tensor sub-filterization $\mathbb{LS}_A = (\mathbb{LS}_A, \tau)$.

$$\text{If } j = 0, \text{ then } \mathcal{X}_0^a \text{ has the asymptotic semicircular law in } \mathbb{LS}_A. \tag{101}$$
$$\text{If } j > 0, \text{ then } \mathcal{X}_j^a \text{ has its asymptotic free distribution, the zero free distribution, in } \mathbb{LS}_A. \tag{102}$$
$$\text{If } j < 0, \text{ then the asymptotic free distribution of } \mathcal{X}_j^a \text{ is undefined in } \mathbb{LS}_A. \tag{103}$$

Proof. The proof of (101) is done by (99).

By (96), if $j > 0$, then, for any $T = \tfrac{1}{\psi(a)} Q_{p,j}^a \in \mathcal{X}_j^a$, one has that

$$\tau_0(T^n) = \tfrac{1}{\psi(a^n)} \tau_0 \left(\left(Q_{p,j}^a \right)^n \right) = 0,$$

for all $n \in \mathbb{N}$. Thus, the asymptotic free distribution of $\mathcal{X}_j^a$ is the zero free distribution in $\mathbb{LS}_A$, as $p \to \infty$ in $\mathcal{P}$. Thus, the statement (102) holds.

Similarly, by (96), if $j < 0$, then the asymptotic free distribution $\mathcal{X}_j^a$ is undefined in $\mathbb{LS}_A$ over $\mathcal{P}$, equivalently, the statement (103) is shown. □

Motivated by (101), (102) and (103), we study the asymptotic semicircular law (over $\mathcal{P}$) on $\mathbb{LS}_A$ more in detail in Section 10 below.

10. Asymptotic Semicircular Laws on $\mathbb{LS}_A$ over $\mathcal{P}$

We here consider asymptotic semicircular laws on the A-tensor sub-filterization $\mathbb{LS}_A = (\mathbb{LS}_A, \tau)$. In Section 9.3, we showed that the asymptotic free distribution of a family

$$\mathcal{X}_0^a = \{\tfrac{1}{\psi(a)} Q_{p,0}^a : p \in \mathcal{P}\} \tag{104}$$

is the semicircular law in $\mathbb{LS}_A$ as $p \to \infty$ in $\mathcal{P}$, for a fixed self-adjoint free random variable $a \in (A, \psi)$ satisfying

(i) $\psi(a) \in \mathbb{R}^\times$, and
(ii) $\psi(a^{2n}) = \psi(a)^{2n}$, for all $n \in \mathbb{N}$.

As an example, the family

$$\mathcal{X}_0^{1_A} = \{Q_{p,0}^{1_A} : p \in \mathcal{P}\} \tag{105}$$

follows the asymptotic semicircular law in $\mathbb{LS}_A$ over $\mathcal{P}$.

We now enlarge such asymptotic behaviors on $\mathbb{LS}_A$ up to certain $*$-isomorphisms.

Define bijective functions g_+ and g_- on $\mathbb{Z}$ by

$$g_+(j) = j+1, \text{ and } g_-(j) = j-1, \tag{106}$$

for all $j \in \mathbb{Z}$.

By (106), one can define bijective functions $g_\pm^{(n)}$ on $\mathbb{Z}$ by

$$g_\pm^{(n)} \stackrel{def}{=} \underbrace{g_\pm \circ g_\pm \circ g_\pm \circ \cdots \circ g_\pm}_{n\text{-times}}, \tag{107}$$

satisfying $g_\pm^{(1)} = g_\pm$ on $\mathbb{Z}$, with axiomatization:

$$g_\pm^{(0)} = id_\mathbb{Z}, \text{ the identity function on } \mathbb{Z},$$

for all $n \in \mathbb{N}_0 = \mathbb{N} \cup \{0\}$. For example,

$$g_\pm^{(n)}(j) = j \pm n, \tag{108}$$

for all $j \in \mathbb{Z}$, for all $n \in \mathbb{N}_0$.

From the bijective functions $g_\pm^{(n)}$ of (107), define the bijective functions $(g_\pm^o)^{(n)}$ on the generator set $\Omega(\mathcal{U}_A)$ of (72) of the A-tensor sub-filterization $\mathbb{LS}_A$ by

$$(g_+^o)^{(n)} (Q_{p,j}^a) = Q_{p, g_+^{(n)}(j)}^a = Q_{p, j+n}^a,$$

$$(g_-^o)^{(n)} (Q_{p,j}^a) = Q_{p, g_-^{(n)}(j)}^a = Q_{p, j-n}^a, \tag{109}$$

with

$$(g_\pm^o)^{(1)} = g_\pm^o, \text{ and } (g_\pm^o)^{(0)} = id,$$

by (108), for all $p \in \mathcal{P}$ and $j \in \mathbb{Z}$, for all $n \in \mathbb{N}_0$, where id is the identity function on $\Omega(\mathcal{U}_A)$.

By the construction (73a) of the generator set $\Omega(\mathcal{U}_A)$ of $\mathbb{LS}_A$ under (73b),

$$\Omega(\mathcal{U}_A) = \bigsqcup_{p \in \mathcal{P}} \{Q_{p,j}^a : a \in A, j \in \mathbb{Z}\},$$

the functions $\left(g_\pm^o\right)^{(n)}$ of (109) are indeed well-defined bijections on $\Omega(\mathcal{U}_A)$, by the bijectivity of $g_\pm^{(n)}$ of (107).

Now, define bounded $*$-homomorphisms $G_\pm$ on $\mathbb{LS}_A$ by the bounded multiplicative linear transformations on $\mathbb{LS}_A$ satisfying that:

$$G_+\left(Q_{p,j}^a\right) = g_+^o\left(Q_{p,j}^a\right) = Q_{p,j+1}^a,$$

(110)

$$G_-\left(Q_{p,j}^a\right) = g_-^o\left(Q_{p,j}^a\right) = Q_{p,j-1}^a,$$

in $\mathbb{LS}_A$, by using the bijections $g_\pm^o$ of (109), for all $Q_{p,j}^a \in \Omega(\mathcal{U}_A)$.

More precisely, the morphisms $G_\pm$ of (110) satisfy that

$$G_\pm\left(\prod_{l=1}^N \left(Q_{p_l,j_l}^{a_l}\right)^{n_l}\right) = \prod_{l=1}^N g_\pm^o\left(\left(Q_{p_l,j_l}^{a_l}\right)^{n_l}\right)$$

$$= \prod_{l=1}^N \left(Q_{p_l,j_l\pm 1}^{a_l}\right)^{n_l}.$$

(111a)

By (111a), one can get that

$$G_\pm\left(\left(\prod_{l=1}^N \left(Q_{p_l,j_l}^{a_l}\right)^{n_l}\right)^*\right) = G_\pm\left(\prod_{l=1}^N \left(Q_{p_{N-l+1},j_{N-l+1}}^{a_{N-l+1}^*}\right)^{n_{N-l+1}}\right)$$

$$= \prod_{l=1}^N \left(\left(Q_{p_{N-l+1},(j_{N-l+1})\pm 1}^{a_{N-l+1}}\right)^{n_{N-l+1}}\right)^*$$

$$= \left(\prod_{l=1}^N \left(Q_{p_l,j_l\pm 1}^{a_l}\right)^{n_l}\right)^*$$

$$= \left(G_\pm\left(\prod_{l=1}^N Q_{p_l,j_l}^{n_l}\right)\right)^*$$

(111b)

for all $Q_{p_l,j_l}^{a_l} \in \Omega(\mathcal{U}_A)$, for $l = 1, ..., N$, for $N \in \mathbb{N}$.

The formula (111a) are obtained by (110) and the multiplicativity of $G_\pm$. The formulas in (111b), obtained from (111a), show that indeed $G_\pm$ are $*$-homomorphisms on $\mathbb{LS}_A$, since

$$G_+(T^*) = (G_\pm(T))^*, \forall T \in \mathbb{LS}_A.$$

By (110) and (111a),

$$G_\pm^n\left(\prod_{l=1}^N \left(Q_{p_l,j_l}^{a_l}\right)^{n_l}\right) = \prod_{l=1}^N \left(Q_{p_l,j_l\pm n}^{a_l}\right)^{n_l},$$

(112)

$$G_\pm^n\left(\left(\prod_{l=1}^N \left(Q_{p_l,j_l}^{a_l}\right)^{n_l}\right)^*\right) = \left(G_\pm^n\left(\prod_{l=1}^N \left(Q_{p_l,j_l}^{a_l}\right)^{n_l}\right)\right)^*,$$

for all $Q_{p_l,j_l}^{a_l} \in \Omega(\mathcal{U}_A)$, for $l = 1, ..., N$, for $N \in \mathbb{N}$, for all $n \in \mathbb{N}_0$.

Definition 16. *We call the bounded $*$-homomorphisms $G_\pm^n$ of (110), the n-($\pm$)-integer-shifts on $\mathbb{LS}_A$, for all $n \in \mathbb{N}_0$.*

Based on the integer-shifting processes on $\mathbb{LS}_A$, one can get the following asymptotic behavior on $\mathbb{LS}_A$ over $\mathcal{P}$.

Theorem 8. *Let $\mathcal{X}_j^a$ be a family (100) of the A-tensor sub-filterization $\mathbb{LS}_A$, for any $j \in \mathbb{Z}$, where a is a fixed self-adjoint free random variable of (A, ψ) satisfying the additional conditions (i) and (ii) above. Then, there exists a $(-j)$-integer-shift G_{-j} on $\mathbb{LS}_A$, such that*

$$G_{-j} = \begin{cases} G_{-}^{|j|} = G_{-}^{j} & \text{if } j \geq 0 \text{ in } \mathbb{Z}, \\ G_{+}^{|j|} = G_{+}^{-j} & \text{if } j < 0 \text{ in } \mathbb{Z}, \end{cases} \tag{113}$$

and

$$\tau_0\left(G_j(T)\right) = \omega_n c_{\frac{n}{2}}, \forall n \in \mathbb{N}, \tag{114}$$

for all $T \in \mathcal{X}_j^a$, where $G_{\mp}^{\pm j}$ on the right-hand sides of (113) are the $|j|$-($\mp$)-integer shifts (110) on $\mathbb{LS}_A$, and where $\tau_0 = \tau \circ \pi$ is the linear functional (95) on $\mathbb{LS}_A$.

Proof. Let $\mathcal{X}_j^a = \left\{ \frac{1}{\psi(a)} Q_{p,j}^a : p \in \mathcal{P} \right\}$ be a family (100) of $\mathbb{LS}_A$, for a fixed $j \in \mathbb{Z}$, where a fixed self-adjoint free random variable $a \in (A, \psi)$ satisfies the above additional conditions (i) and (ii).

Assume first that $j \geq 0$ in $\mathbb{Z}$. Then, one can take the $(-j)$-$(-)$-integer-shift G_{-}^j of (110) on $\mathbb{LS}_A$, satisfying

$$G_{-}^j\left(Q_{p,j}^a\right) = Q_{p,j-j}^a = Q_{p,0}^a \text{ in } \mathbb{LS}_A,$$

for all $Q_{p,j}^a \in \Omega(\mathcal{U}_A)$.

Second, if $j < 0$ in $\mathbb{Z}$, then one can have the $|j|$-(+)-integer shift G_{+}^{-j} of (110) on $\mathbb{LS}_A$, satisfying that

$$G_{+}^{-j}\left(Q_{p,j}^a\right) = Q_{p,j+(-j)}^a = Q_{p,0}^a \text{ in } \mathbb{LS}_A,$$

for all $Q_{p,j}^a \in \Omega(\mathcal{U}_A)$.

For example, for any $Q_{p,j}^a \in \Omega(\mathcal{U}_A)$, we have the corresponding $(-j)$-integer-shift G_{-j},

$$G_{-j} = \begin{cases} G_{-}^j & \text{if } j \geq 0, \\ G_{+}^{-j} & \text{if } j < 0, \end{cases}$$

on $\mathbb{LS}_A$ in the sense of (113), such that

$$G_{-j}\left(Q_{p,j}^a\right) = Q_{p,0}^a \text{ in } \mathbb{LS}_A,$$

for all $p \in \mathcal{P}$.

Then, for any $X_{p,j}^a = \frac{1}{\psi(a)} Q_{p,j}^a \in \mathcal{X}_j^a$, we have that

$$\tau_0\left(G_{-j}\left(\left(X_{p,j}^a\right)^n\right)\right) = \tau_0\left(\frac{1}{\psi(a)^n}\left(G_{-j}(Q_{p,j}^a)\right)^n\right),$$

since G_{-j} is a $*$-homomorphism (113) on $\mathbb{LS}_A$

$$= \tau_0\left(\frac{1}{\psi(a^n)}\left(Q_{p,0}^a\right)^n\right) = \omega_n c_{\frac{n}{2}},$$

by (96) and (98), for all $n \in \mathbb{N}$. Therefore, formula (114) holds true. □

By the above theorem, we obtain the following result.

Corollary 4. *Let $\mathcal{X}_j^a$ be a family (100) of the A-tensor sub-filterization $\mathbb{LS}_A$, for $j \in \mathbb{Z}$, where a self-adjoint free random variable $a \in (A, \psi)$ satisfies the conditions (i) and (ii). Then, the corresponding family*

$$\mathcal{G}_j^a = \left\{ G_{-j}(X) : X \in \mathcal{X}_j^a \right\} \tag{115}$$

has its asymptotic free distribution, the semicircular law, in $\mathbb{LS}_A$ over $\mathcal{P}$, where G_{-j} is the $(-j)$-integer shift (113) on $\mathbb{LS}_A$, for all $j \in \mathbb{Z}$.

Proof. The asymptotic semicircular law induced by the family $\mathcal{G}_j^a$ of (115) in $\mathbb{LS}_A$ is guaranteed by (114) and (45), for all $j \in \mathbb{Z}$. □

By the above corollary, the following result is immediately obtained.

Corollary 5. Let $\mathcal{X}_j^{1_A}$ be in the sense of (100) in $\mathbb{LS}_A$, where 1_A is the unity of (A, ψ), and let

$$\mathcal{G}_j^{1_A} = \left\{ G_{-j}(X) : X \in \mathcal{X}_j^{1_A} \right\}$$

be in the sense of (115), for all $j \in \mathbb{Z}$. Then, the asymptotic free distributions of $\mathcal{G}_j^{1_A}$ are the semicircular law in $\mathbb{LS}_A$ over $\mathcal{P}$, for all $j \in \mathbb{Z}$.

Proof. The proof is done by Corollary 4. Indeed, the unity 1_A automatically satisfies the conditions (i) and (ii) in (A, ψ). □

More general to Theorem 8, we obtain the following result too.

Theorem 9. Let $a \in (A, \psi)$ be a self-adjoint free random variable satisfying the conditions (i) and (ii), and let $p_0 \in \mathcal{P}$ be an arbitrarily fixed prime. Let

$$\mathcal{G}_j^a[\geq p_0] \stackrel{def}{=} \left\{ G_{-j}(X_{p,j}) \;\middle|\; \begin{array}{c} X_{p,j}^a \in \mathcal{X}_j^a \text{ and} \\ p \geq p_0 \text{ in } \mathcal{P} \end{array} \right\},$$

where $\mathcal{X}_j^a$ is the family (100), and $\mathcal{G}_j^a$ is the family (115), for $j \in \mathbb{Z}$. Then, the asymptotic free distribution of the family $\mathcal{G}_j^a[\geq p_0]$ is the semicircular law in $\mathbb{LS}_A$.

Proof. The proof of this theorem is similar to that of Theorem 8. One can simply replace

$$\text{"} p \to \infty \text{"} \equiv \text{"} \lim_{n \to \infty} h^n(2); 2 \in \mathcal{P} \text{,"}$$

in the proof of Theorem 8 to

$$\text{"} p \to \infty \text{"} \equiv \text{"} \lim_{n \to \infty} h^n(p_0); p_0 \in \mathcal{P} \text{,"}$$

where ($\equiv$) means "being symbolically same". □

Funding: This research received no external funding.

Conflicts of Interest: The author declares no conflict of interest.

References

1. Cho, I. Adelic Analysis and Functional Analysis on the Finite Adele Ring. *Opusc. Math.* **2017**, *38*, 139–185. [CrossRef]
2. Cho, I.; Jorgensen, P.E.T. Krein-Space Operators Induced by Dirichlet Characters, Special Issues: Contemp. Math.: Commutative and Noncommutative Harmonic Analysis and Applications. *Math. Amer. Math. Soc.* **2014**, 3–33. [CrossRef]
3. Alpay, D.; Jorgensen, P.E.T.; Levanony, D. On the Equivalence of Probability Spaces. *arXiv* **2016**, arXiv:1601.00639.
4. Alpay, D.; Jorgensen, P.E.T.; Kimsey, D. Moment Problems in an Infinite Number of Variables. *Infinite Dimensional Analysis, Quantum Probab. Relat. Top.* **2015**, *18*, 1550024. [CrossRef]
5. Alpay, D.; Jorgensen, P.E.T.; Salomon, G. On Free Stochastic Processes and Their Derivatives. *Stoch. Process. Their Appl.* **2014**, *124*, 3392–3411. [CrossRef]
6. Cho, I. *p*-Adic Free Stochastic Integrals for *p*-Adic Weighted-Semicircular Motions Determined by Primes *p*. *Lib. Math.* **2016**, *36*, 65–110.
7. Albeverio, S.; Jorgensen, P.E.T.; Paolucci, A.M. Multiresolution Wavelet Analysis of Integer Scale Bessel Functions. *J. Math. Phys.* **2007**, *48*, 073516. [CrossRef]
8. Gillespie, T. Superposition of Zeroes of Automorphic *L*-Functions and Functoriality. Ph.D. Thesis, University of Iowa, Iowa City, IA, USA, 2010.
9. Gillespie, T. Prime Number Theorems for Rankin-Selberg *L*-Functions over Number Fields. *Sci. China Math.* **2011**, *54*, 35–46. [CrossRef]

10. Jorgensen, P.E.T.; Paolucci, A.M. q-Frames and Bessel Functions. *Numer. Funct. Anal. Optim.* **2012**, *33*, 1063–1069. [CrossRef]
11. Jorgensen, P.E.T.; Paolucci, A.M. Markov Measures and Extended Zeta Functions. *J. Appl. Math. Comput.* **2012**, *38*, 305–323. [CrossRef]
12. Radulescu, F. Random Matrices, Amalgamated Free Products and Subfactors of the C^*-Algebra of a Free Group of Nonsingular Index. *Invent. Math.* **1994**, *115*, 347–389. [CrossRef]
13. Radulescu, F. Free Group Factors and Hecke Operators, notes taken by N. Ozawa. In *Proceedings of the 24th International Conference in Operator Theory, Timisoara, Romania, 2–7 July 2012*; Theta Series in Advanced Mathematics; Theta Foundation: Indianapolis, IN, USA, 2014.
14. Radulescu, F. Conditional Expectations, Traces, Angles Between Spaces and Representations of the Hecke Algebras. *Lib. Math.* **2013**, *33*, 65–95. [CrossRef]
15. Speicher, R. Combinatorial Theory of the Free Product with Amalgamation and Operator-Valued Free Probability Theory. *Mem. Am. Math. Soc.* **1998**, *132*, 627–627. [CrossRef]
16. Speicher, R. Speicher A Conceptual Proof of a Basic Result in the Combinatorial Approach to Freeness. *Infin. Dimens. Anal. Quantum Probab. Relat. Top.* **2000**, *3*, 213–222. [CrossRef]
17. Speicher, R. Multiplicative Functions on the Lattice of Non-crossing Partitions and Free Convolution. *Math. Ann.* **1994**, *298*, 611–628. [CrossRef]
18. Speicher, R.; Neu, P. Physical Applications of Freeness. In *Proceedings of the XII-th International Congress of Mathematical Physics (ICMP '97), Brisbane, Australian, 13–19 July 1997*; International Press: Vienna, Austria, 1999; pp. 261–266.
19. Vladimirov, V.S. p-Adic Quantum Mechanics. *Commun. Math. Phys.* **1989**, *123*, 659–676. [CrossRef]
20. Vladimirov, V.S.; Volovich, I.V.; Zelenov, E.I. *p-Adic Analysis and Mathematical Physics*; Series on Soviet and East European Mathematics; World Scientific: Singapore, 1994; Volume 1, ISBN 978-981-02-0880-6.
21. Voiculescu, D. Aspects of Free Analysis. *Jpn. J. Math.* **2008**, *3*, 163–183. [CrossRef]
22. Voiculescu, D.; Dykema, K.; Nica, A. *Free Random Variables*; CRM Monograph Series; Published by American Mathematical Society: Providence, RI, USA, 1992; Volume 1.
23. Cho, I.; Jorgensen, P.E.T. Semicircular Elements Induced by p-Adic Number Fields. *Opusc. Math.* **2017**, *35*, 665–703. [CrossRef]
24. Cho, I. Semicircular Families in Free Product Banach $*$-Algebras Induced by p-Adic Number Fields over Primes p. *Complex Anal. Oper. Theory* **2017**, *11*, 507–565. [CrossRef]
25. Cho, I. Asymptotic Semicircular Laws Induced by p-Adic Number Fields over Primes p. *Complex Anal. Oper. Theory* **2019**. [CrossRef]
26. Albeverio, S.; Jorgensen, P.E.T.; Paolucci, A.M. On Fractional Brownian Motion and Wavelets. *Complex Anal. Oper. Theory* **2012**, *6*, 33–63. [CrossRef]
27. Alpay, D.; Jorgensen, P.E.T. Spectral Theory for Gaussian Processes: Reproducing Kernels. *Random Funct. Oper. Theory* **2015**, *83*, 211–229.
28. Connes, A. Hecke Algebras, Type III-Factors, and Phase Transitions with Spontaneous Symmetry Breaking in Number Theory. *Sel. Math.* **1995**, *1*, 411–457.
29. Connes, A. Trace Formula in Noncommutative Geometry and the Zeroes of the Riemann Zeta Functions. Available online: http://www.alainconnes.org/en/download.php (accessed on 15 March 2019).
30. Alpay, D.; Jorgensen, P.E.T. Spectral Theory for Gaussian Processes: Reproducing Kernels, Boundaries, & L_2-Wavelet Generators with Fractional Scales. *Numer. Funct. Anal. Optim.* **2015**, *36*, 1239–1285.
31. Jorgensen, P.E.T. *Operators and Representation Theory: Canonical Models for Algebras of Operators Arising in Quantum Mechanics*, 2nd ed.; Dover Publications: Mineola, NY, USA, 2008; ISBN 978-0586466651.
32. Connes, A. *Noncommutative Geometry*; Academic Press: San Diego, CA, USA, 1994; ISBN 0-12-185860-X.

 © 2019 by the author. Licensee MDPI, Basel, Switzerland. This article is an open access article distributed under the terms and conditions of the Creative Commons Attribution (CC BY) license (http://creativecommons.org/licenses/by/4.0/).

MDPI
St. Alban-Anlage 66
4052 Basel
Switzerland
Tel. +41 61 683 77 34
Fax +41 61 302 89 18
www.mdpi.com

Symmetry Editorial Office
E-mail: symmetry@mdpi.com
www.mdpi.com/journal/symmetry

www.ingramcontent.com/pod-product-compliance
Lightning Source LLC
LaVergne TN
LVHW070730100526
838202LV00013B/1207